Caudata
(Salamanders)

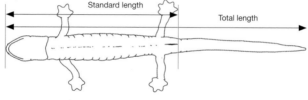

Anura
(Frogs & Toads)

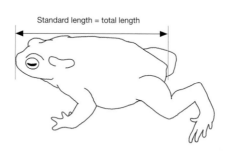

Illustrations

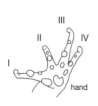

Species accounts for most anura are accompanied by schematic drawings showing diagnostic features of their hands and feet, including the amount of webbing and the shape of finger and toe disks (when present). Note that specific fingers and toes are identified throughout the text by roman numerals; digits are numbered from those closest to the body axis outward, as shown here.

The hands and feet ⌐ ...
very similar in tern ...
and the amount of ...
that salamander ha ...
four digits, wherea ...
Therefore, only a f ...
salamander accou ...

In toads (family Bu ...
and size of the par ...
as the presence ar ...
crests, are more di ...
hands and feet; sp ...
family are therefore ...
illustration of their F ...

Amphibians of Costa Rica

A FIELD GUIDE

TWAN LEENDERS

A Zona Tropical Publication

FROM

Comstock Publishing Associates

a division of

Cornell University Press

Ithaca and London

Text copyright © 2016 by Twan Leenders
Illustrations and range maps copyright © 2016 by Twan Leenders
Photographs copyright © 2016 by respective photographers

All rights reserved. Except for brief quotations in a review, this book, or parts
thereof, must not be reproduced in any form without permission in writing
from the publishers. For information within Costa Rica, visit Zona Tropical at
www.zonatropical.net. For information in the rest of the world, address Cornell
University Press, Sage House, 512 East State Street, Ithaca, New York 14850, or
visit www.cornellpress.cornell.edu.

First published 2016 by Cornell University Press

First printing, Cornell Paperbacks, 2016
Printed in China

Library of Congress Cataloging-in-Publication Data

Names: Leenders, Twan, author.
Title: Amphibians of Costa Rica : a field guide / Twan Leenders.
Description: Ithaca : Comstock Publishing Associates, a division of Cornell University Press,
    2016. | "A Zona Tropical publication." | Includes bibliographical references and index.
Identifiers: LCCN 2016021844 | ISBN 9781501700620 (pbk. : alk. paper)
Subjects: LCSH: Amphibians—Costa Rica—Identification.
Classification: LCC QL656.C783 L43 2016 | DDC 597.8097286—dc23
LC record available at https://lccn.loc.gov/2016021844

Zona Tropical Press ISBN 978-0-9894408-3-7

Cornell University Press strives to use environmentally responsible suppliers
and materials to the fullest extent possible in the publishing of its books. Such
materials include vegetable-based, low-VOC inks and acid-free papers that are
recycled, totally chlorine-free, or partly composed of nonwood fibers. For further
information, visit our website at www.cornellpress.cornell.edu.

Paperback printing 10 9 8 7 6 5 4 3 2 1

Book concept design: Dave Huth and Gabriela Wattson
Book design: Gabriela Wattson

For those with a curious mind, a love of nature, and a desire to learn: share that passion widely so that future generations will be able to enjoy the stars of this book!

Contents

Foreword

If there were a pub quiz on Central American amphibians, I would feel pretty confident about winning if I were on Twan Leenders' team.

I have been fortunate to accompany Twan into the rainforest habitat in which he is so at home, bearing witness to his eagle eyes, fast hands, and astounding understanding of natural history. Quite where he stores his encyclopedic knowledge is beyond me, but it's hard to spend time with him in the field without hoping that at least some of it is absorbed like water through a frog's skin.

But what impresses me even more than his knowledge is his passion for sharing it. Few people are capable of doing this so graciously and with such good cheer. This new field guide illuminates the natural history of Costa Rica's rich amphibian fauna through range maps and detailed but engaging descriptions that will transport you to the forest floor, rolling a mossy log in the hopes of finding an elusive caecilian.

But it is the brilliant photographs of animals on luminous white backgrounds that draw the eye, page after page. These crisp images reveal features—the eardrums on a poison dart frog, for example—not easily seen in the field. It is this combination of beautiful photographs and fascinating text that makes this book such an invaluable resource.

Understanding is the first step to caring, and what excites me most about this book is the possibility that it will grow the community of people who are fascinated by amphibians—and eager to protect them. And we can all do something. Whether you are a hiker recording an important species sighting on your phone or a conservationist deciding which forests to protect and manage, this book will be an invaluable and inspiring resource.

I grew up looking for amphibians in the moors of Scotland, where I had the chance of finding four amphibian species. I thought this was diversity. The moment I entered the humid forests of Costa Rica I was left mesmerized as every swing of the headlamp illuminated something different. Such an explosion of diversity in sizes, colors, and forms is as daunting as it is exciting. In order to understand that diversity, and truly appreciate the richness of amphibians in Costa Rica, I would highly recommend having Twan by your side. Failing that, this guide is the next best option.

Dr. Robin Moore
Author, *In Search of Lost Frogs: The Quest to Find the World's Rarest Amphibians*

Acknowledgments

This book is the result of nearly 25 years of studying the amphibians of Costa Rica, dozens of trips to the region, countless nights spent hiking rainforest trails or wading through streams and swamps in search of frogs and salamanders, and hundreds of hours wrangling jumpy and squirmy creatures to obtain photographs. Although I shudder at how much time and effort went into this pursuit, all of this pales in comparison to the cumulative exploits of many, many other students of Costa Rican amphibians whose work—published or shared in person—forms the foundation of this book. I would like to express my appreciation to all those people whose love of nature and natural history study, focused on some of my favorite creatures in the world, helped shape this book in many ways.

Furthermore, I would like to acknowledge the support of the people most affected by my work: my wife and kids. Casey, Madeleine, and Jason tolerate my absence several times a year for field work and, lately, suffered the loss of considerable family time while I was writing this book. No other people have invested as much in this book, and I am exceedingly grateful for their love and enthusiasm.

Many people have helped make this project possible, and there simply is not enough room here to list and thank them all. I would be remiss, however, not to extend special appreciation to Tim Paine and Alex Shepack, with whom I have shared many memorable times in the field over the years. Others who have generously contributed time, effort, experience, and insight over the years, in ways that have improved this project, include Juan Abarca, Victor Acosta Chaves, Abel Batista, Clay Bolt, Peter Dahlquist, Don Filipiak, Sean Graesser, Ann Gutierrez, Andreas Hertz, Dave Huth, Robin Moore, Juan Pineda, Michael Roy, Stanley Salazar, Angel Solís, José Solís, Javier Sunyer, and David Wake. A special thanks goes to all the students who have joined me on a variety of research trips, and to Wendy Welshans and Laurie Doss, who, through their efforts, continue to introduce young people to the marvels of tropical rainforest ecology. Greg Watkins-Colwell and the rest of the staff at the Division of Vertebrate Zoology of Yale University's Peabody Museum of Natural History helped support aspects of the research for this book and facilitated access to specimens.

Not all of the images and materials provided by the following people made it into the final design of this book, but their kind contributions were nevertheless extremely valuable, and I acknowledge them here: César Barrio-Amorós, Eduardo Boza Oviedo, David Cannatella, Luis Coloma, William Duellman, Cesar Jaramillo, Karl-Heinz Jungfer, Karen Lips, Stefan Lotzkat, Piotr Naskrecki, Kenji Nishida, Tobias Palmberg, Todd Pierson, Marcos Ponce, Robert Puschendorf, Ignacio de la Riva, Sean Rovito, Paddy Ryan, Rob Schell, Jen Stabile, Steven Whitfield, and Brad Wilson. (Detailed photo credits are provided on page 525.)

Ultimately, it was John McCuen, of Zona Tropical Press, who made sure that this book became a reality—which is truly an accomplishment!

Thank you all!
Twan Leenders

Introduction

This book is first and foremost a field guide, meant to help readers identify the country's frogs and toads, salamanders, and caecilians—and to give them information about the natural history of these fascinating animals. It is also an introduction to the amazing diversity of the herpetofauna of Costa Rica, home to 147 species of frog and toad, 53 species of lungless salamander, and 7 species of caecilian, ancient bizarre creatures.

Each species account provides a brief description of the animal; information on distribution (within Costa Rica and other countries), including a range map and an ecoregion bar that indicates the habitats in which the species occurs; natural history information, including a description of the conservation status of the species; key characteristics, with accompanying photographs and illustrations of hands, feet, and, in some instances, head; and a description of similar species.

Geography, Climate, and Weather

Costa Rica is a small country of roughly 19,730 square miles (51,100 km²). Its Caribbean and Pacific coasts are separated by four mountain chains (called *cordilleras* in Spanish) that run roughly in a northwest to southeast direction: the Guanacaste, Tilarán, Central, and Talamanca Mountain Ranges. These mountains include one of the highest peaks in Central America, Cerro Chirripó (12,530 ft/3,820 m), and several active volcanoes. The Central Valley, where the majority of Costa Ricans live, is a large highland valley enclosed on all sides by the mountains of the Central Mountain Range and the northern part of the Talamanca Mountain Range. The Caribbean side of the country consists mainly of a large, flat coastal plain that is widest near the Nicaraguan border and narrows toward the southern border with Panama. The Pacific side of the country can be divided into two climate regions: the dry northern region, which encompasses the lowland plains of Guanacaste and the Nicoya Peninsula, and the wet Osa Peninsula and adjacent lowlands in the south.

Costa Rica lies within the tropics and is about halfway between the Tropic of Cancer and the Equator. At any given place in the country, the average temperature in the hottest month of the year does not exceed the average temperature in the coolest month by more than 9 °F (5 °C). This small variance is strikingly different from the large changes in temperature that can occur locally on a daily basis. In some areas it is invariably hot during the day, but can turn cool at night. With temperatures relatively constant, seasons in the tropics are not defined by temperature differences but rather by variation in rainfall.

All of Costa Rica experiences a dry season and a wet season each year, but the duration and timing of each vary significantly depending on the region. Generally, the dry season begins toward the end of November or early December and lasts until April or May, when the rainy season begins. However, there is a marked difference in the amount of rainfall received by the Pacific and Caribbean slopes. The Pacific slope has an average annual precipitation of around 98 in (2,500 mm), while the Caribbean foothills may receive more than twice as much. Note that even in the dry season it is not uncommon to see daily precipitation in many parts of the country, particularly on the Caribbean slope. Another well-known phenomenon is the occurrence of a *veranillo*, or little summer, a short dry spell during the rainy season that lasts a few weeks and usually takes place in July or August. The regional variation in rainfall patterns explains why the timing of the dry and wet season indicated in different species accounts varies depending on where the species occurs.

The northwestern part of Costa Rica, north of the Tarcoles River and including the Nicoya Peninsula, is very hot, and substantially drier than the rest of the country. Annual precipitation there ranges from 51 to 91 in (1,300 to 2,300 mm). The forest in this part of the country has a low canopy, with trees usually less than 49 ft (15 m) tall. The trees generally lose their leaves during the dry season, which is particularly harsh and may last up to eight months, starting around October. Many of the amphibian species in this region retreat into burrows or tree holes and are neither seen nor heard during this time.

The highest density and diversity of amphibians is found in wet lowland areas, especially in the Caribbean region and in the area on, and surrounding, the Osa Peninsula on the southern Pacific coast. In most of these evergreen forests, it rains almost year-round, and some areas may receive more than 236 in (6,000 mm) of rain annually. In most of the remaining lowland areas, the rainfall is much less extreme, with 98 to 157 in (2,500 to 4,000 mm) of rain per year.

At elevations above roughly 4,900 ft (1,500 m), a cooler climate prevails, with temperatures from 50 to 61 °F (10 to 16 °C). Near the summit of some of the higher peaks, at elevations above 11,500 ft (3,500 m), there is occasional frost. Because most highland regions are frequently covered in clouds, the resulting ground-level mist creates the very humid conditions that are preferred by many amphibians. Whereas the lowland regions are home to the greatest number of species and individuals, the highlands, above roughly 4,900 ft (1,500 m), are home to more elusive, and often endemic, species. Many of Costa Rica's endemic species are restricted to one or a few mountains, isolated from neighboring peaks by intolerably hot and/or dry lowlands.

Seasonal changes in rainfall can significantly affect both the size of amphibian populations and species diversity. During year-round sampling of the leaf-litter herpetofauna at La Selva Biological Station, in the Caribbean lowlands near Puerto Viejo de Sarapiquí, the highest density and diversity of amphibians was recorded in February, and the lowest in August. These findings indicate that the greatest diversity and abundance of herpetofauna occurs between the end of the dry season and the onset of the rainy season. Generally an increase in leaf litter depth, as a result of trees dropping their leaves during dry periods, tends to increase habitat and prey availability for amphibians that inhabit leaf litter. Conversely, a recent study indicated that during an unusually wet climatic event (a recent occurrence of the La Niña Southern Oscillation, which results in extraordinarily high precipitation levels), leaf litter amphibian populations were significantly reduced, but rebounded to normal levels within 12 months.

Ecoregions

Within Costa Rica, there are several ecoregions, each defined by specific environmental conditions and the assemblage of species they contain. Eight such ecoregions are formally recognized in Costa Rica, but three (two types of coastal mangrove forests and the Cocos Island Moist Forest ecoregion) harbor no amphibians and are therefore not included in this book. The remaining five, however, provide important habitat and ecological resources for the 207 species described here. The geographic distribution range of many amphibians in Costa Rica is defined by the same

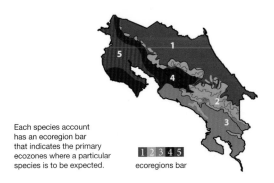

Each species account has an ecoregion bar that indicates the primary ecozones where a particular species is to be expected.

ecoregions bar

environmental parameters that help define these ecoregions. Therefore, each distribution map in this book is accompanied by an ecoregion bar that indicates the numerical code of the ecoregion/s where a given species is most likely encountered.

1. Isthmian-Atlantic Moist Forests. This ecoregion spans the Atlantic lowlands and foothills of Nicaragua, Costa Rica, and Panama. Most amphibians that are endemic to this assemblage occur below 1,640 ft (500 m). This ecoregion represents the archetypical wet, evergreen tropical forest most people conjure up when imagining what a rainforest looks like. It is a hot and humid environment with tall trees, a closed canopy, and dense understory vegetation. Much of this ecoregion has been severely affected by logging, agriculture, and fragmentation; remaining large blocks of pristine habitat now only exist in protected areas such as Barro del Colorado Wildlife Refuge, Tortuguero National Park, Gandoca Manzanillo Wildlife Refuge, and the lower reaches of Braullio Carrillo National Park.

2. Talamancan Montane Forests. This ecoregion is confined to the highlands of Costa Rica and western Panama, and is characterized by relatively cool temperatures and high levels of rainfall. Locally, habitats and species unique to this ecoregion can be found as low as 2,460 ft (750 m), but generally this assemblage occupies higher elevations, ranging from roughly 4,925 ft (1,500 m) to the summits of the highest peaks in Costa Rica. It includes habitats such as cloud forest, pine-oak forest, and, above the tree line, high elevation páramo. Many

sections of this ecoregion are protected in national parks and other preserves in the Central Mountain Range (e.g., Poás National Park and Irazú National Park) and the Talamanca Mountain Range, which includes Chirripó and Tapantí National Parks and La Amistad International Park, a World Heritage Site in southern Costa Rica and western Panama.

3. Isthmian-Pacific Moist Forests. Occupies the Pacific lowlands and foothills of southern Costa Rica and Panama, primarily below 1,640 ft (500 m) but extending into the foothills, where some overlap with ecoregions 2 and 4 may occur. This ecoregion is characterized by tropical evergreen forest habitats, warm average temperatures, and relatively high amounts of rainfall. It is similar to ecoregion 1 in the overall appearance of its prevailing habitat, although Pacific slope evergreen forests often have a higher diversity of palms. The geological history of this region, and its relative isolation be-

cause of the dry ecoregions bordering to the north and high mountains to the east, have led to a high level of endemism. Logging and development have caused significant loss of the original forest cover in this ecoregion, and habitat remnants of any size are currently restricted to protected areas such as Corcovado, Manuel Antonio, and Piedras Blancas National Parks.

4. Costa Rican Seasonal Moist Forests. This ecoregion is located on the Pacific slope of Costa Rica and Nicaragua, and characterized by a distinct seasonality. Its dry season is less intense and shorter than that in the nearby dry forests (ecoregion 5), and the cooling influence from the Central and Tilarán Mountain Ranges, which form the ecoregion's eastern border, provide seasonal increases in moisture levels. The forest

types in this region are primarily deciduous; typically, many trees drop their leaves during the dry season. This ecoregion includes much of the Central Valley, foothills of the Pacific northwest, as well as the higher elevation forests on the southern end of the Nicoya Peninsula. It includes Rincón de la Vieja National Park, Cabo Blanco National Park, and the Guanacaste Conservation Area.

5. Central American Dry Forests. Extends along the Pacific coast from southern Mexico to northwestern Costa Rica. This ecoregion is highly fragmented and few relatively undisturbed areas of dry forest remain; it is one of the most endangered ecosystems in Latin America. Dry forests typically experience a prolonged dry season that may last from five to eight months. The habitat is often characterized by relatively low trees, thorny scrub, and plants like cacti and agave that are adapted to arid conditions. Many of the species that inhabit this area are unique to it. Santa Rosa National Park is a notable example of this ecoregion.

Classification and Scientific Names

At present, more than 7,500 species of amphibian are known to walk, hop, and crawl on our planet. Taxonomy and phylogeny, the branches of biology concerned with the naming and classification of living things, offer the framework humans use to categorize and discuss individual species, as well as the evolutionary groupings into which each species seems to best fit. Historically, classification has been based primarily on morphological traits and the shared characteristics that groups of species inherited from a common ancestor. However, in the past decades analysis of molecular-level characteristics such as DNA sequences have been providing a parallel system of classification that can be applied in conjunction with comparative anatomy to better categorize the known amphibians. Broad application of such techniques on some of the more complex groups of amphibians has lately spurred several major reclassifications of amphibian families, genera, and species. Since some of those reclassifications are primarily based on molecular traits rather than morphological ones, this poses challenges for field identification (and field guides) because there may not be any clearly distinguishable, morphological features available to easily assign certain species to their respective genus or family.

The scientific names, genus, and family allocations used to describe the amphibians in this book differ considerably from previously used classifications in older literature. As more information becomes available and our understanding increases, the currently accepted classification of Costa Rica's amphibian fauna will likely see additional changes in years to come. Phylogenetic analysis is only as good as the data that drives it, and more is better. Given that taxonomic decisions are inherently based on imperfect information, a good deal of interpretation is involved, and it is possible for two experts who agree on classification criteria to disagree on specific phylogenetic placement of organisms, or groups of organisms.

Each formally described species is given a unique, binomial scientific name that consists of a genus and species label, a system that traces back to the nested hierarchical classification system implemented by Linnaeus in the mid-1700s. A genus forms the hierarchical level above a species and can contain a single, unique species that is sufficiently different from other forms that it warrants its own taxonomic group, or (more commonly) forms a taxonomic unit that contains multiple closely related species, which (as a group) share characteristics that set them apart from other species. Multiple genera usually form a family, multiple families are contained within an order, and the three orders contained in this book (caecilians, salamanders, and frogs) together comprise the class amphibia.

The unique label that is a scientific name is indispensable for communication in scientific circles but admittedly can be awkward for the nonspecialist. Even though scientific names for particular species may change as a result of reclassification, these nomenclatural changes are carefully documented and remain associated with the species. Common names, on the other hand, are not formally assigned to a particular species—at least not in the case of tropical amphibians. Several well-known Costa Rican amphibian species have commonly used vernacular names associated with them, but most do not. For this book, an attempt was made to provide a common name for each species in the hope that they will eventually become established.

Observing and Identifying Amphibians

Most amphibians are furtive, many are well camouflaged, and the majority active only at night. Observing these animals obviously requires some familiarity with their biology in order to know where and when to look—and you also need a good deal of patience. With a little preparation, finding amphibians is not terribly hard, however.

Finding amphibians is one thing, but identifying them is an entirely different matter. Some species require close inspection since their distinguishing features are very subtle. Success in this endeavor may require a hand lens, keen eyes, and a bit of practice. Following are pointers that will improve chances for success.

Amphibians generally shy away from direct exposure to sunlight. Their thin, permeable skin makes them vulnerable to rapid water loss in a warm and dry climate, resulting in dehydration and ultimately death. Most Costa Rican amphibians are active at night, when cool and humid climate conditions are much more favorable. Those amphibian species active during the day are mostly found in water or in moist, heavily shaded habitats.

Poison dart frogs like this strawberry poison dart frog (*Oophaga pumilio*) are often seen on daytime hikes.

Areas with "disturbed" plant life—forests in which trees have been felled, pastures, and plantations, for example—often still harbor amphibians. In fact, herpetofauna sometimes occurs in higher densities in disturbed areas than in primary rainforests. However, the number of species living in an undisturbed forest, with its very large trees and understory of palms, lianas, epiphytes, and other plant life, is invariably higher than in any secondary forest or cultivated area. Population patterns of Costa Rican amphibians are typical of the New World tropics in that there are only a few abundant species and a lot of uncommon ones. It is the abundant species, which tend to be more adaptable to different environmental conditions, that are found in large numbers in disturbed areas; many of the species that inhabit the forest interior do not survive extensive habitat alteration.

There are certain habitats in Costa Rica where, with some careful observation, you are likely to find common amphibians. During the day, while hiking in a lowland rainforest, you may be able to see a few species of rain frog (genus *Craugastor*), poison dart frogs, as well as a toad or ranid frog. But nighttime is usually the better time to look. On night walks, even the smallest creature can be spotted with the aid of a powerful flashlight. While walking slowly along a rainforest trail, carefully scan the forest floor, tree trunks, and foliage. You may encounter night-active frogs of the families Craugastoridae, Eleutherodactylidae, and Strabomantidae on foliage along forest trails; glass frogs on (or underneath) leaves that overhang streams; and tree frogs in vegetation near streams and ponds. Toads and leptodactylid frogs may be seen on the forest floor and in streambeds, where it is also possible to find true frogs and rain frogs. Lowland marshes and ponds prove best for observing several types of tree frog, and on nights when their breeding choruses are in full swing, the sounds produced by hundreds of frogs can be heard from far away (and be quite deafening up close!).

A flashlight, a pair of rubber boots, and a raincoat are the essential tools for observing amphibians. You should be able to find quite a few species just about anywhere with remaining natural habitat. Please note that small amphibians are easily hurt and may overheat if held in your hand for too long. In general it is better to avoid handling these animals and to enjoy them where you find them. However, if you do handle one for closer examination or identification, make sure your hands are free of sunscreen and insect repellent; they contain chemicals that can harm or kill amphibians. And, amphibian skin contains all sorts of chemicals that may affect an observer; avoid touching your mouth, nose, or eyes, and wash your hands carefully with soap and water after handling any amphibian.

Declining Amphibian Populations

Across the globe, loss of natural habitat due to logging, agriculture, mining, and construction is the biggest cause of amphibian declines. The same is true for Costa Rica. In addition, climate change, water and air pollution, use of agrochemicals, erosion and road runoff, and the spread of non-native invasive plants and animals, all symptoms of an expanding human population, take a significant toll.

In the past few decades, amphibian populations have also been affected by disease. The fungal pathogen *Batrachochytrium dendrobatidis*, generally referred to as *Bd* or chytrid fungus (because it is a member of a huge family of otherwise benign soil fungi, the chytrid fungi), has been implicated in amphibian declines and extinctions around the globe. This fungus inhabits relatively cool, moist habitats and produces mobile zoospores that are aquatic. It infects adult amphibians and inhabits their skin. Infected

Feared extinct, the variable harlequin toad (*Atelopus varius*) has recently reappeared in a few areas.

amphibians can develop a disease, called chytridiomycosis, which induces the sloughing of layers of epidermis and causes sections of the skin to thicken. The pathogen particularly affects areas of the skin that frogs and toads use to absorb water from their environment and can fairly quickly lead to death of the host because the fluid and gas exchange in these animals is disrupted. When amphibians are infected during their larval phase, the chytrid fungus has the exact opposite effect: rather than hardening a tadpole's skin, it instead causes loss of the hard, keratinized mouthparts and hinders the ability of tadpoles to feed.

Not all species of amphibian are equally susceptible to this pathogen. While some species die quickly upon its arrival, others survive, though in reduced numbers. Since the fungus only survives on amphibian skin, it is critical for its survival to have some hosts persist in a given area. As a result, arrival of *Bd* generally causes massive die-offs among local amphibians, with losses of up to 60% of the local species in a relatively short period, while the few infected survivors then aid in keeping the pathogen present in the habitat, thus reducing the chances of recolonization.

Bd and the disease it causes arrived in Costa Rica in the late 1980s. The first indication that something was amiss came when the famous golden toad (*Incilius periglenes*), endemic to the Monteverde area and one of few Costa Rican amphibians being monitored at the time, suddenly disappeared in 1989. Over the course of several years, into the first half of the 1990s, additional species disappeared from other parts of the country, as far south as the Panamanian border, and the fungal culprit was eventually identified in 1998. Although several other theories explaining the possible cause of these amphibian declines were investigated, irrefutable evidence came when researchers established a baseline of local amphibian diversity and density in the central Panamanian preserve of El Copé, west of where the *Bd* front was thought to be, and waited for its arrival. When it did, in 2004, dead and dying frogs were observed, and within five months after the chytrid fungus arrived in El Copé *Bd* had extirpated 50% of the local amphibian species, and 80% of individuals.

In Costa Rica, many species were affected, but the most severely impacted groups include all highland toads (*Incilius fastidiosus*, *I. holdridgei*, and

I. periglenes), all *Atelopus* species, all middle and high elevation *Craugastor* species of the stream-dwelling *rugulosus*-species group, several middle elevation tree frogs, as well as a highland ranid, *Rana vibicaria*. All of these species had disappeared completely by the early 1990s, whereas other frogs and toads persisted in greatly reduced numbers. Due to the difficulty of finding any salamanders or caecilians, the impact on these groups is poorly understood, but many have not been observed for decades. In reality, so little was known about healthy population sizes of local anura that it was difficult to fully comprehend the impact chytridiomycosis had on even common Costa Rican frogs and toads.

Some of the species that had disappeared upon the arrival of *Bd* were once common or abundant. The variable harlequin toad (*Atelopus varius*), for example, was known to exist in more than 100 populations throughout the mountains of Costa Rica; and in Monteverde, their density reportedly could reach almost two frogs per meter of stream bank. It was last observed in Monteverde in 1988, and no more *A. varius* were found anywhere in Costa Rica after 1996. Surprisingly, in late 2003 a small isolated population of *A. varius* was discovered in the Pacific lowlands, where the chytrid fungus is present. Since then, a few more relict populations of *A. varius* have been found, and other species of frog and toad that were thought to be extinct have recently reappeared in very small numbers as well. Much more information and careful monitoring of these so-called Lazarus species is needed to better assess their population dynamics during these tenuous recoveries in the presence of *Bd*.

Note that the distribution map in each species account indicates the current distribution range in yellow, while a species' historic range is indicated in gray.

Citizen Science

The author encourages readers to share their observations with other naturalists, researchers, and conservationists. Even though many dedicated people spend a lot of time and effort studying these animals, the reality remains that very little is known about many of these animals. In recent decades, many Costa Rican amphibian populations have seen very dramatic declines, and some species are possibly extinct, while others are seemingly making a comeback in areas from which they were long absent. There simply are not enough resources to send observers into the field to find all missing amphibian populations, to monitor increasing or declining populations, to develop conservation strategies, and to gather the natural history information that is needed to protect even the most common of species.

At the same time, people are undoubtedly making valuable observations of amphibians that researchers are desperately trying to study or locate; but for a variety of reasons—including the inability to correctly identify such species or the fact that casual observers may not be aware of the critical population status of a given species—this information often does not reach interested parties. This book will hopefully help increase communication between observers, researchers, and conservationists by providing citizen scientists with the information they need to identify amphibian species in Costa Rica and to learn their conservation status. In the age of the cell phone, virtually everyone carries at all times a GPS-enabled communication device with a camera.

By documenting potentially valuable sightings with a quick cell phone picture and sharing that with local researchers, reserve managers, park rangers, or with the author (anfibiosCR@rtpi.org), your observations can contribute to a greater understanding of the amazing amphibians of Costa Rica.

Caecilians
(Order Gymnophiona)

Approximately 205 species of caecilian comprise the order Gymnophiona. Caecilians are limbless, elongate amphibians that occur throughout tropical regions of the world. Their taxonomy has changed drastically in recent years, and those changes have affected our understanding of species in Costa Rica. The seven species known to occur in the country are now placed in two families and three genera; and what was once thought to be a single widespread species, *Dermophis mexicanus*, has been split up into several endemic forms only found in Costa Rica.

Caecilians look like giant earthworms—indeed both animals appear to have segmented bodies and generally lack externally visible features. Caecilians have a clearly visible mouth, but no external ear openings, and their eyes are vestigial. In all species, the eye sockets are covered with skin and in some the eye sockets are covered over with both bone and skin. The latter condition occurs in Costa Rican *Gymnopis* and *Oscaecilia,* and these creatures appear eyeless, while the remaining caecilians in the country (in the genus *Dermophis*) have eyes that are discernible as a dark spot on each side of the head.

Caecilians are perfectly adapted to the particulars of their unique life-style, although the morphological features that facilitate that lifestyle are not easy to discern. Since they spend much of their life underground, external ear openings and eyes with moveable eyelids would only create problems for these animals in an environment where such features are of very limited use. Instead, caecilians have developed unique features and biological traits that are finely attuned to the specifics of their preferred habitat, feeding strategies, and reproductive needs. Since these animals are difficult to study systematically—they are difficult to find reliably in large enough numbers to gather sufficient data—much remains to be learned about them. The limited information that is available is summarized here and should provide a glimpse into their highly specialized nature.

Caecilians, like salamanders and frogs, are amphibians. These three groups share common traits, including a vertebrate skeleton; caecilians have a fully articulated vertebral column that is associated with ribs and a highly adapted skull. Fossil remains indicate that some species once had limbs. All three groups also have a permeable and highly glandular skin that allows for the exchange of gases and moisture.

Nonetheless, caecilians possess several biological traits that are unknown (or highly unusual) among other amphibians. For example, fertilization takes place inside the female's body; the male inseminates the female by inserting a penislike structure (phallodeum) into the female's cloaca. Generally, copulation is very uncommon among amphibians; in most species, fertilization takes place outside the female's body. Also highly unusual is the fact that most caecilians bear live young (as opposed to laying eggs). Although some caecilian species reproduce by laying eggs, as far as we know, all Costa Rican species are live-bearing. The larvae develop within the mother's body and feed on a highly nutritious substance that is secreted from the lining of the female's oviducts. The unborn young have special fetal teeth that are used to scrape food from the oviduct walls; this scraping, in turn, stimulates the continued production of nutritious secretions. Shortly after birth, the fetal teeth are replaced by regular adult teeth suitable for eating invertebrate prey. Some species of egg-laying caecilians feed their recently hatched young with rapidly generating layers of skin that the young consume, using specialized teeth. This behavior has not been observed in any Costa Rican species, however.

A caecilian creates tunnels by bracing its body against the earth and pushing its bullet-shaped head forward into soft soil. The head is covered with smooth skin that contains numerous tiny glands. These glands produce a slimy coating that helps reduce friction as the animal burrows (and makes it exceedingly difficult to restrain these animals when handled). Caecilians typically have a recessed mouth, tiny nostrils, eyes covered by skin and/or bone, and no external ear openings, all adaptations to a fossorial life. In spite of the seeming absence of an auditory organ, caecilians are still capable of detecting low-frequency sounds; sound waves are transmitted through the lower jaw structure to remnants of an inner ear structure. Likewise, in spite of the seeming absence of eyes, most species seem to retain the retina and a vestigial lens; even with the protective covering over the eyes, caecilians may still be capable of detecting differences in light intensity.

The most prominent feature on a caecilian head is a pair of short, retractable tentacles, one on each side of the head. These tentacles, which protrude slightly beyond the streamlined head and can be moved with specialized muscles, serve to detect chemical cues that aid in locating prey, mates, and possibly suitable habitat features.

The caecilian diet includes a variety of ground-dwelling or burrowing invertebrates such as worms, termites, and insect larvae. Prey is not restricted to invertebrates, however, and records exist of caecilians eating eggs, small- to medium-sized lizards, and other small vertebrates. In turn, caecilians fall prey to a variety of predators, including birds and snakes. Coral snakes, in particular, are known to eat the occasional caecilian.

Caecilians spend most of their time hunting for prey underground or hiding beneath logs, rocks, or leaf litter. Chance encounters with these animals generally occur when flipping over objects that they happen to be hiding under. Occasionally, during or after heavy rains, caecilians are found on the surface of the forest floor. This may indicate that these animals sometimes forage on the surface, but perhaps they go to the surface simply to escape their flooded burrows. Recent observations of two Costa Rican species indicate that caecilians may also become more active on the surface during periods of extreme drought, possibly because the drying, and thus hardening, of the top soil makes it more difficult for these animals to burrow, causing them to move and forage on the surface more than they would normally do.

All caecilians have annular skin folds (annuli) that encircle the body. The so-called primary annuli appear at more or less regular intervals. Their number is related to the skeletal structure of the animal: there is one primary fold for each vertebra. Since the number of vertebrae is variable within each species, so is the number of primary annuli. Secondary folds are those skin folds that are located between two primary skin folds; they often do not completely encircle the body. The number of primary and secondary skin folds on individuals of a given species varies, and the upper and lower limits define a range that may aid in species identification. Note that several of these species are known only from a very small number of individuals and discovery of additional specimens may expand the known variation in the number of annular folds. Features that distinguish one genus from another are the location of the tentacle relative to the eye or nostril and the presence (or absence) of minute dermal scales in the annular skin folds.

Costa Rica has two families of caecilian: Caeciliidae and Dermophiidae.

Costa Rican Caecilians
at a Glance

Costa Rica is home to two families and three genera of caecilian. This is a summary of key features that should help you identify these very similar looking and rarely seen amphibians.

Oscaecilia. Slender body and head. Eyes not visible (eye sockets covered with bone), tentacle located near nostril. Has only primary annuli.

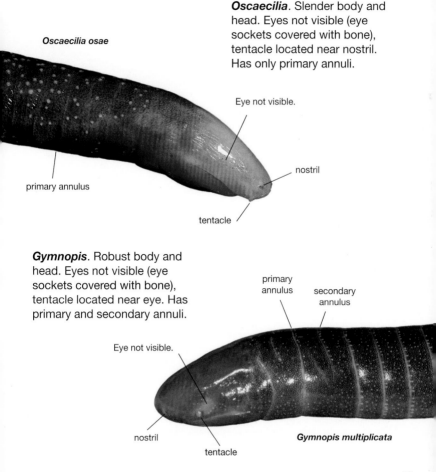

Oscaecilia osae

Eye not visible.

nostril

primary annulus

tentacle

Gymnopis. Robust body and head. Eyes not visible (eye sockets covered with bone), tentacle located near eye. Has primary and secondary annuli.

Eye not visible.

primary annulus

secondary annulus

nostril

tentacle

Gymnopis multiplicata

Dermophis. Eyes visible, tentacle located between eye and nostril. Has primary and secondary annuli.

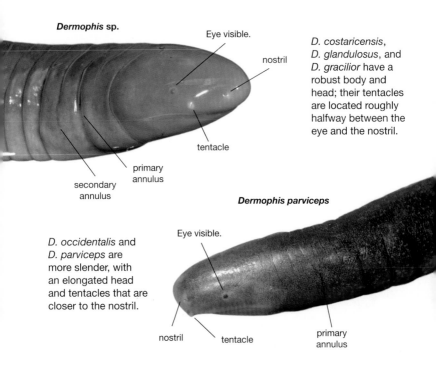

Dermophis sp.

Eye visible.

nostril

tentacle

primary annulus

secondary annulus

D. costaricensis, *D. glandulosus*, and *D. gracilior* have a robust body and head; their tentacles are located roughly halfway between the eye and the nostril.

Dermophis parviceps

D. occidentalis and *D. parviceps* are more slender, with an elongated head and tentacles that are closer to the nostril.

Eye visible.

nostril

tentacle

primary annulus

Table 1.1. Caecilian annular skin fold count

Species	Primary annuli	Secondary annuli	Total annular folds
Oscaecilia osae	232	0	232
Dermophis costaricense	107-117	74-96	186-208
Dermophis glandulosus	91-106	37-60	132-159
Dermophis gracilior	91-102	65-78	159-176
Dermophis occidentalis	95-112	29-37	126-149
Dermophis parviceps	85-102	11-26	97-126
Gymnopis multiplicata	112-133	84-107	201-250

Perhaps the surest way to identify a caecilian is with reference to the number of primary and secondary annuli and the total number of annular skin folds. Note that individual caecilians with the highest (or lowest) primary annuli count do not necessarily have the highest (or lowest) secondary annuli count, therefore the total annular fold count range does not conform to the added values of the highest and lowest primary and secondary annuli counts.

Family Caeciliidae

Until recently, all Costa Rican caecilians were included in this family, but evaluation of the world's gymnophiona has led to a redefinition of the various families, based primarily on morphological features that are beyond the scope of this field guide. Currently, the family Caeciliidae contains 42 species in 2 genera (*Caecilia* and *Oscaecilia*). Most species occur in South America, but some range into southern Central America. Only the genus *Oscaecilia* (barely) reaches Costa Rica.

Genus *Oscaecilia*

The caecilians of the genus *Oscaecilia* inhabit southern Central America and northern South America. Nine species are recognized in the genus, one of which reaches extreme southwestern Costa Rica. *Oscaecilia* are slender to very slender caecilians whose eyes are not visible externally because the eye sockets are covered with bone. A short tentacle is located directly below each nostril. Some *Oscaecilia* lack secondary annuli; all may have minute dermal scales in their annular folds, which is a highly unusual trait for amphibians.

Oscaecilia osae
Osa Caecilian   **Data deficient**

15.04 in (382 mm) in total length. A skinny, pinkish-gray caecilian without visible eyes and with a short tentacle below each nostril.

Ecoregions
1 2 **3** 4 5

Oscaecilia osae is endemic to southwestern Costa Rica and only known from low elevations on the Osa Peninsula and the lowlands surrounding the Golfo Dulce; near sea level to about 330 ft (100 m). It likely occurs in adjacent Panama but has not yet been detected there.

Natural history

This species was originally discovered in 1984, in southwestern Costa Rica, on the flooded grassy airstrip at the Sirena Ranger Station in Corcovado National Park. It was still known from only a single individual when it was formally described in 1992, but additional observations since then indicate that this species may not be rare and is certainly more widespread than first thought.

In 2001, at La Gamba Biological Station, an Allen's coral snake (*Micrurus alleni*) was observed eating an *O. osae*, representing the first record of this species outside the Osa Peninsula, on the Costa Rican mainland. Similar observations of coral snakes eating these caecilians have been made in several locations in the region in recent years, and a 2014 report of coatis consuming *O. osae* near Playa San Pedrillo on the north side of the Osa Peninsula adds another known predator and another confirmed locality. All sites where this species has been found seem to involve relatively open areas surrounded by

O. osae is the northernmost representative of this primarily South American genus; it is endemic to southwestern Costa Rica.

tropical lowland wet forest. One locality was near a stream, another in a dry, rocky streambed, while a recent sighting was made in sandy soil along a beach.

Due to its secretive nature and fossorial lifestyle, not much is known about *O. osae*. Like other caecilians, it is considered a nocturnal fossorial species that inhabits humid forest habitats. Deforestation and the subsequent drying of the soil likely render habitat unsuitable. *O. osae* is known from protected areas such as Corcovado National Park (on the Osa Peninsula) and the La Gamba-Las Esquinas Preserves near Golfito; other preserves in the area undoubtedly provide suitable habitat for this species.

Description

O. osae lacks secondary annuli; the total number of (primary) annular folds in the single studied specimen is 232.

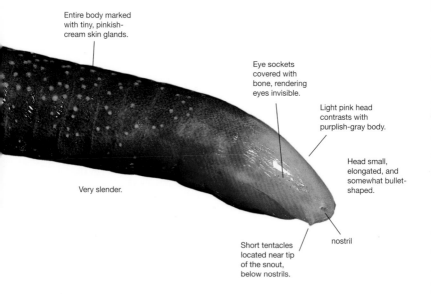

Entire body marked with tiny, pinkish-cream skin glands.

Eye sockets covered with bone, rendering eyes invisible.

Light pink head contrasts with purplish-gray body.

Head small, elongated, and somewhat bullet-shaped.

Very slender.

Short tentacles located near tip of the snout, below nostrils.

nostril

Similar species

Dermophis occidentalis (p. 25) likely co-occurs with this species but differs in having visible eyes and tentacles that are located about halfway between eye and nostril. *Gymnopis multiplicata* (p. 30) may co-occur with *O. osae*; it is a much more robust caecilian and its tentacles are associated with its eye sockets.

Family Dermophiidae

A recent revision of the world's caecilians has led
to the removal of 9 species of Central and South
American species from the family Caeciliidae
and their placement in Dermophiidae. This family
currently contains the 9 New World forms, as
well as an additional 5 species from tropical
Africa. Together, these 14 species represent the
only caecilians in the world that are live-bearing
(viviparous), have both primary and secondary
annuli, and have minute dermal scales in their
annular folds (not visible without magnification).
Two genera of this family occur in Costa Rica:
Dermophis and *Gymnopis*.

Genus *Dermophis*

This genus contains seven species, all native to Central America and adjacent regions of South America; five of these occur in Costa Rica. Caecilians in the genus *Dermophis* are medium-sized and relatively robust. The members of this genus share two diagnostic features: the eyes are not covered by bone (and are thus visible) and each tentacle is located between the eye and the nostril. In some species it may be situated about halfway, while in others it may be closer to the nostril.

Dermophis costaricense
Atlantic Montane Caecilian

Data deficient

Maximum known length 15.24 in (387 mm). A stocky caecilian with visible eyes and a tentacle located halfway between each eye and nostril; found at middle elevations on the Atlantic slope. This species has the highest annuli count of any Costa Rican *Dermophis*.

Ecoregions

Dermophis costaricense is endemic to Costa Rica. It occurs at middle elevations on the Atlantic slopes of the Tilarán, Central, and northern Talamanca Mountain Ranges; 3,300 to 4,300 ft (1,000 to 1,300 m).

Natural history

D. costaricense is fossorial and is usually found underground. It is only rarely seen when individuals hiding under logs, rocks, or surface debris are uncovered. Its hidden lifestyle makes it difficult to study and little is known of this species' biology, preferred habitat, or population status. *D. costaricense* inhabits relatively undisturbed premontane rainforest and occurs in protected areas such as Braulio Carrillo National Park and Tapantí National Park.

Description

The number of annular folds ranges from 186 to 208 (107 to 117 primary annuli and 74 to 96 secondary annuli), the highest number of annuli of any *Dermophis* species (only *Gymnopis multiplicata* has a higher count).

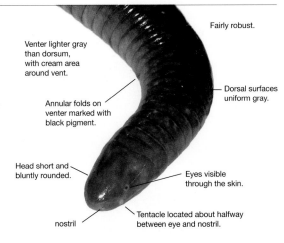

Venter lighter gray than dorsum, with cream area around vent.

Annular folds on venter marked with black pigment.

Fairly robust.

Dorsal surfaces uniform gray.

Head short and bluntly rounded.

nostril

Eyes visible through the skin.

Tentacle located about halfway between eye and nostril.

Similar species

D. parviceps (p. 27) occurs at lower elevations along the Atlantic slope, is much more slender, and has far fewer annuli. *Gymnopis multiplicata* (p. 30) is similar in appearance but its tentacles are associated with the eye socket. *D. glandulosus* (p. 22), *D. gracilior* (p. 24), and *D. occidentalis* (p. 25) all look similar but they only occur on the Pacific slope.

Dermophis glandulosus
Glandular Caecilian

Data deficient

Maximum known length is 15.95 in (405 mm). A robust, large caecilian with a relatively long head, visible eyes, and a tentacle located roughly halfway between each eye and nostril; identifiable only by number of annuli. Found at low and middle elevations on the Pacific slope.

Ecoregions
1 2 **3** 4 5

Dermophis glandulosus is a poorly known species, presumed to be wide-ranging. Records attributed to this species exist from northern Colombia and the Darién region in southwestern Panama, through eastern Panama to the southern and central Pacific slope of Costa Rica. Not enough information exists, however, to assess whether these records all represent a single species. In Costa Rica it is known primarily from low and moderate elevations on the central and southern Pacific slope; from near sea level to at least 3,900 ft (1,200 m).

Natural history

This mostly fossorial species is rarely encountered. It is presumed to feed underground on insects and other soft-bodied invertebrates. However, a recent finding of a lizard egg (*Norops* sp.) in the stomach of a *D. glandulosus* may indicate that these animals actively hunt immobile prey—and hunt on the surface as well. Caecilians are most often uncovered when turning over objects such as logs, rocks, or leaf litter piles. *D. glandulosus* inhabits humid lowland and premontane areas but has also been encountered in relatively disturbed areas such as plantations and clearings in forested habitat. Extensive deforestation and opening of the canopy will cause soil to dry and harden, making it unsuitable for caecilians. This species is known from protected areas such as Las Cruces Biological Station and the La Amistad International Park, where its habitat is secure. However, insufficient information is available to assess its conservation status.

Many caecilians show scars from old injuries, possibly indicating predation attempts or signs of physical damage sustained while burrowing through substrates with abrasive materials.

Description

Annular folds range from 132 to 159 (91 to 106 primary annuli and 37 to 60 secondary annuli).

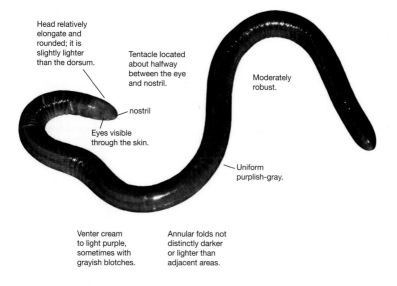

Head relatively elongate and rounded; it is slightly lighter than the dorsum.

Tentacle located about halfway between the eye and nostril.

Moderately robust.

nostril

Eyes visible through the skin.

Uniform purplish-gray.

Venter cream to light purple, sometimes with grayish blotches.

Annular folds not distinctly darker or lighter than adjacent areas.

Similar species

D. gracilior (p. 24) co-occurs with *D. glandulosus* at higher elevations; these two species can only be reliably distinguished based on annular counts. *D. occidentalis* (p. 25) likely co-occurs at lower elevations; it has a more elongate head that is lighter in color than the dorsum, but it may be difficult to distinguish based on coloration alone. *D. costaricense* (p. 21) is similar in appearance but inhabits Atlantic slope foothills; *D. costaricense* has a much higher number of annuli surrounding its body. *Gymnopis multiplicata* (p. 30) is similar in appearance but its tentacles are associated with the eye socket. *Oscaecilia osae* (p. 17) may co-occur with *D. glandulosus* in southwestern Costa Rica, but it is much more slender and its tentacles are located close to the nostrils.

Dermophis gracilior
Talamanca Caecilian

Data deficient

Maximum known length 13.58 in (345 mm). A dark gray, relatively robust caecilian with visible eyes and a tentacle located halfway between each eye and nostril; often has a distinctly light venter. Only found at moderate elevations in extreme southern Costa Rica.

Dermophis gracilior is known from scattered lowland and foothill locations on the Pacific slope of central and extreme southwestern Costa Rica, as well as adjacent western Panama; 1,300 to 6,550 ft (400 to 2,000 m).

Ecoregions
1 2 **3** 4 5

Natural history

This species inhabits humid premontane forests, where it has been found under logs and surface debris. Not enough records are available to adequately assess its taxonomic or conservation status. Fortunately, it occurs in protected areas such as Las Cruces Biological Station (Costa Rica) and in La Amistad International Park (Costa Rica and Panama).

Description

Annular folds range from 159 to 176 (91 to 102 primary annuli and 65 to 78 secondary annuli). The primary annuli count in this species broadly overlaps with that of *D. glandulosus*, but the secondary and total annuli count in *D. gracilior* is consistently higher than in the latter species.

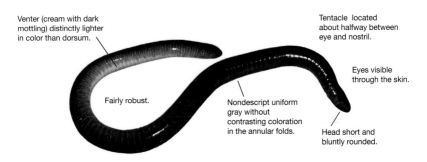

Venter (cream with dark mottling) distinctly lighter in color than dorsum.

Tentacle located about halfway between eye and nostril.

Eyes visible through the skin.

Fairly robust.

Nondescript uniform gray without contrasting coloration in the annular folds.

Head short and bluntly rounded.

Similar species

D. gracilior may overlap geographically with *D. glandulosus* (p. 22); these two species can only be reliably distinguished by annular counts. *D. occidentalis* inhabits lower elevation habitats and has a head that is more elongate and that is lighter in color than the dorsum. *D. costaricense* (p. 21) is similar in appearance but there is no range overlap. *D. costaricense* has a much higher number of annuli surrounding its body. *Gymnopis multiplicata* (p. 30) is similar in appearance but its tentacles are associated with the eye socket.

Dermophis occidentalis
Pacific Slender Caecilian

Data deficient

Maximum known total length of 9.25 in (235 mm). A small and fairly slender caecilian with a somewhat elongate head, visible eyes, a tentacle located halfway between each eye and nostril, and a relatively low number of annuli; occurs in Pacific lowlands only.

Ecoregions
1 2 **3** 4 5

Dermophis occidentalis is endemic to the central and southern Pacific lowlands and the Pacific slopes of the western Central Valley; from near sea level to 3,200 ft (970 m).

Natural history

Like other caecilians, *D. occidentalis* is a rarely seen subterranean species and not much is known about its biology or population status. It has been found under logs and in leaf litter in lowland and premontane rainforest. Parts of its range lie within protected areas such as Corcovado National Park.

D. occidentalis is one of several *Dermophis* species in the region. These forms are poorly defined and superficially quite similar, making it difficult to identify them.

Description

Annular folds range from 126 to 149 (95 to 112 primary annuli and 29 to 37 secondary annuli).

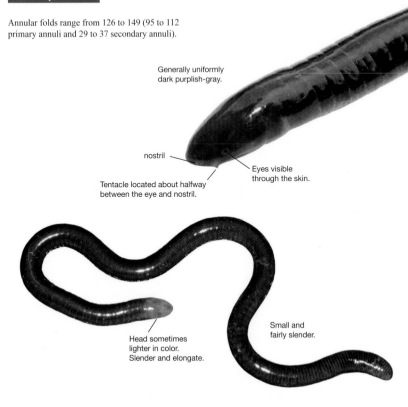

Generally uniformly dark purplish-gray.

nostril

Eyes visible through the skin.

Tentacle located about halfway between the eye and nostril.

Head sometimes lighter in color. Slender and elongate.

Small and fairly slender.

Similar species

This species is most similar to *D. parviceps* (p. 27) but has more annuli and does not co-occur. *D. gracilior* (p. 24) is more robust and occurs at higher elevations in extreme southern Costa Rica. *D. glandulosus* (p. 22) overlaps in range but is more robust. *Gymnopis multiplicata* (p. 30) is similar in appearance but its tentacles are associated with the eye socket. *Oscaecilia osae* (p. 17) co-occurs with *D. occidentalis* in southwestern Costa Rica but differs in having its tentacles located close to the nostrils.

Dermophis parviceps
Atlantic Slender Caecilian

Least concern

Maximum known total length 16.73 in (425 mm). A very slender caecilian with visible eyes and a tentacle located halfway between each eye and nostril. It has a uniform purplish-gray body and a distinctly lighter, pinkish head. Occurs in Atlantic lowlands and foothills only.

Ecoregions
1 2 3 4 5

Dermophis parviceps is known from scattered lowland locations along the Atlantic coast of Costa Rica, up to 2,300 ft (700 m); it has also been found at an isolated premontane site in eastern Panama, at 4,000 ft (1,220 m).

Natural history

D. parviceps inhabits undisturbed lowland and premontane rainforests. It is a rarely encountered, secretive animal that spends much of its time underground or under logs, rocks, or leaf litter. Surface activity is restricted to nighttime forays, presumably to hunt for invertebrate prey, or during heavy rains, when flooding burrows may bring these animals to the surface. Underground cavity nests belonging to *D. parviceps* were discovered recently. Each baseball-sized cavity was lined with mud and located at a depth of 20 to 25 in (50 to 65 cm) below the surface of a secondary forest; each contained a single caecilian. It is unclear whether these cavities were used as nests for reproduction.

Description

Annular folds range from 97 to 126 (85 to 102 primary annuli and 11 to 26 secondary annuli). Secondary annuli are found on the posterior section of the body, near the vent; this species has fewer secondary annuli than any other Costa Rican caecilian (except *Oscaecilia osae*, which has none).

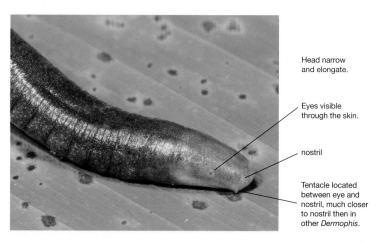

Head narrow and elongate.

Eyes visible through the skin.

nostril

Tentacle located between eye and nostril, much closer to nostril then in other *Dermophis*.

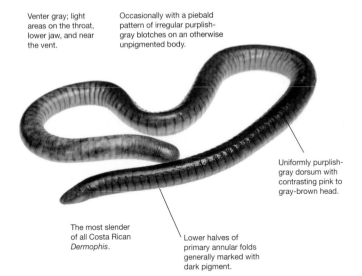

Venter gray; light areas on the throat, lower jaw, and near the vent.

Occasionally with a piebald pattern of irregular purplish-gray blotches on an otherwise unpigmented body.

Uniformly purplish-gray dorsum with contrasting pink to gray-brown head.

The most slender of all Costa Rican *Dermophis*.

Lower halves of primary annular folds generally marked with dark pigment.

Similar species

Gymnopis multiplicata (p. 30) co-occurs with *D. parviceps*, but it is much more robust and its tentacles are associated with the eye sockets. *D. occidentalis* (p. 25) is most similar morphologically to *D. parviceps* but is only found on the southern Pacific slope of Costa Rica. *D. costaricense* (p. 21) inhabits higher elevations on the Atlantic slope, is much more robust, and has a much higher annuli count.

Genus *Gymnopis*

The genus *Gymnopis* consists of two species, both native to Central America. One of those, *Gymnopis multiplicata*, is widespread and occurs throughout much of Costa Rica. These are large and robust caecilians whose eyes are covered with bone and not visible externally. Their chemosensory tentacle is located immediately in front of the eye and is indicated by a pale area on each side of the head. Small dermal scales are present in most annular folds, though the scales are only visible under a microscope.

Gymnopis multiplicata
Purple Caecilian   **Least concern**

Maximum total length to at least 27.55 in (700 mm). A large, robust purplish caecilian; the only species in Costa Rica whose eyes are not visible and whose tentacles are located adjacent to the eye sockets.

Ecoregions
1 2 3 4 5

Gymnopis multiplicata is widespread in Central America, ranging from extreme eastern Guatemala along the Atlantic slope of Honduras, Nicaragua, and Costa Rica to extreme western Panama; from sea level to 2,950 ft (900 m). It also occurs along the Pacific slope of Costa Rica, from sea level to 4,600 ft (1,400 m), and on several islands in the Bocas del Toro archipelago, Panama, as well as on Isla del Caño in southwestern Costa Rica.

Natural history

G. multiplicata is likely the most commonly observed caecilian in Costa Rica but is still not often seen by the casual observer. Like other caecilians, this species is nocturnal and lives a mostly subterranean life. Individuals are found when turning over rocks, logs, leaf piles, or other cover objects. During heavy nighttime rains, they may get driven out of flooded burrows and emerge above ground. Occasionally caecilians hunt on the surface for soft-bodied invertebrate prey such as earth worms and termites. Recent sightings of this species active on the surface on warm, dry nights during extreme drought conditions may indicate that the drying and subsequent hardening of the soil sometimes pushes these animals to the surface.

Several caecilians are known to engage in surprisingly sophisticated parental care behavior. On being discovered, this female G. *multiplicata* curled her body into a protective coil around her recently born young.

This is predominantly a lowland species, but it occasionally ranges into the foothills. It occupies a wide variety of habitats and appears relatively well-adapted to life in disturbed, secondary habitats such as gardens, pastures, and plantations, although it is also found in undisturbed dry, humid, and wet forest types.

G. multiplicata gives birth to live young, producing litters of 2 to 10 young; each measures from 4.92 to 5.16 in (125 to 131 mm). Gestation period is about 11 months and females reproduce every 2 years. These caecilians display some very involved and unique parental care. While inside the female's body, embryos initially receive nutrition from their yolk supply; once the yolk supply is depleted, the young use specialized fetal teeth to scrape the inside of the oviducal walls and stimulate the release of specialized nutritional substances. Most likely nutrition comes in the form of fat-rich cellular materials that grow rapidly in response to the scraping action of the young. During this phase of their development, the young possess gills, which are lost before birth. After giving birth, females have been observed guarding their offspring and forming their body into a protective, tight ball around the young.

Even though *G. multiplicata* has long been considered the largest of the Costa Rican caecilians, historical measurements indicated that it was just under 19.5 in (500 mm), not that much larger than the other large caecilians in the country. But a 2014 discovery of an individual in excess of 27.5 in (700 mm) in length and with a circumference of 4.1 in (105 mm) caused the maximum possible size for this species to be adjusted significantly upward.

Annular folds range from 201 to 250 (112 to 133 primary annuli and 84 to 107 secondary annuli); *G. multiplicata* has the highest number of annular folds of any Costa Rican caecilian.

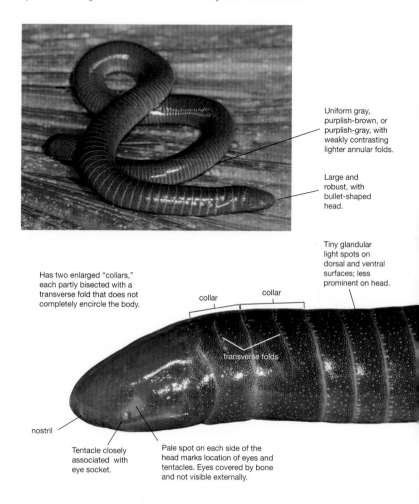

Uniform gray, purplish-brown, or purplish-gray, with weakly contrasting lighter annular folds.

Large and robust, with bullet-shaped head.

Tiny glandular light spots on dorsal and ventral surfaces; less prominent on head.

Has two enlarged "collars," each partly bisected with a transverse fold that does not completely encircle the body.

collar

collar

transverse folds

nostril

Tentacle closely associated with eye socket.

Pale spot on each side of the head marks location of eyes and tentacles. Eyes covered by bone and not visible externally.

Similar species

Several species of *Dermophis* occur within the range of *G. multiplicata*. All differ in having visible eyes and the tentacle located about halfway between each eye and nostril. *Oscaecilia osae* (p. 17) is very slender and its tentacles are located near the nostrils.

Salamanders
(Order Caudata)

Salamanders are also referred to as the tailed amphibians, for obvious reasons. Anura (frogs and toads) only have a tail during their larval (tadpole) stage, while caecilians tend to lack one altogether or only have a very short tail that is barely recognizable as such. The order Caudata currently has 676 species in 10 families and includes a very wide range of body shapes and sizes. The general form and structure of a salamander's body is very similar to that of the first prehistoric amphibians. The fact that their overall body plan has not changed significantly in more than 150 million years is testament to the success of this animal group.

Salamanders have distinctly separated body segments (head, body, and tail), unlike frogs, toads, and caecilians, whose head appears to be a mere extension of the body. Usually, front and hind limbs are present, projecting at right angles from the body; but in many species, the limbs may be reduced in size or absent altogether. Most salamanders have four digits on each front limb and five digits on each hind limb, though some species have fewer digits or digits that have been fused together.

Salamanders are often mistaken for lizards because their body shapes are superficially similar. However, salamanders, which are amphibians, have moist, glandular, permeable skin, and they lack the scaly skin, claws, and external ear openings seen in lizards.

In temperate climates, one can find salamander species that dwell on land (terrestrial), live in water (aquatic), or inhabit both environments (semi-aquatic). However, even those species that spend a lot of time on land usually begin life in a body of water, first as an egg, then as a free-swimming larva. In contrast, almost all salamanders that inhabit tropical regions are terrestrial and do not use ponds or streams for reproduction. All salamanders in Costa Rica have developed a completely terrestrial life and have even lost the free-swimming larval stage. The female deposits her eggs in a moist place (e.g., leaf litter, wet moss, or under a log or rock), but never in water. The larvae develop within the watery confines of the egg, undergo metamorphosis there, and hatch as miniature replicas of their parents.

Tropical humidity and rainfall provide adult salamanders with sufficient moisture to prevent dehydration. However, warm temperatures and intense solar radiation increase evaporation, which may put tropical amphibians at risk of excessive water loss. Salamanders are particularly sensitive to dehydration, which is why in the tropics many species occur at higher altitudes, where prevailing cool climates are more similar to those in temperature zones. In Costa Rica, it is possible to find salamanders near the summit of some of the country's highest peaks, at elevations exceeding those where frogs and toads occur. Another adaptation to prevent excessive water loss is a behavioral one: all tropical salamanders have adopted an entirely nocturnal activity pattern and hide in moist microclimates during the warmer parts of the day and the drier times of the year.

Salamanders do not copulate. Males have special glands in their cloaca that produce a gelatinous, cone-shaped structure called a spermatophore, which is capped with a small packet of sperm. During mating, a pair of sexually active salamanders initiates a courtship ritual in which the male places a spermatophore on the ground and then coerces the female to move over it. The female then picks up the packet of sperm with the lips of her cloaca and stores it there. When the environmental conditions are suitable, the female finds a place to lay her eggs; as the eggs pass through the female's cloaca, the stored sperm is released onto them and fertilization takes place. In some species of tropical salamander, sperm can remain within the female's body for more than a year prior to egg-laying. Courtship patterns, as well as the size and shape of the male's spermatophore, differ from species to species, thus preventing crossbreeding between species.

Costa Rican salamanders are fairly closely related, and all are placed in the family Plethodontidae, the lungless salamanders.

Costa Rican Salamanders at a Glance

Members of the three genera of Costa Rican plethodontids can be distinguished as follows:

Genus *Bolitoglossa*

Bolitoglossa lignicolor

Webfoot salamanders are robust, with well-developed limbs and broad (width ≥ length) hands and feet. They have a relatively short tail. There are always less than 15 costal grooves between the insertion of the front and hind limbs.

Genus *Nototriton*

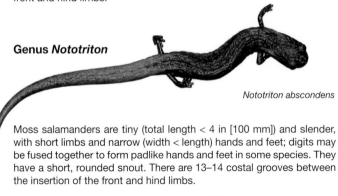

Nototriton abscondens

Moss salamanders are tiny (total length < 4 in [100 mm]) and slender, with short limbs and narrow (width < length) hands and feet; digits may be fused together to form padlike hands and feet in some species. They have a short, rounded snout. There are 13–14 costal grooves between the insertion of the front and hind limbs.

Genus *Oedipina*

Oedipina gracilis

Worm salamanders are exceedingly slender and very long-tailed, with tiny limbs and miniscule hands and feet (often padlike, with fused digits). There are 17–20 costal grooves between the insertion of the front and hind limbs.

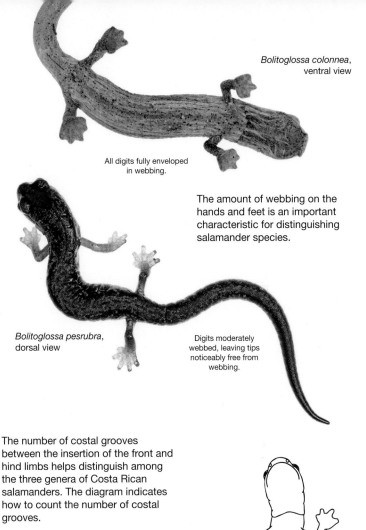

Bolitoglossa colonnea,
ventral view

All digits fully enveloped
in webbing.

The amount of webbing on the
hands and feet is an important
characteristic for distinguishing
salamander species.

Bolitoglossa pesrubra,
dorsal view

Digits moderately
webbed, leaving tips
noticeably free from
webbing.

The number of costal grooves
between the insertion of the front and
hind limbs helps distinguish among
the three genera of Costa Rican
salamanders. The diagram indicates
how to count the number of costal
grooves.

Female salamanders tend to have a
longer body than do males, which
results in having a greater distance
separating their adpressed limbs
(held flat against the body). The
standard way in which this distance
is expressed is in the number of
costal folds visible between the tips
of the digits on adpressed limbs, as
indicated in this diagram. Note that
costal grooves lie on either side of
the costal folds.

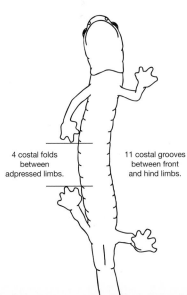

4 costal folds
between
adpressed limbs.

11 costal grooves
between front
and hind limbs.

Family Plethodontidae
(Lungless Salamanders)

Plethodontidae, the largest family of salamanders, currently contains nearly 450 species (about two-thirds of all the world's salamanders). This number is in part explained by the unique adaptations and abilities of lungless salamanders that have allowed them to colonize habitats that are inaccessible to other salamanders. As a result, they have an extensive geographic distribution; although the epicenter of plethodontid diversity is the Appalachian region of eastern North America, numerous species have radiated out from there across the continent and into Central and South America. Isolated pockets of lungless salamanders still persist in outlying regions as far away as northern Italy and the Mediterranean island of Sardinia, as well as South Korea.

Salamanders are primarily found in northern temperate zones and it is only members of the family Plethodontidae that range into the tropical latitudes of Central America and northern South America. Costa Rica harbors only three salamander genera: *Bolitoglossa*, *Nototriton*, and *Oedipina*. Each occupies a slightly different ecological niche. Within the constraints of their geographical, habitat, and biological preferences, each genus has diversified into a significant number of species, and at least 53 salamander species are currently known to exist in Costa Rica or just outside its borders. Several additional forms are still awaiting formal description and more research is needed to better define several of the known species.

In tropical and subtropical regions, lungless salamanders seem to display a preference for cool, humid habitats like those found in forested mountain ranges. As their common name indicates, plethodontid salamanders have no lungs. Gas exchange takes place through the skin, and is facilitated by continuous skin moisture. Humid tropical air not only prevents adult salamanders from dehydrating, but also protects their terrestrial egg clutches from drying out, which would eventually kill the embryos.

All Costa Rican salamanders presumably reproduce through direct development, which means that an embryo completes its larval development entirely within the confines of the egg. There is no free-swimming tadpole stage, and the transition from larva to salamander takes place before hatching. Compared to frogs and toads, which can sometimes lay egg clutches with thousands of eggs, female salamanders produce relatively small egg clutches, with usually between 10 and 30 eggs. The eggs have a large yolk supply that nourishes the developing embryo. The time between egg-laying and hatching can be very long by amphibian standards; an egg clutch of the species *Bolitoglossa compacta* was reported to hatch about 250 days after being laid, a period of over 8 months.

Some tropical salamanders display parental care. The female (sometimes the male) coils around the egg clutch during embryonic development. The parent protects the eggs against predators, and contact between parent and eggs helps regulate incubation temperature and humidity. At times, the parent will agitate the eggs to ensure they receive sufficient oxygen. Skin secretions from the parent also seem to reduce fungal and bacterial growth that may harm the eggs. Observations of abandoned egg clutches of *Bolitoglossa pesrubra* indicate that clutches not attended by a parent invariably die. However, a report on the successful reproduction in captivity of *Bolitoglossa mexicana* (a non-Costa Rican species) showed that parental attendance of an egg clutch is not critical in all *Bolitoglossa* species, since those eggs developed into perfectly healthy young without adult attention. Egg guarding behavior has been reported for several Costa Rican *Bolitoglossa*, but has not been observed in any *Nototriton* or *Oedipina*.

The life expectancy of salamanders, especially those living at high elevation locations with a cool climate, commonly exceeds 10 years; most frogs and toads do not live longer than 4 to 5 years. A Japanese species, the Japanese giant salamander (*Andrias japonicus*), lived 55 years in captivity. Little information exists about longevity in tropical salamanders, but *Bolitoglossa pesrubra* reaches an estimated maximum age of 18 years. In a population study of this species, males were found to reach sexual maturity after 6 years, whereas females reached sexual maturity after 12 years. Most species of lungless salamander, however, are thought to first reproduce 0.5 to 3.5 years after hatching. It is likely that highland species that have limited surface activity due to the prevailing cold conditions in their preferred habitat develop more slowly and live longer than their warm-climate lowland counterparts.

Salamanders lack an outer and middle ear, but do possess inner ear structures. Although they are thought to have a poorly developed sense of hearing, their capacity to hear remains largely unstudied. With very few exceptions, salamanders are voiceless. Most have acute vision and see especially well at night or in dark burrows.

Chemoreception, the perception of chemical cues in the environment, is probably the most important sense for salamanders. It helps them locate food, identify potential mates, and avoid predators. All

Nasolabial groove, connecting tip of nasolabial protuberance with nostril.

Adult ridge-headed webfoot salamander (*Bolitoglossa colonnea*) showing well-developed nasolabial protuberances. The associated nasolabial grooves carry chemical cues from the environment to sensitive chemosensory organs located in the nostrils.

plethodontids have nasolabial grooves, visible on each side of the snout as a furrow connecting the nostril and the upper lip. These organs appear to play an important role in the reception of chemical traces. Crawling salamanders can often be seen tapping their nose on the substrate. Experiments have shown that the waterborne chemicals present in the film of moisture that covers everything in the salamander's humid environment are carried through the nasolabial grooves by capillary action to a highly sensitive sensory organ. Most Costa Rican salamanders have prominent nasolabial protuberances, short, tentacle-like bumps below each nostril that extend downward from the front edge of the upper jaw and improve their ability to pick up chemical cues from their environment. In many species, these nasolabial protuberances are more prominently developed in males than in females. The exact reason for this phenomenon is not known but may reflect a stronger reliance on chemical cues in males than in females, and it possibly aids in finding a mate in the complex matrix that makes up a tropical habitat.

Several salamanders have noxious skin secretions that are bitter and can burn the eyes or the lining of the mouth of would-be predators. In some species of *Bolitoglossa* these secretions are sufficiently toxic to cause a predator such as a snake to regurgitate its prey. In severe cases, the secretions can impair coordination, or even kill the predator.

A specialized means of escaping predation that is widespread in the family Plethodontidae is pseudoautotomy, or tail breakage after the tail has been grasped. After the animal loses its tail, a replacement grows back completely, including vertebrae and all other tissues (unlike lizards, which regrow a rodlike, secondary tail that does not have vertebrae). Many salamander species have a slight constriction at the tail base immediately behind the cloaca where tail breakage occurs most often. A number of species have an unusually long or heavy tail, and they can use the weight contained in this section of their body as leverage to flip their body off vegetation or to change direction quickly when moving rapidly in an attempt to escape predators. Some species, in particular those that

Constriction in tail base where tail breakage takes place if salamander is restrained.

Bolitoglossa tica

live in trees, have a prehensile tail that acts as an additional anchor when climbing through dense vegetation.

Due to the ability of these animals to lose and regrow their tail, the most reliable measure of their size is the combined length of the head and body, referred to as the standard length (measured from the tip of the snout to the posterior edge of the vent). Salamander sizes are therefore indicated as such in this book. Where possible, an approximate total length (which includes a complete tail) is indicated, although this measure is sometimes less reliable. Bear in mind that encounters with tropical salamanders are quite rare and little quantitative data exists for many of the species described here. Therefore, most indicated sizes should be considered best possible estimates that are subject to change as additional individuals are encountered and measured.

Tropical salamanders are rarely seen because of their nocturnal and secretive lifestyle and seemingly low population densities. In addition, several formerly common species are now increasingly rare due to various environmental pressures. Because of the difficulty of observing these animals, little information exists on their natural history. Observations on breeding behavior, interactions between individuals, diet, and/or habitat use are noteworthy and should be documented and shared to further our understanding of these fascinating creatures.

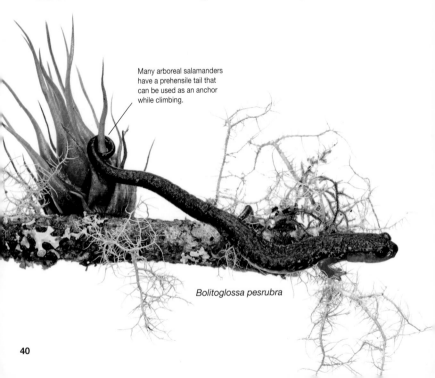

Many arboreal salamanders have a prehensile tail that can be used as an anchor while climbing.

Bolitoglossa pesrubra

Genus *Bolitoglossa*
Webfoot Salamanders

With 130 species currently recognized, *Bolitoglossa* constitutes the largest salamander genus in the world. Even though salamanders are primarily a temperate zone group, all *Bolitoglossa* are restricted to the neotropics.

Webfoot salamanders are small to large salamanders with a relatively stocky build, robust limbs, and broad hands and feet. Usually a substantial amount of webbing is present between the digits; in many species the tips of the digits are free from webbing, but in other species all fingers and toes may be completely enveloped in skin, making the hands and feet look like fleshy pads. The webbing provides additional traction when the salamanders climb, and the amount of webbing connecting the fingers and toes generally correlates with the preferred habitat of each species. Many of the *Bolitoglossa* with fully webbed hands and feet are arboreal and expertly navigate wet leaf surfaces and branches, often aided by a prehensile tail; more terrestrial species tend to have less interdigital webbing.

As far as is currently known, all *Bolitoglossa* reproduce through direct development and are not dependent on bodies of water. They lay their eggs in a moist microhabitat (under moss, in leaf litter, or in a leaf axil of a bromeliad, for example), and the embryos develop into tiny salamanders inside the egg, skipping a free-swimming larval stage. In some species parental care occurs, and a parent can be seen near an egg clutch.

Costa Rica is home to 27 species of *Bolitoglossa*; 2 more species are known from extreme western Panama, within a few miles of the border, and are included in this section. The preferred habitat of those 2 species, *B. anthracina* and *B. pygmaea*, extends into Costa Rica, and they very likely occur there as well.

Bolitoglossa alvaradoi
Alvarado's Webfoot Salamander   **Endangered**

Standard length to 3.11 in (79 mm); females generally larger than males. A robust salamander with fully webbed hands and feet, dark (black, dark brown, or dark olive) coloration, and a pattern of discrete black dots and irregularly shaped blotches of light tan, copper, or pinkish.

Ecoregions
1 2 3 4 5

Bolitoglossa alvaradoi is endemic to the Atlantic slope of Costa Rica, where it inhabits relatively undisturbed humid lowland and premontane forest habitats, from near sea level to at least 3,600 ft (1,100 m).

Natural history

This species displays the striking ability to change color, and at night its dark base color becomes significantly lighter, turning a light bluish-gray or olive-gray. The black spots and dark outlines of the costal folds are prominently visible at night but may be obscured during the day, when the salamander's coloration darkens.

B. alvaradoi is a rarely seen species that is invariably found on vegetation. It is thought to be a canopy dweller, and its extensively webbed hands and feet provide a good grip when climbing. In addition, this species uses its prehensile tail to grip leaves and branches as it moves across vegetation at night. Individuals have been extracted from bromeliads and from leaf axils of small palms during the day; they may retreat there during periods of inactivity. The perceived rarity of *B. alvaradoi* may be mostly due to its secretive habits, arboreal lifestyle, and the difficulty of locating it in densely vegetated forest habitat.

B. alvaradoi is found only in relatively undisturbed rainforest and does not appear to survive in degraded habitat. The main threat to this species is likely the loss of habitat through deforestation.

In April and May, courting pairs of *B. alvaradoi* have been observed on low vegetation at night. Presumably males and females locate each other using chemical cues that are detected by their sensitive chemosensory organs. This species has well-developed nasolabial protuberances that serve an important function in detecting chemical cues.

The few sightings of this species—and its variable coloration and pattern—have led to considerable confusion in defining *B. alvaradoi*. A recently described Nicaraguan species, *Bolitoglossa indio*, is now thought to occur in Costa Rica as well, within the currently known range of *B. alvaradoi*. These two species share several characteristics and may be closely related, or even constitute examples of a single, variable species. Alternatively, both may be valid species, but there are too few individuals of each species available to determine whether their geographic or altitudinal distribution ranges overlap. Additional research is needed to clarify this situation.

During the day, near black with visible light blotches.

At night, dark base color lightens, revealing a pattern of dark spots and dark-outlined costal grooves. Light blotches not as prominently visible.

Adpressed limbs are separated by 3 to 4 costal folds in females and 1.5 to 2 in males.

Irregularly shaped light blotches (yellowish or pinkish-tan to copper) on head. These may extend to form short, paired dorsolateral bands that appear past the insertion of the front limbs.

Dorsum black, olive-brown, or grayish-green, marked with discrete black spots.

Isolated light blotches usually present on the base of the tail, but can be more widespread.

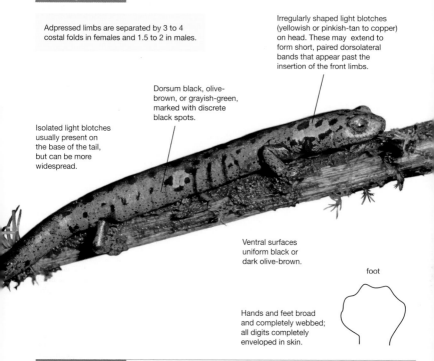

Ventral surfaces uniform black or dark olive-brown.

foot

Hands and feet broad and completely webbed; all digits completely enveloped in skin.

Similar species

B. indio (p. 60) is most similar to *B. alvaradoi*; both are poorly understood species that are difficult to distinguish in the field. As currently understood, *B. indio* is a smaller species, with brown dorsal coloration and a pair of irregularly shaped lighter (cream, tan) dorsolateral bands; it is known only from low elevations (below 230 ft [70 m]). *B. colonnea* (p. 50) has a prominent transverse fleshy ridge across the head, between the eyes. *B. striatula* (p. 89) has a unique color pattern of irregular dark brown, longitudinal dorsal and lateral stripes on a tan or light brown background color.

Bolitoglossa anthracina
Coal Black Webfoot Salamander

Data deficient

Standard length to 2.79 in (71 mm). One of several large black salamanders that occurs in the uplands of the Talamanca Mountain Range of Costa Rica and adjacent Panama. Though it may be difficult to identify in the field based on morphology or coloration, *Bolitoglossa anthracina* is the only large, entirely black salamander found at middle elevations on the Atlantic slope of western Panama; it may also occur in nearby Costa Rica.

Ecoregions
1 2 3 4 5

Bolitoglossa anthracina has been found at middle elevations on the north slope of Cerro Pando, Bocas del Toro Province, Panama, and in a few nearby areas. This species is not currently known from Costa Rica but should be expected in adjacent southern sections of the Talamanca Mountain Range; 3,600–4,750 ft (1,100–1,450 m).

Natural history

The description of this poorly known species is based on three individuals collected in the early 1960s and a fourth individual observed in 1976. It occurs in mostly inaccessible, undisturbed cloud forest habitats that have not received herpetological attention since that time. No additional *B. anthracina* have been reported in the last 40 years. Information on the life history of this species is limited to the observation that this is a nocturnal and highly arboreal species. Two of the known individuals were observed at night climbing on vegetation, 3–6 ft (1–2 m) above the ground. Another was found inside a bromeliad about 59 ft (18 m) above the forest floor. No natural history information is available for the remaining individual.

All known localities for this species are between 3,600 and 4,750 ft (1,100 and 1,450 m), but it may occur at still higher elevations. The type locality on Cerro Pando is mere miles from Costa Rica, and the premontane forest habitat that predominates there continues across the border.

Description

- Coloration in this species is uniform black except for the soles of its hands and feet, which are somewhat lighter gray.
- Head robust and distinctly wider than neck, with small eyes and small nostrils.
- Snout rounded; bears only weakly developed nasolabial protuberances.
- Large and moderately robust body.
- Limbs long and slender, with bluntly truncated, long fingers and toes.
- Tail slender and long, exceeds standard length.

One to 1.5 costal folds separate adpressed limbs.

foot

Hands and feet moderate in size; bear only basal webbing.

Similar species

B. anthracina likely co-occurs with *B. robusta* (p. 80); the latter species has a light colored ring around its tail base. The other large black salamanders in Costa Rica in the *Bolitoglossa nigrescens*-complex (*B. nigrescens* [p. 70], *B. obscura* [p. 72], and *B. sombra* [p. 84]) are very similar in morphology and coloration, but their respective distribution ranges do not overlap with *B. anthracina.*

Bolitoglossa aureogularis
Yellow-throated Webfoot Salamander

Not assessed

Standard length 1.93 in (49 mm) in the only known adult, a female. A slender salamander with partially webbed hands and feet and a uniquely yellow chin, throat, and chest.

Ecoregions

Bolitoglossa aureogularis is endemic to Costa Rica. It is known only from a relatively small area, the headwaters of the Río Coén, close to the continental divide and to Cerro Arbolado, on the Atlantic slope, Limón Province; 5,510–6,890 ft (1,680–2,100 m).

Natural history

Recently discovered in 2007, *B. aureogularis* is a relatively new addition to the Costa Rican herpetofauna. It is only known from a single adult female and a handful of juveniles. All individuals were collected in a small area on the Atlantic slope of the Talamanca Mountain Range, in mature, undisturbed cloud forest. A few individuals were found active at night, on low vegetation and in leaf litter on the forest floor. During the day, other individuals were discovered hiding and sleeping in arboreal bromeliads located at heights 9–20 ft (2.8–6.2 m) above the ground. One *B. aureogularis* reportedly used its prehensile tail to grab stems while climbing through vegetation.

Although only a single adult is known so far, there appears to be no significant difference in coloration between juveniles and adults.

Based on recent molecular analysis, the apparent closest relative to *B. aureogularis* is *B. robinsoni* (p. 78), a larger and more robust species from extreme southeastern Costa Rica and adjacent Panama that lacks the yellow throat and chest. Together, these two species form a sister group to the widespread *Bolitoglossa subpalmata*-species group (p. 91) of montane salamanders, several of which occur in Costa Rica.

The dorsal coloration of *B. aureogularis* is inconspicuous, but its yellow throat is unique among Costa Rican salamanders.

Three to 4 costal folds between adpressed limbs, which are relatively short.

Dorsal coloration ranges from yellowish-tan on the head and body to reddish-brown on the dorsal surfaces of the tail; small white specks may be present on the body and tail.

Relatively small head with a broadly rounded snout and small eyes.

Dark brown lateral band separates the dorsal coloration from the light venter; band extends onto the head and tail.

Upper surfaces of limbs yellowish-tan.

Relatively slender, small to moderate-sized.

Ventral surfaces cream to dirty white, marked with a central, irregularly edged, dark band.

Color pattern unique among Costa Rican salamanders. *Aureogularis* means golden throat, which refers to the bright yellow chin, throat, and chest.

Tail in only known adult is incomplete; intact tail length presumably about the same length as the standard length.

foot

Hands and feet partially webbed; tips of its fingers and toes protrude from the webbing.

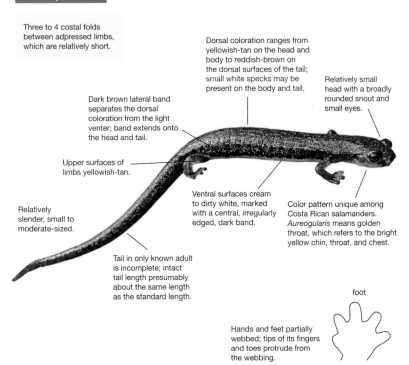

Other salamanders in the Talamanca Mountain Range that are similar in size and coloration to *B. aureogularis* include *B. pesrubra* (p. 73), *B. subpalmata* (p. 91), and *B. gomezi* (p. 57). None of these display the diagnostic yellow throat and chest.

Bolitoglossa bramei
Brame's Webfoot Salamander

Data deficient

Standard length to 1.61 in (41 mm) in females; males usually somewhat smaller than females. A small salamander in the *Bolitoglossa subpalmata*-species group (p. 91). Under field conditions, it may be difficult to distinguish it from related species. *Bolitoglossa bramei* has a very slender build, long limbs, and a long tail that typically has a reddish tip.

Ecoregions
1 2 3 4 5

Bolitoglossa bramei occurs in the southern part of the Talamanca Mountain Range, on both the Atlantic and Pacific slopes. In Costa Rica it is known from near the Panamanian border, between Las Tablas and Cerro Pando, as well as from the Valle del Silencio and Cerro Frantzius; 6,250–10,500 ft (1,900–3,200 m). Other populations occur in adjacent Panama.

Natural history

B. bramei is a nocturnal salamander that has been found both on the ground and up to a few meters above the ground in vegetation. Like several other webfoot salamanders, it uses its prehensile tail extensively as it climbs across vegetation at night. An individual of this species was discovered in its daytime retreat, located under a clump of moss on the ground. Recent observations have significantly expanded the known distribution and elevational range of this species. Nevertheless, very little is known about the biology and population status of *B. bramei.*

Description

Adpressed limbs separated by 0–0.5 costal folds in males and 0.5–1.0 folds in females.

Posterior half of tail dark brown, the same color as the flanks and head; tail typically ends with a reddish tip. Underside of tail orange with dark blotches.

Head narrow but distinctly wider than neck; dark brown, punctuated by clusters of black spots that extend to the shoulders.

Tail length generally exceeds standard length.

Snout bluntly rounded, with small nostrils and weakly developed nasolabial protuberances.

Dorsum often shows an irregularly edged reddish-brown dorsal field that extends from behind the head along the dorsum and onto the tail.

Limbs lighter than dorsum; yellowish-brown with bright orange or white marks.

Ventral surface either black or gray with light lines, mottled with irregular dark and light spots.

Fingers and toes robust and truncated, extending significantly beyond webbing.

foot

Similar species

B. gomezi (p. 57), *B. kamuk* (p. 62), *B. robinsoni* (p. 78), and other members of the *Bolitoglossa subpalmata*-species group (p. 91) are not reliably distinguished in the field. *B. bramei* is primarily distinguished by details of its dentition and DNA, though its small size and slender morphology may also aid in identification. *B. colonnea* (p. 50) has fully webbed feet and a distinctive fleshy ridge across the top of its head. *B. minutula* (p. 68) is very small and stocky, with fully webbed feet. *B. compacta* (p. 52) is black with bright orange-red blotches on its dorsum.

Bolitoglossa cerroensis
Talamanca Highland Webfoot Salamander   **Least concern**

Standard length to 2.91 (74 mm); maximum known total length 6.14 in (156 mm). Females generally larger than males. *Bolitoglossa cerroensis* is a highland endemic. It is medium-sized, with moderately webbed hands and feet. Has dense yellowish-cream mottling on a brown to purplish-brown base color; its light markings often form a pair of irregular dorsolateral lines.

Ecoregions
1 2 3 4 5

This endemic species is found solely in Costa Rica, in the northern Talamanca Mountain Range; 8,300–9,800 ft (2,530–2,990 m).

Natural history

B. cerroensis inhabits rock slides and talus at the base of steep, forest-covered slopes. During the day, individuals are occasionally found under rocks. At night they surface and hunt small invertebrates such as flies, springtails, and mites. *B. cerroensis* is uncommonly encountered due to its nocturnal habits and the general inaccessibility of its habitat. Based on its presence in altered habitats such as road cuts and power-line clearings, this species may be somewhat adaptable to habitat alteration; however, it likely does not survive in severely degraded forest. It is still observed (albeit infrequently) in its appropriate habitat, and its population appears stable.

Up close, the yellow speckling on *B. cerroensis* can be seen to form irregular dorsolateral lines.

Limbs fairly short and slender; when adpressed, limbs separated by 0.5–1 costal folds in males and 1–2 in females.

Upper eyelids may be darker in coloration than the rest of the head and body.

Nasolabial protuberances are weakly developed.

Coloration distinctive; dorsum light brown to dark purplish-brown, suffused with yellow or cream stippling and blotching.

Head narrow (not wider than neck), with a short, truncated snout and large eyes.

Light stippling, particularly pronounced on the flanks and undersurfaces, often forms irregular dorsolateral stripes or transverse bands that align with the costal grooves.

Moderate-sized and slender, with a fairly long tail that is about equal in length to, or slightly exceeds, the standard length and that has a noticeable constriction at its base.

Feet moderately webbed. Fingers and toes with rounded tips that extend significantly beyond the interdigital webbing; but inner toe and inner finger entirely enveloped in webbing.

foot

Tail often darker than body, can be almost black at its tip.

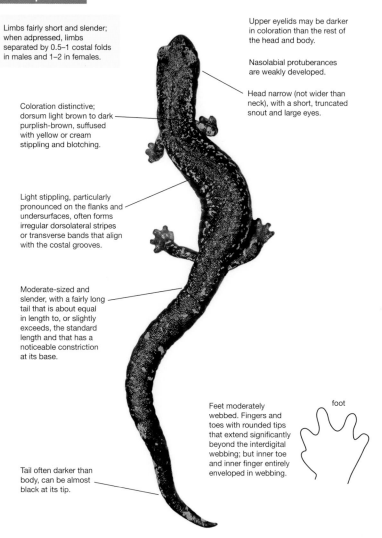

B. pesrubra (p. 73) is commonly found in the same range as *B. cerroensis*. It displays a quite variable color pattern, but most individuals have reddish-orange or yellowish-orange limbs and a dark head, body, and tail. *B. sooyorum* (p. 86) is a rare, co-occurring species with longer limbs, larger hands and feet, and a pattern of large yellowish spots or flecks (not fine yellowish-cream stippling) on a dark background.

Bolitoglossa colonnea
Ridge-headed Webfoot Salamander   **Least concern**

Standard length to 2.13 in (54 mm); total length to 4.45 in (113 mm). No known sexual size difference. A smallish salamander with fully webbed hands and feet and a unique, raised, fleshy ridge between its eyes.

Ecoregions
1 2 3 4 5

Bolitoglossa colonnea is found in the lowlands and foothills of the Atlantic slope of Costa Rica and adjacent Panama; it also occurs at middle elevations of Pacific southwestern Costa Rica, near the border with Panama. Additional reports of this species exist from isolated Pacific lowland localities such as the Osa Peninsula; near sea level to 5,250 ft (1,600 m).

Natural history

Like several other webfoot salamanders, *B. colonnea* is capable of significant color change. This species also undergoes a striking difference in skin texture between night and day. At night, it has smooth skin and uniform salmon to grayish-tan coloration, sometimes marked by small dark specks. During the day, the skin texture becomes rugose with pronounced longitudinal skin folds, and its coloration changes to a dark brown mottling.

B. colonnea is nocturnal and is mostly seen at night perched low on broad-leaved plants. These salamanders seem most active on warm, humid nights, when they hunt for invertebrate prey or look for a mate. Examination of the reproductive organs in this species showed that there is no apparent seasonality in reproductive activity. It is not uncommon to find one of these salamanders active at night and discover a second one close by, usually of the opposite sex. Whether this is simply the result of ideal environmental conditions that induce many individuals to become active simultaneously is not clear, but active individuals possibly attract potential mates, likely through chemical cues. *B. colonnea* deposits small clutches of eggs in moist environments away from water; their larvae complete their development entirely inside the egg.

During the day, these salamanders have been found hiding inside large, rolled-up dead leaves in the leaf litter layer on the forest floor. One individual was found under the loose bark of a dead tree. Sleeping *B. colonnea* tend to be tightly curled, with the tail wrapped snugly around their body.

B. colonnea moves extremely slowly at night. Movement is sped up considerably when the surrounding vegetation is moving in the wind, presumably because this masks the salamander's movements and thus makes it less cautious. When disturbed during the day, *B. colonnea* twitches its body and occasionally even hops small distances by flipping the heavy tail around. Individuals placed on low vegetation during the day display a similar behavior. They launch themselves off the leaf surface and drop to the ground, where they rapidly disappear into the leaf litter. Another defensive behavior observed in this species involves holding the body rigid with the hands and feet folded flat against the body. This behavior, combined with the rugose, textured skin and mottled brown daytime coloration, makes these animals resemble a dead twig.

This is one of the more frequently observed salamanders in the humid forests of Costa Rica's Atlantic slope. Its populations seem stable in most mid-elevation areas with relatively intact forest cover, and this species is not considered threatened, although apparently its numbers have decreased some at low-elevation sites.

Coloration highly variable, with shades of gray, tan, brown, or reddish-brown; sometimes with spots, blotches, or irregular streaks.

Description

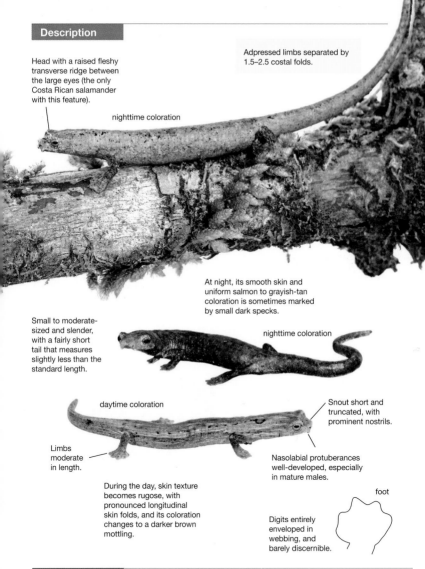

Head with a raised fleshy transverse ridge between the large eyes (the only Costa Rican salamander with this feature).

nighttime coloration

Adpressed limbs separated by 1.5–2.5 costal folds.

At night, its smooth skin and uniform salmon to grayish-tan coloration is sometimes marked by small dark specks.

Small to moderate-sized and slender, with a fairly short tail that measures slightly less than the standard length.

nighttime coloration

daytime coloration

Snout short and truncated, with prominent nostrils.

Limbs moderate in length.

Nasolabial protuberances well-developed, especially in mature males.

During the day, skin texture becomes rugose, with pronounced longitudinal skin folds, and its coloration changes to a darker brown mottling.

foot

Digits entirely enveloped in webbing, and barely discernible.

Similar species

No other Costa Rican salamander has a fleshy transverse ridge across the head. *B. striatula* (p. 89) most resembles this species but has a characteristic color pattern of irregular, dark brown, longitudinal dorsal and lateral stripes on a tan or light brown background. *B. alvaradoi* (p. 42) is larger and has a dark brown to black dorsal coloration marked with a few large, irregularly shaped reddish-brown to tan blotches. *B. indio* (p. 60) has a dark brown dorsum with a pair of irregularly shaped light dorsolateral bands. *B. minutula* (p. 68) co-occurs at Las Tablas but is smaller and lacks the fleshy ridge across the head.

Bolitoglossa compacta
Red-spotted Webfoot Salamander

Endangered

Standard length to 2.91 in (74 mm); reaches a maximum total length of 5.63 in (143 mm). Males generally considerably smaller than females. A strikingly colored, robust salamander with a unique series of large, irregularly shaped orange or red blotches on a very dark background.

Ecoregions

Found in humid lower montane areas along the Costa Rica-Panama border. In Costa Rica it has been found sporadically on both the Atlantic and the Pacific slopes, 5,400–8,200 ft (1,650–2,500 m); in adjacent Panama it occurs at elevations up to 9,125 ft (2,780 m).

Natural history

B. compacta is a rarely encountered and poorly studied salamander that inhabits humid montane forests, where it has been found at night on the ground, under logs, and on low vegetation. Individuals have been collected from moist forested habitats with significant moss growth and an understory of palms and ferns. Due to the small number of individuals known and the disappearance and degradation of much of the suitable habitat within its range, this species is considered endangered, although further surveys are needed to adequately assess its status. *B. compacta* is known from at least one protected area, La Amistad International Park, which straddles the border between Costa Rica and Panama.

It undergoes direct development; egg deposition occurs during the dry season and young hatch at the beginning of the wet season. An adult female of this species laid 39 eggs in a tight clump shortly after it was collected from the wild. This salamander did not exhibit any parental care behavior and the eggs were incubated in a laboratory at 55 °F (13 °C), comparable to the ambient temperature of the capture site. Two of these eggs hatched 249 and

251 days, respectively, after oviposition. The very long development period of more than 8 months seems consistent with the few observations made on other tropical salamander egg clutches that have been successfully hatched in captivity.

B. compacta appears most closely related to the montane salamanders of the *Bolitoglossa subpalmata*-species group (p. 91).

B. compacta is a strikingly colored salamander with a unique pattern formed by orange spots.

Description

Adpressed limbs separated by 1.5–2.5 costal folds.

Moderate-sized and robust species with a fairly short tail that is slightly compressed laterally.

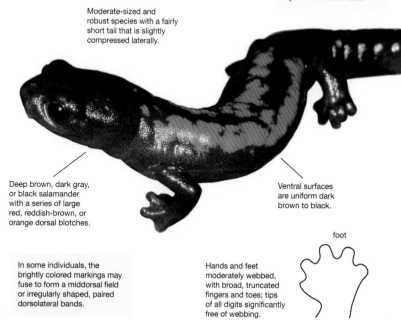

Deep brown, dark gray, or black salamander with a series of large red, reddish-brown, or orange dorsal blotches.

Ventral surfaces are uniform dark brown to black.

foot

In some individuals, the brightly colored markings may fuse to form a middorsal field or irregularly shaped, paired dorsolateral bands.

Hands and feet moderately webbed, with broad, truncated fingers and toes; tips of all digits significantly free of webbing.

Similar species

Its distinctive color pattern distinguishes *B. compacta* from other salamanders in its range (*B. bramei*, *B. colonnea*, *B. marmorea*, *B. minutula*, *B. pesrubra*, and *B. robinsoni*); of these, only *B. pesrubra* (p. 73) displays a dark body with contrasting red markings, but in that species the red coloration is generally restricted to the limbs.

Bolitoglossa diminuta
Diminutive Webfoot Salamander

Vulnerable

Standard length to 1.26 in (32 mm); maximum known total length only 2.60 in (66 mm). An extremely small salamander with moderate interdigital webbing. Has a unique color pattern: a broad, bronze to light brown middorsal band, bordered on either side by a wavy, dark chocolate-brown band. Its ventral surfaces are pale, without a midventral dark stripe.

Ecoregions
1 2 3 4 5

Bolitoglossa diminuta is endemic to Costa Rica and has only been found at Quebrada Valverde, a steep stream near Tapantí, Cartago Province, on the Atlantic slope of the Talamanca Mountain Range; elevation 5,100 ft (1,555 m).

Natural history

This is a very rare salamander of undisturbed premontane rainforest habitat. Only two adult females and two egg clutches have ever been observed. The first sample of this species ever discovered was a female in a mat of arboreal liverworts attached to a branch that had recently fallen from the canopy. The same mat of vegetation contained a clutch of 7 eggs that later, in a lab, hatched out into tiny salamanders with a standard length of 0.31 in (8 mm). The discovery of the female and eggs in a fallen mat of arboreal vegetation led to the assumption that this species lives an arboreal life. It was also assumed that the female salamander was associated with these eggs and therefore sexually mature, in spite of her very small size.

Due to the lack of additional individuals and the difficulty of discerning clear, diagnostic characters on an animal this small, the taxonomic position of *B. diminuta* is uncertain. It is currently thought to be a member of the *Bolitoglossa*

subpalmata-species group (p. 91), but its exact affiliation with other members of the genus remains unknown. This species may be conspecific with *B. gracilis* (p. 59), another exceedingly rare and co-occurring species that differs subtly in the extent of interdigital webbing and coloration. Additional research and more samples of both species are needed to resolve this issue.

In spite of its perceived rarity (no individuals of this species have been observed since it was formally described in 1976), its natural habitat is relatively secure and *B. diminuta* may continue to live hidden from view in the forest canopy in the area where it was first identified. Due to its very small known distribution range, it remains at risk of extinction, either as a result of any events impacting its preferred habitat or, possibly, disease. Therefore, while the conservation status assigned to this species is "vulnerable," that label may not reflect its true population status.

One of the smallest species in the genus *Bolitoglossa*; presumably males are slightly smaller than females.

Body fairly robust, with 13 costal grooves; adpressed limbs separated by 2 costal folds.

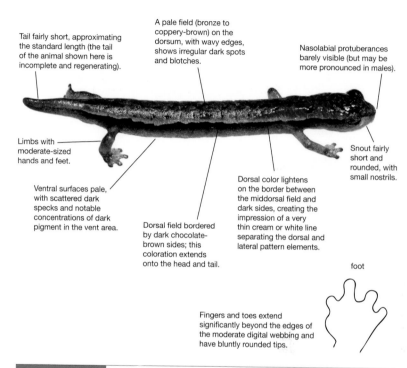

Tail fairly short, approximating the standard length (the tail of the animal shown here is incomplete and regenerating).

A pale field (bronze to coppery-brown) on the dorsum, with wavy edges, shows irregular dark spots and blotches.

Nasolabial protuberances barely visible (but may be more pronounced in males).

Limbs with moderate-sized hands and feet.

Snout fairly short and rounded, with small nostrils.

Ventral surfaces pale, with scattered dark specks and notable concentrations of dark pigment in the vent area.

Dorsal color lightens on the border between the middorsal field and dark sides, creating the impression of a very thin cream or white line separating the dorsal and lateral pattern elements.

Dorsal field bordered by dark chocolate-brown sides; this coloration extends onto the head and tail.

foot

Fingers and toes extend significantly beyond the edges of the moderate digital webbing and have bluntly rounded tips.

B. gracilis (p. 59) is very similar in size and overall appearance but has a yellowish ground color with irregular dark brown lines and a dark midventral stripe. *B. colonnea* (p. 50) has fully-webbed feet and a raised, fleshy ridge across the top of the head. *B. subpalmata* (p. 91) and *B. pesrubra* (p. 73) are considerably larger, have less interdigital webbing, and differ in coloration; both are generally found at higher elevations.

Bolitoglossa epimela
Black-headed Webfoot Salamander

Data deficient

Standard length to 1.81 in (46 mm), maximum known total length 3.82 in (97 mm). Females generally somewhat larger than males. A small, slender dark salamander with very large hands and feet, extensive finger and toe webbing, and rounded toe tips.

Ecoregions
1 2 3 4 5

Bolitoglossa epimela is endemic to Costa Rica and known only from the Río Chitaria, near Turrialba, and from Tapantí National Park, Cartago Province; 2,550–5,100 ft (775–1,550 m). Both localities are on the Atlantic slope of central Costa Rica.

Natural history

This rare salamander is only known from 10 individuals and has not been observed in recent years. Consequently, not much is known about this species, and it is possible that individuals from the two known localities actually represent two different species.

B. epimela is a nocturnal species that has been observed active at night, moving across logs and low vegetation up to 3 ft (1 m) above the forest floor. Its large webbed hands and feet provide increased adhesion to wet leaf surfaces, and

individuals have been observed walking upside down on the undersurfaces of leaves.

One of the known localities for *B. epimela* is now incorporated into Tapantí National Park, which is adequately protected. However, the other site, Río Chitaria, has been subject to significant small-scale logging, and little natural habitat remains. Intensive searches in recent years failed to produce any additional individuals and continued surveys are urgently needed to assess the status of this species.

Description

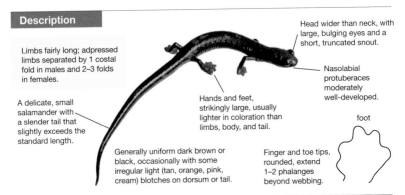

Limbs fairly long; adpressed limbs separated by 1 costal fold in males and 2–3 folds in females.

A delicate, small salamander with a slender tail that slightly exceeds the standard length.

Head wider than neck, with large, bulging eyes and a short, truncated snout.

Nasolabial protuberaces moderately well-developed.

foot

Hands and feet, strikingly large, usually lighter in coloration than limbs, body, and tail.

Generally uniform dark brown or black, occasionally with some irregular light (tan, orange, pink, cream) blotches on dorsum or tail.

Finger and toe tips, rounded, extend 1–2 phalanges beyond webbing.

Similar species

B. colonnea (p. 50) has fully-webbed feet and a raised, fleshy ridge across the top of the head. *B. diminuta* (p. 54) is very small, with moderately webbed hands and feet, but with a yellowish dorsum. *B. gracilis* (p. 59) is also small, with moderately webbed hands and feet, but it has yellowish dorsal coloration and a dark midventral stripe. *B. subpalmata* (p. 91) and *B. pesrubra* (p. 73) are considerably larger, have less interdigital webbing, and differ in coloration; both are generally found at higher elevations.

Bolitoglossa gomezi
Gomez's Webfoot Salamander

Data deficient

Standard length to 2.17 in (55 mm); generally males are smaller than females. A small to moderate-sized member of the *Bolitoglossa subpalmata*-species group (p. 91) that, under field conditions, may not be readily distinguishable from other species in the group. *Bolitoglossa gomezi* has relatively short limbs, large hands and feet, and distinctly robust, bluntly truncated digits that extend significantly beyond the moderate webbing.

Ecoregions
1 2 3 4 5

Bolitoglossa gomezi occurs on the Pacific slope, in a relatively small area in southern Costa Rica and adjacent western Panama; 3,850–7,875 ft (1,170–2,400 m).

Natural history

B. gomezi is a nocturnal and arboreal salamander, found in premontane and montane tropical rainforest and cloud forest habitats. Individuals have been found in arboreal bromeliads up to 10 ft (3 m) above the ground. At higher elevations, this species finds daytime retreats in arboreal and terrestrial bromeliads in oak forests, and in moss carpets in sites with páramo habitat. In spite of extensive searches in those habitats, no individuals were ever seen under rocks, logs, or in leaf litter. One individual was reportedly found at night actively walking on a large-leaved heliconia plant.

Not much is known about the biology of this species. *B. gomezi* is apparently quite common in a variety of habitat types and may be somewhat adaptable. However, it likely cannot tolerate significant habitat alteration such as logging and clear cutting for agriculture. Recent surveys indicate that the distribution range of *B. gomezi* is more extensive than originally thought. It occurs in at least one protected area, at Las Cruces Biological Station.

B. gomezi is highly variable in coloration and not readily identified based on morphological features.

57

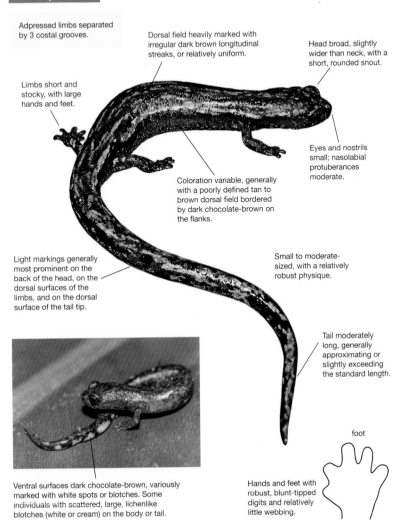

Adpressed limbs separated by 3 costal grooves.

Dorsal field heavily marked with irregular dark brown longitudinal streaks, or relatively uniform.

Head broad, slightly wider than neck, with a short, rounded snout.

Limbs short and stocky, with large hands and feet.

Eyes and nostrils small; nasolabial protuberances moderate.

Coloration variable, generally with a poorly defined tan to brown dorsal field bordered by dark chocolate-brown on the flanks.

Light markings generally most prominent on the back of the head, on the dorsal surfaces of the limbs, and on the dorsal surface of the tail tip.

Small to moderate-sized, with a relatively robust physique.

Tail moderately long, generally approximating or slightly exceeding the standard length.

foot

Ventral surfaces dark chocolate-brown, variously marked with white spots or blotches. Some individuals with scattered, large, lichenlike blotches (white or cream) on the body or tail.

Hands and feet with robust, blunt-tipped digits and relatively little webbing.

Similar species

B. bramei (p. 47), *B. kamuk* (p. 62), *B. robinsoni* (p. 78), and other members of the *Bolitoglossa subpalmata*-species group (p. 91) may not be easily distinguished in the field. *B. robinsoni* generally occurs at higher elevations. *B. colonnea* (p. 50) has fully webbed feet and a distinctive fleshy ridge across the top of its head. *B. minutula* (p. 68) is smaller and stocky, with fully webbed feet. *B. compacta* (p. 52) is black with bright orange-red blotches.

Bolitoglossa gracilis
Slender Webfoot Salamander

Vulnerable

Standard length to 1.65 in (42 mm), maximum known total length 3.58 in (91 mm). Females likely somewhat larger than males. A small, slender salamander with moderately webbed hands and feet. Has a reddish-brown midventral stripe.

Ecoregions

1 2 3 4 5

Bolitoglossa gracilis is endemic to the Atlantic slope of Costa Rica and known only from two localities, Río Quirí and Quebrada Casa Blanca, in the vicinity of Tapantí, Cartago Province; 4,025–4,200 ft (1,225–1,280 m).

Natural history

Not much is known about this uncommon species. The few individuals of *B. gracilis* ever collected were found in undisturbed premontane rainforest habitat. One of the two previously known localities (Río Quirí) has been lost to large-scale coffee plantations. However, the other locality is now protected as part of Tapantí National Park.

Like other *Bolitoglossa*, this is likely a nocturnal species. It is not clear whether it is an arboreal or terrestrial species; one individual was collected

from a moss mat on a tree, located close to the ground.

Comparison of DNA from *B. gracilis* with DNA from other salamanders showed that it is most closely related to *B. subpalmata* and *B. tica* and is therefore assigned to the *Bolitoglossa subpalmata*-species group (p. 91). Another closely related species, *B. diminuta*, was not included in this analysis because no genetic material is available for that rare species.

Description

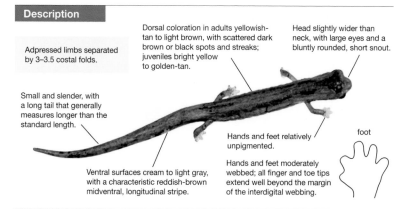

Adpressed limbs separated by 3–3.5 costal folds.

Dorsal coloration in adults yellowish-tan to light brown, with scattered dark brown or black spots and streaks; juveniles bright yellow to golden-tan.

Head slightly wider than neck, with large eyes and a bluntly rounded, short snout.

Small and slender, with a long tail that generally measures longer than the standard length.

Ventral surfaces cream to light gray, with a characteristic reddish-brown midventral, longitudinal stripe.

Hands and feet relatively unpigmented.

Hands and feet moderately webbed; all finger and toe tips extend well beyond the margin of the interdigital webbing.

foot

Similar species

B. diminuta (p. 54) is very similar in size and overall appearance, but has a different color pattern: a broad, bronze to light brown middorsal band, bordered on either side by a wavy, dark chocolate-brown band. It lacks a midventral brown stripe. *B. colonnea* (p. 50) has fully webbed feet and a raised, fleshy ridge across the top of the head. *B. subpalmata* (p. 91) and *B. pesrubra* (p. 73) are considerably larger, have less interdigital webbing, and differ in coloration; both species are mostly found at higher elevations.

59

Bolitoglossa indio
Río Indio Webfoot Salamander

Not assessed

Maximum known standard length 1.85 in (47 mm); maximum known total length approaching 3.54 in (90 mm). A robust, small to moderate-sized salamander with fully webbed hands and feet. Has a dark (black, dark brown, dark olive) base color; a pair of irregularly edged light brown to tan dorsolateral stripes are sometimes broken up into a linear series of light blotches.

Ecoregions

1 2 3 4 5

Originally known from a single individual discovered at 82 ft (25 m) elevation in Dos Bocas del Río Indio, a lowland site in the Río San Juan area of southeastern Nicaragua. A second specimen, collected in 1890 in Boca de Arenal in Alajuela Province (elevation 223 ft [68 m]) and currently housed in the collection of the British Museum of Natural History, was recently assigned to this species and now represents the only record for Costa Rica.

Natural history

In 2008, *B. indio* was described as a new species, based on a single individual. It is similar in general appearance to the poorly known species *B. alvaradoi*, and more information is needed to better define both species and their respective distribution ranges. Based on the recent discovery of a historic specimen from northeastern Costa Rica, it is possible that salamanders from nearby areas (e.g., Rara Avis Rainforest Reserve, adjacent to Braulio Carrillo National Park, at 710 m [2,329 ft]) that are currently assigned to *B. alvaradoi* may actually represent *B. indio*. If this is the case, it would extend the latter species' elevation and distribution range considerably, but only the discovery of additional individuals will help clarify this situation.

As currently understood, *B. alvaradoi* is larger than *B. indio*; in *B. alvaradoi*, females reach 3.11 in (79 mm) in standard length, males reach 2.56 in (65 mm). It also occurs at higher elevations. While both species have completely webbed hands and feet, some differences in color pattern exist between the two species, though the extent of interspecific variation in each is poorly known due to lack of observations. The color pattern of *B. alvaradoi* is quite variable and changes dramatically between day and night. Limited observations of the only living *B. indio* ever documented

indicate that it shows no significant color change under different light levels.

The holotype of *B. indio* was encountered in leaf litter on the forest floor during early afternoon, at the hottest time of day and in full sun. However, as a group webfoot salamanders are generally nocturnal and found active on cool, humid nights. It is possible, therefore, that the holotype of this species had been dislodged from a canopy retreat, though no real firsthand observations exist on the life history of this enigmatic species.

B. indio is an enigmatic species whose distribution and taxonomic status require additional study.

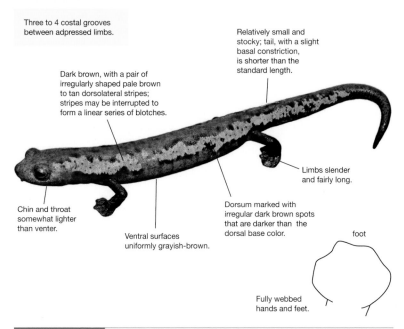

Three to 4 costal grooves between adpressed limbs.

Relatively small and stocky; tail, with a slight basal constriction, is shorter than the standard length.

Dark brown, with a pair of irregularly shaped pale brown to tan dorsolateral stripes; stripes may be interrupted to form a linear series of blotches.

Limbs slender and fairly long.

Chin and throat somewhat lighter than venter.

Dorsum marked with irregular dark brown spots that are darker than the dorsal base color.

foot

Ventral surfaces uniformly grayish-brown.

Fully webbed hands and feet.

Similar species

B. alvaradoi (p. 42) is larger, usually black, with irregular salmon, tan, or copper blotches. *B. striatula* (p. 89) is light tan or brown, with dark brown longitudinal stripes and fine, irregular cream or white lines. *B. colonnea* (p. 50) has a fleshy transverse ridge across the top of the head, between the eyes.

Bolitoglossa kamuk
Kamuk Webfoot Salamander

Not assessed

Maximum standard length about 1.38 in (35 mm); total length to 2.76 in (70 mm). A small, slender salamander with partly webbed hands and feet. Has no easily observed morphological characteristics that would aid in reliable identification. Due to its small distribution range, however, geographic provenance may aid in identification.

Ecoregions
1 2 3 4 5

A Costa Rican endemic, only known from Cerro Apri in the Kamuk Massif of the Talamanca Mountain Range, Limón Province; 10,255 ft (3,126 m).

Natural history

Three individuals were discovered in 2007; they were formally described in 2012 based on DNA characteristics. Morphologically, no field marks easily distinguish *B. kamuk* from related species, and the small sample size does not allow adequate assessment of interspecific variation in color and pattern. Interestingly, one of the three known *B. kamuk* individuals is a juvenile with notably different coloration; its orange dorsum gradually darkens from bright orange to orange-gray on the tail, and is marked with gray patches. Its undersurfaces range from pale yellow-orange on the throat to golden-yellow with black markings on the venter.

B. kamuk inhabits sub-páramo habitat on the continental divide. Its high-elevation habitat is characterized by deep moss-pillows and low vegetation, including temperate zone herbaceous plants and low woody vegetation. Terrestrial bromeliads are present and arboreal epiphytes adorn the few trees found at this elevation. Individual *B. kamuk* were reportedly found under moss and in arboreal bromeliads. Additional surveys of the grassy, treeless páramo on the summit of the Kamuk Massif did not reveal any salamanders. The exact distribution range of *B. kamuk* is not well understood, but it may co-occur at lower elevations with another member of the *Bolitoglossa subpalmata*-species group (p. 91), *B. gomezi*.

Description

Only known juvenile with reddish-orange dorsum that gradually darkens to orange-gray on the tail; marked with gray patches.

juvenile

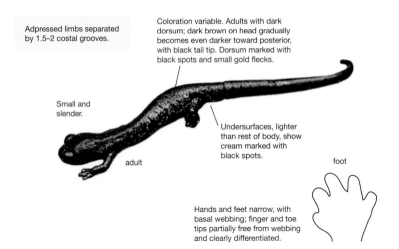

Adpressed limbs separated by 1.5–2 costal grooves.

Coloration variable. Adults with dark dorsum; dark brown on head gradually becomes even darker toward posterior, with black tail tip. Dorsum marked with black spots and small gold flecks.

Small and slender.

Undersurfaces, lighter than rest of body, show cream marked with black spots.

adult

foot

Hands and feet narrow, with basal webbing; finger and toe tips partially free from webbing and clearly differentiated.

Similar species

B. gomezi (p. 57) is slightly larger, also with variable coloration, and may be difficult to distinguish from *B. kamuk* in the field. *B. splendida* (p. 87) and *B. pesrubra* (p. 73) have a black body with bright red markings.

Bolitoglossa lignicolor
Wood-colored Webfoot Salamander

Vulnerable

Standard length to 3.19 in (81 mm), total length to at least 6.30 in (160 mm). Females significantly larger than males. The only moderate-sized, brown webfoot salamander in the Pacific lowlands and foothills that has completely webbed hands and feet and lacks a transverse ridge across the head.

Ecoregions
1 2 **3** 4 5

Bolitoglossa lignicolor is found in humid lowlands and in foothills of southwestern Costa Rica; from near sea level to at least 3,450 ft (1,050 m). In Panama, it occurs in regions across the border from Costa Rica, as well as on the Azuero Peninsula and Isla Coiba.

Natural history

This uncommon species is usually encountered at night, when it moves across the branches and foliage of large-leaved plants, 2–7 ft (0.6–2 m) above the ground. It uses extensively webbed hands and feet to adhere to wet leaves, and often secures its position by grabbing a leaf or branch with its prehensile tail. During the day, individuals have been found hidden between the axils of arboreal bromeliads, under the bark of standing trees, and under logs on the ground.

The cryptic coloration of *B. lignicolor* shows considerable variation. The dorsal field tends to be relatively uniform in color in juveniles, but becomes increasingly marked with dark streaks or irregular blotches in older individuals. Likewise,

the dark ventral surfaces are relatively unmarked in juveniles but can become suffused with discrete yellowish-cream spots in adults. In addition, *B. lignicolor* is capable of significant color change between day and night; at night, the dark flanks and venter can change to a grayish-tan, becoming lighter than the dorsal field.

Extensive development of Pacific lowland habitats, primarily for agriculture, has greatly reduced the amount of lowland moist and wet forest and premontane rainforest habitat that is suitable for *B. lignicolor*. This species does not seem to tolerate habitat degradation, and its occurrence is increasingly limited to undisturbed habitats, often in protected areas such as Corcovado National Park.

Coloration cryptic with varying shades of brown (*lignicolor* means color of wood).

The prehensile tail and fully webbed hands and feet provide added traction when climbing through vegetation.

64

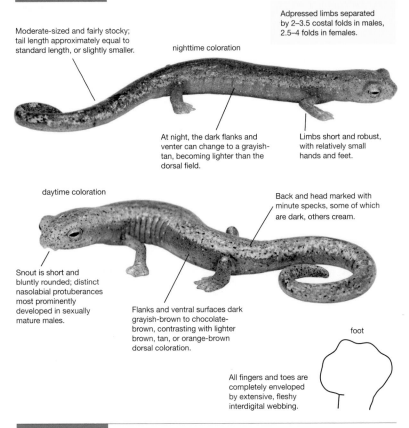

Moderate-sized and fairly stocky; tail length approximately equal to standard length, or slightly smaller.

Adpressed limbs separated by 2–3.5 costal folds in males, 2.5–4 folds in females.

nighttime coloration

At night, the dark flanks and venter can change to a grayish-tan, becoming lighter than the dorsal field.

Limbs short and robust, with relatively small hands and feet.

daytime coloration

Back and head marked with minute specks, some of which are dark, others cream.

Snout is short and bluntly rounded; distinct nasolabial protuberances most prominently developed in sexually mature males.

Flanks and ventral surfaces dark grayish-brown to chocolate-brown, contrasting with lighter brown, tan, or orange-brown dorsal coloration.

foot

All fingers and toes are completely enveloped by extensive, fleshy interdigital webbing.

Similar species

B. colonnea (p. 50) co-occurs with *B. lignicolor* in Corcovado and other locations in southwestern Costa Rica. It also has a brown coloration and fully webbed hands and feet but is instantly distinguished by the presence of a raised, fleshy ridge across its head. *B. striatula* (p. 89) occurs locally in the Pacific lowlands (e.g., Barú area, Dominical) and has fully webbed feet, but has a distinctive color pattern of longitudinal brown bands and light stripes. *B. minutula* (p. 68) also occurs on the Pacific slope and has fully webbed hands and feet, but is much smaller and occurs at higher elevations.

Bolitoglossa marmorea
Marbled Webfoot Salamander

Endangered

Standard length to 2.83 in (72 mm), maximum known total length 5.28 in (134 mm). Females likely larger than males. A moderate-sized salamander with long limbs; large, moderately webbed hands and feet; and digits that clearly extend beyond the margins of the webbing. Its base color is dark chestnut-brown to purplish-brown, heavily marked with irregular cream, yellow, or yellow-orange flecks that create a marbled appearance. In some individuals, light flecks are fused to form large irregular blotches.

Ecoregions
1 2 3 4 5

Known from the slopes of the Talamanca Mountain Range and Barú Volcano region in extreme southwestern Costa Rica and western Panama; 6,300–11,300 ft (1,920–3,444 m). There are only a few records of *Bolitoglossa marmorea* from Costa Rica, primarily from the western slope of Cerro Pando in the border region with Panama.

Natural history

B. marmorea was first collected under rocks at the crater of Barú Volcano, Panama. At night, it apparently emerges from its rocky daytime retreat and forages on low vegetation and on mossy tree trunks close to the ground. Although its primary habitat is humid montane forest, this species can withstand some degree of habitat alteration. Several individuals were recently found near the summit of Barú Volcano, suggesting that

B marmorea still persists there in significant numbers. However, surveys are needed to determine the current population status of this species elsewhere, especially in Costa Rica, where it has not been detected for several years.

B. marmorea occurs in at least two protected areas: La Amistad International Park, which straddles the border between Panama and Costa Rica, and Barú Volcano National Park, in Panama.

Usually 13 costal grooves between insertion of front and hind limbs. When adpressed, limbs nearly touch in males and are separated by 1.5–2.5 costal grooves in females.

Relatively slender and medium-sized; tail length approximates or slightly exceeds the standard length.

Coloration variable; generally dark chestnut-brown to purplish-brown, heavily marked with irregular cream, yellow, or yellow-orange flecks that create a marbled appearance.

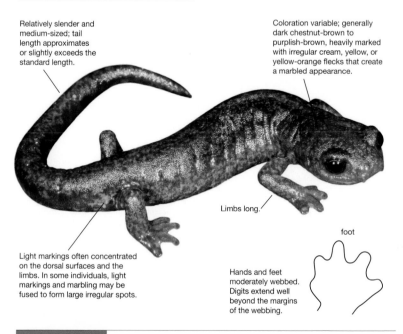

Limbs long.

foot

Light markings often concentrated on the dorsal surfaces and the limbs. In some individuals, light markings and marbling may be fused to form large irregular spots.

Hands and feet moderately webbed. Digits extend well beyond the margins of the webbing.

Similar species

B. robinsoni (p. 78) has narrower hands and feet but is similar in coloration and may be difficult to distinguish in the field. *B. compacta* (p. 52) has a striking pattern of red-orange blotches or stripes on a jet-black base color. *B. marmorea* is morphologically similar to *B. sooyorum* (p. 86) except for minor details in dentition and the amount of webbing; the two species are geographically separated.

Bolitoglossa minutula
Minute Webfoot Salamander

Males and female standard length to 1.46 in (37 mm); maximum known total length 2.99 in (76 mm). A very small montane salamander found in extreme southern Costa Rica. It has fully webbed hands and feet, relatively rugose skin, and a raised parotoid-like structure on each side of the neck.

Ecoregions
1 2 3 4 5

Bolitoglossa minutula occurs on both the Atlantic and Pacific slopes of the Talamanca Mountain Range, in southern Costa Rica and adjacent western Panama; 5,475–8,725 ft (1,670–2,660 m).

Natural history

This is a nocturnal species of undisturbed, humid montane forests. *B. minutula* forages at night on low vegetation (usually less than 3 ft [1 m] above the ground) and retreats into moist hiding places, such as the leaf axils of arboreal bromeliads, during the day. Apparently, this species is still relatively common in appropriate habitat within its small geographic range. Stable populations have been reported from the Las Tablas region in Costa Rica as recently as 2008, and from Barú Volcano National Park and La Amistad International Park in adjacent Panama in 2010.

Description

Females are stockier than males. Adult males tend to be more slender, with longer limbs and more prominent nasolabial protuberances.

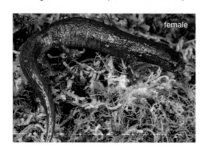

female

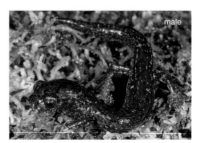

male

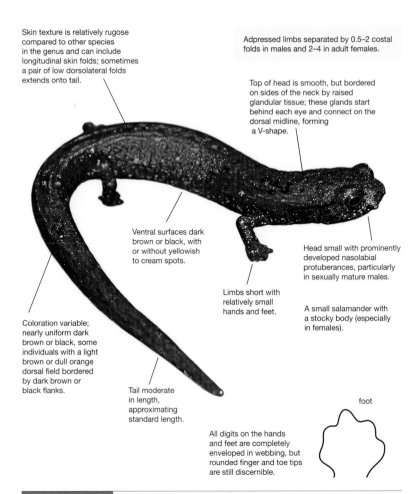

Skin texture is relatively rugose compared to other species in the genus and can include longitudinal skin folds; sometimes a pair of low dorsolateral folds extends onto tail.

Adpressed limbs separated by 0.5–2 costal folds in males and 2–4 in adult females.

Top of head is smooth, but bordered on sides of the neck by raised glandular tissue; these glands start behind each eye and connect on the dorsal midline, forming a V-shape.

Ventral surfaces dark brown or black, with or without yellowish to cream spots.

Head small with prominently developed nasolabial protuberances, particularly in sexually mature males.

Limbs short with relatively small hands and feet.

A small salamander with a stocky body (especially in females).

Coloration variable; nearly uniform dark brown or black, some individuals with a light brown or dull orange dorsal field bordered by dark brown or black flanks.

Tail moderate in length, approximating standard length.

foot

All digits on the hands and feet are completely enveloped in webbing, but rounded finger and toe tips are still discernible.

Similar species

B. colonnea (p. 50) co-occurs with *B. minutula* in the Las Tablas area; both have fully webbed feet, but *B. colonnea* has a raised, fleshy ridge across the top of its head. *B. pygmaea* (p. 76) is the same size and occurs in the same region (though at higher elevations); it differs in having only moderately webbed hands and feet. *B. bramei* (p. 47), *B. gomezi* (p. 57), and *B. robinsoni* (p. 78) are generally larger and have only slightly webbed hands and feet.

Bolitoglossa nigrescens
Webfoot Salamander

Endangered

Maximum known standard length 3.70 in (94 mm). Males likely smaller than females. One of several black salamanders that occurs in the uplands of the Talamanca Mountain Range in Costa Rica and adjacent Panama. *Bolitoglossa nigrescens* may be difficult to distinguish from other species in this group based on its morphology or coloration alone. However, this is the only large, entirely black salamander found at middle elevations on the Pacific slope of the northern Talamanca Mountain Range.

Ecoregions

1 2 3 **4** 5

Bolitoglossa nigrescens is endemic to Costa Rica and known only from a few scattered localities in the northern and central Talamanca Mountain Range, roughly between the Cerros de Escazú, south of San José, to Villa Mills near Cerro de la Muerte; 5,400–9,850 ft (1,650–3,000 m).

Natural history

B. nigrescens is a rare species that has not been observed for many years. It apparently inhabits the forest floor of humid premontane and montane zones. Individuals have been collected from holes under logs and rocks. It is active at night, but surface activity may be limited in the cool, high-elevation sections of its range. This species has been observed in areas where habitat quality has been somewhat compromised, and it therefore seems to be relatively adaptable.

Until 2005, several large black salamanders from Costa Rica and Panama, collectively called the *Bolitoglossa nigrescens*-complex, were lumped under the name *B. nigrescens*. Detailed morphological comparison of the various populations captured under this name led to the identification and description of the Costa Rican species *B. obscura* and *B. sombra*, as well as a Panamanian form not within the scope of this book. Based on tooth counts, proportional differences in morphology, and the amount of interdigital webbing, these different taxa could be separated into geographically isolated species. However, a recent molecular study suggests that *B. nigrescens* and *B. sombra* may in fact be the same species. Future research, ideally supported by additional observations, will hopefully clarify this situation.

Description

Adpressed limbs separated by 2.5–4 costal folds.

This is an entirely dark brown to black salamander, both dorsally and ventrally. Only the soles of the hands and feet are somewhat lighter in color.

Head very broad, wider than neck; snout very short and truncated.

eye

nasolabial protuberance

A large, very robust salamander; relatively short tail is considerably shorter than its standard length.

Limbs stocky and short, with large hands and feet.

Eyes and nostrils small; nasolabial protuberances only weakly developed.

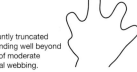

foot

Digits bluntly truncated and extending well beyond margins of moderate interdigital webbing.

Similar species

B. robusta (p. 80) is a large black salamander but typically has a light ring around its tail base. *B. pesrubra* (p. 73) generally has a black body, head, and tail, but with contrasting red, orange, or yellowish limbs. Entirely black morphs of this species are smaller and much more slender in morphology. *B. cerroensis* (p. 48) is dark brown with cream or yellowish mottling. *B. sooyorum* (p. 86) is smaller and exceedingly slender, with long limbs and very large hands and feet; it is never solid black.

The other large black salamanders in Costa Rica (*B. anthracina*, *B. obscura*, and *B. sombra*) are very similar in morphology and coloration, but their respective distribution ranges do not overlap with that of *B. nigrescens*.

Bolitoglossa obscura
Tapantí Black Webfoot Salamander **Vulnerable**

The only known individual measures 3.46 in (88 mm) in standard length. One of several large black sala-manders that occurs in the uplands of the Talamanca Mountain Range in Costa Rica and adjacent Panama. *Bolitoglossa obscura* may not be identifiable in the field based on appearance alone, but it is the only large, entirely black salamander found at middle elevations on the Atlantic slope of the northern Talamanca Mountain Range in central Costa Rica.

Ecoregions
1 2 3 4 5

Bolitoglossa obscura is endemic to Costa Rica and known only from a single adult female, collected in 1969 from Quebrada Valverde in Tapantí National Park, Cartago Province; 5,100 ft (1,555 m).

Natural history

Distinguished from the other large, black salaman-ders without light markings mainly by its dental characteristics, minor variations in digital web-bing, and proportional differences in morphology.

This species inhabits undisturbed premontane rainforest habitat. The single adult female known was collected right after she laid 31 eggs, each measuring 0.27 in (7 mm) in diameter. There is insufficient information to accurately assess its conservation status, but the type locality of *B. obscura* is protected in a national park that seems to contain appropriate habitat. Nevertheless, no additional specimens have been found since its original discovery in 1969.

Description

Adpressed limbs are separated by 2.5 costal folds.

- A large and robust salamander; its tail is shorter than its standard length.
- Limbs short and stout.
- Head fairly robust, with a short, truncated snout and moderately developed nasolabial protuberances.
- Uniform dark gray to black with minimal light speckling on the head. The soles of its hands and feet are unpigmented.

Hands and feet large, with bluntly truncated finger and toe tips that protrude significantly beyond the moderate interdigital webbing.

foot

Similar species

B. robusta (p. 80) is easily distinguished by the presence of a light colored ring at the base of its tail. *B. obscura* is broadly sympatric with several other species (*B. diminuta* [p. 54], *B. epimela* [p. 56], *B. gracilis* [p. 59], and *Nototriton picadoi* [p. 106]) in the salamander-rich Tapantí area; all are much smaller and not solid black.

The other large black salamanders in Costa Rica (*B. anthracina* [p. 44], *B. nigrescens* [p. 70], and *B. sombra* [p. 84]) are very similar in morphology and coloration, but their respective distribution ranges do not overlap with *B. obscura*.

Bolitoglossa pesrubra
Red-legged Webfoot Salamander

Vulnerable

Standard length to 2.64 in (67 mm), maximum recorded total length 5.16 in (131 mm). Males considerably smaller than females. A small to medium-sized highland salamander with moderate digital webbing and clearly exposed finger and toe tips; coloration varies, but often with a dark head, body, and tail and contrasting red or orange limbs.

Ecoregions
1 2 3 4 5

As currently understood, *Bolitoglossa pesrubra* is a Costa Rican endemic, restricted to the higher elevations of the Talamanca Mountain Range; 6,150–11,875 ft (1,870–3,620 m).

Natural history

B. pesrubra shows great variation in color and pattern. Future research may show that some of the color morphs in fact represent cryptic species that warrant distinct species status.

Due to its extensive elevational range, *B. pesrubra* occupies a variety of habitats in the Talamanca Mountain Range. At lower elevations below the tree line, it is semi-arboreal and often inhabits arboreal bromeliads and moss-cushions on branches. At higher elevations, it occurs in páramo, where it is terrestrial and lives in leaf litter, in or under logs, under rocks, in clumps of terrestrial moss, and in terrestrial bromeliads.

Active salamanders have been observed on nights when temperatures were 44–55 °F (6.4–12.8 °C); at lower temperatures, they remain in hiding. During the drier months of the year, this species restricts its surface activity, as dry windy nights significantly increase the risk of dehydration. Whenever temperature and humidity conditions are suitable at night, individuals actively explore crevices for insect prey. Prey, which includes a wide variety of insects and other small invertebrates, is captured using a projectile tongue.

Due to its low metabolism and cold living conditions, these salamanders are estimated to grow very slowly. Males reach sexual maturity at age 6, while females may take up to 12 years; *B. pesrubra* may live as long as 18 years. Like other Costa Rican salamanders, this species reproduces independent of water, and deposits its eggs in moist micro-sites on land (under logs or rocks), or in

arboreal bromeliads at lower elevations. The eggs develop directly into fully formed young and skip a free-swimming larval stage. *B. pesrubra* produces clutches of 13–38 eggs, which are guarded by an adult, usually the female. The parent attends the clutch continuously for the 4- to 5-month incubation period, coiling around the eggs, with its head and throat resting on top of the egg mass. Every few days the eggs are rotated, likely to ensure that all eggs are in regular contact with the adult's skin. Antibiotic and antifungal properties of the amphibian skin may increase egg survival; it has been shown that unattended eggs of this species fail to develop.

Laboratory studies with adult *B. pesrubra* and non-native predators (North American garter snakes) have shown that these salamanders

Small juveniles usually with reddish-brown dorsum and tail and darker sides. Red limbs are diagnostic.

produce potent skin toxins that can cause severe discomfort in at least certain predators. However, very few snakes occupy the high elevation range of these salamanders.

In literature published prior to 2002, *B. pesrubra* is generally reported as *B. subpalmata*. As currently understood, *B. subpalmata* does not inhabit the Talamanca Mountain Range, but rather the Central Mountain Range, Tilarán Mountain Range, and the Guanacaste Mountain Range (Cerro Cacao).

Historically, *B. pesrubra* was abundant in some localities and as a result it became one of the most studied salamanders in Central America. In some areas near the summit of the Cerro de la Muerte, it once reached population densities exceeding 9,000 individuals per hectare (2.5 acres). However, since 1985 or so its numbers have drastically declined;

some populations have declined by more than 90%, though others still remain relatively stable. The reasons for this decline are not clear, but habitat destruction, pollution, climate change, and disease are all likely candidates. *B. pesrubra* appears relatively adaptable to some forms of habitat alteration and is regularly found in disturbed areas such as road sides, power-line cuts, and even garbage dumps, which may indicate that disease is a more likely driver for its decline than habitat quality. However, further study is needed to adequately address this question.

B. pesrubra is still regularly found in many places throughout its range and is ensured protected habitat in several parks and preserves, including Chiripó National Park, Tapantí National Park, and Cerro las Vueltas Biological Reserve.

Description

This species is highly variable in coloration. Two distinct pattern types are recognized: The *pesrubra*-type (*pesrubra* means red-legged) generally has a dark brown to near black head, body, and tail, variously marked with bronze or brown blotches; limbs are red, orange, pink, or yellowish. The *torresi*-type displays a broad, irregularly shaped silver-gray to black middorsal field that extends onto the top of the head and tail and that is variously marked with silver-gray to cream or light reddish-brown blotches. Its limbs may be brightly colored or black.

pesrubra-type

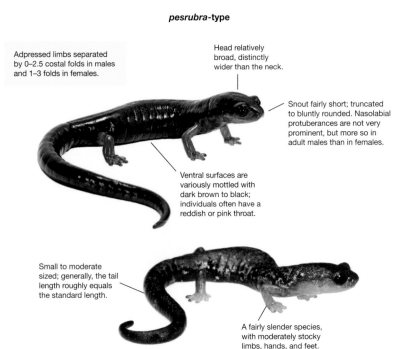

Adpressed limbs separated by 0–2.5 costal folds in males and 1–3 folds in females.

Head relatively broad, distinctly wider than the neck.

Snout fairly short; truncated to bluntly rounded. Nasolabial protuberances are not very prominent, but more so in adult males than in females.

Ventral surfaces are variously mottled with dark brown to black; individuals often have a reddish or pink throat.

Small to moderate sized; generally, the tail length roughly equals the standard length.

A fairly slender species, with moderately stocky limbs, hands, and feet.

torresi-type

Note that on the *torresi*-type the limbs can either be brightly colored or black.

foot

Interdigital webbing leaves a significant portion of the finger and toe tips exposed; digits are bluntly rounded.

At the highest points in its distribution range, above 10,500 ft (3,200 m), individuals are very small and dark, and generally lack red limbs (as seen in these two individuals).

Similar species

B. cerroensis (p. 48) has a color pattern that consists of numerous yellowish-cream speckles that often form a pair of irregular dorsolateral lines; both species may co-occur in certain areas. *B. subpalmata* (p. 91) is similar in appearance, but generally lacks the reddish limbs. There seems to be no range overlap; *B. subpalmata* occupies the mountain ranges north of the Talamanca Mountain Range. *B. sooyorum* (p. 86) has long, skinny limbs; large, moderately webbed hands and feet; and a dark purplish-brown coloration, marked with yellow flecks. It may co-occur with *B. pesrubra* in some areas.

Other salamanders in the *Bolitoglossa subpalmata*-species complex that also occur in the Talamanca Mountain Range (e.g., *B. bramei* [p. 47], *B. gomezi* [p. 57], *B. gracilis* [p. 59], and *B. tica* [p. 94]) may be difficult to distinguish from *B. pesrubra*, although none of those species has reddish limbs.

Bolitoglossa pygmaea
Pygmy Webfoot Salamander

Not assessed

Standard length to 1.46 in (37 mm), total length less than 2.75 in (70 mm). Females larger than males. A very small webfoot salamander, among the smallest in the genus, with a short tail and only moderately webbed hands and feet. The coloration of live individuals is not known, but this species is likely pale, with a pair of darker dorsolateral lines and a distinctive, black-pigmented area on the stomach.

Ecoregions
1 2 3 4 5

As currently understood, this species is endemic to the Fábrega Massif of extreme western Panama. This highland plateau has a mean elevation of more than 9,850 ft (3,000 m) and is located north of the highest peak in the area, Cerro Fábrega (10,940 ft [3,335 m]). The Panama site lies within 3 mi (5 km) of the Costa Rican border, and thus *Bolitoglossa pygmaea* may eventually be found in Costa Rica.

Natural history

Based on its morphological features, *B. pygmaea* is considered to be a member of the *B. subpalmata*-species group (p. 91).

The color of live individuals is not known, but based on the nearly pigmentless preserved specimens it likely has a pale dorsum. Markings include a darker pair of dorsolateral lines that extend from the snout along the sides of the neck and body, and sometimes continues onto the tail. Dark markings appear on top of the eyelids. Most distinctive is a conspicuous black patch of clearly visible internal pigment on the animal's ventral surfaces; this patch covers the stomach and posterior abdomen, near the hind limbs. The function of this black internal pigmentation is unclear, though the same feature has been observed on internal integuments of some Costa Rican highland lizards (e.g., *Sceloporus*

malachiticus and *Mesaspis monticola*); the pigmentation perhaps helps the animals thermoregulate in cool high-elevation habitats by increasing the body temperature around critical organs. Based on the location of the black pigment in *B. pygmaea*, it may benefit digestion of food items in the stomach, or possibly play a role in the development of eggs.

The type series of these salamanders was apparently discovered in grassy tussocks in an open area on the treeless plateau of the Fábrega Massif. Only 25 individuals were collected (during a 10-day period in early March), providing some insight into the active period and population density of this species. These collections and observations were made in 1984, prior to the arrival of chytrid fungal pathogens in the region. The current population status is unknown, and further research is needed.

Description

Adpressed limbs separated by 2.5 costal grooves.

B. pygmaea was identified and described as a new species in 2009 based on a series of preserved specimens collected in 1984. There are no known photographs of a live individual.

- Very small, with a short tail (length never exceeds the standard length).
- Head broad, with a rounded snout and large nostrils; adult males with relatively small nasolabial protuberances.

Limbs short, with fairly large hands and feet; digits extend well beyond the moderate webbing and have bluntly rounded tips.

foot

Bolitoglossa minutula (p. 68), another small webfoot salamander, occurs in the same general region but differs in having fully webbed hands and feet.

   Bolitoglossa robinsoni (p. 78) is a member of the *Bolitoglossa subpalmata*-species group (p. 91). Though significantly larger than *B. pygmaea*, it does share morphological traits. *B. robinsoni*, as currently understood, may include a number of cryptic species. It may not be possible to distinguish between the two species in the field.

Bolitoglossa robinsoni
Robinson's Web-footed Salamander

Not assessed

Standard length to 2.48 in (63 mm), total length to at least 4.84 in (123 mm). Females likely somewhat larger than males. This is one of the larger salamanders within its distribution range; it has a robust body, relatively long and moderately webbed digits, and, in some populations, extensive yellowish spotting on the body, limbs, and tail. Variation in morphology and coloration between known populations suggests that this species may represent an amalgam of as many as four cryptic forms, complicating exact physical description and identification in the field.

Ecoregions
1 2 3 4 5

Isolated populations of salamanders currently assigned to *Bolitoglossa robinsoni* occur on the continental divide, in the border region with Panama. Known localities include the Valle del Silencio, Cerro Biné, Cerro Burú, and Cerro Echandi, in the Talamanca Mountain Range, in southern Costa Rica; this species also occurs in an adjacent area of Panama, on a high plateau of the Fábrega Massif; 8,040–10,940 ft (2,450–3,335 m).

Natural history

Not much is known about the natural history of *B. robinsoni*. It is apparently reasonably common in its appropriate habitat, high-elevation páramo and relatively undisturbed montane forests.

The exact taxonomic placement of the salamanders assigned to *B. robinsoni* requires further study.

As currently understood, these animals are most closely related to *B. aureogularis* (p. 45), which occurs outside of the known range of *B. robinsoni*. Based on molecular analysis, this combined '*robinsoni-aureogularis*' group represents a sister group to the larger *B. subpalmata*-species group (p. 91).

Description

The number of costal folds visible between adpressed limbs varies between populations. In southernmost Costa Rica, 1.5–3.5; in Valle del Silencio, in the northern limit of its range, 0.5–2.0.

Stocky and moderate-sized; fairly short tail generally less than the standard length.

Head slightly wider than neck, with a rounded snout and small nostrils. Nasolabial protuberances small, but likely more developed in adult males than in females.

Limbs robust and fairly long.

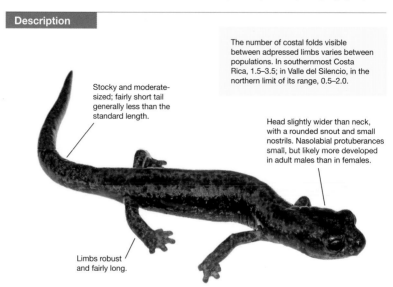

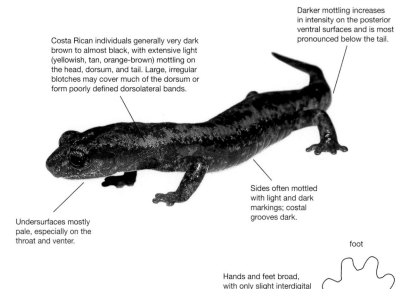

Costa Rican individuals generally very dark brown to almost black, with extensive light (yellowish, tan, orange-brown) mottling on the head, dorsum, and tail. Large, irregular blotches may cover much of the dorsum or form poorly defined dorsolateral bands.

Darker mottling increases in intensity on the posterior ventral surfaces and is most pronounced below the tail.

Sides often mottled with light and dark markings; costal grooves dark.

Undersurfaces mostly pale, especially on the throat and venter.

foot

Hands and feet broad, with only slight interdigital webbing; all fingers and toes are bluntly truncated and protrude well beyond the edges of the webbing.

Similar species

Other potentially co-occurring members of the *Bolitoglossa subpalmata*-species group (e.g., *B. bramei* [p. 47] and *B. gomezi* [p. 57]) occur at slightly lower elevations; all three are similar and may not be easily distinguished in the field.

B. pygmaea (p. 76) co-occurs with *B. robinsoni* on the Fábrega Massif but is much smaller and has different coloration. *B. marmorea* (p. 66) has even broader hands and feet but is similar in coloration and may be difficult to distinguish in the field. *B. sombra* (p. 84) has narrower hands and feet and is uniform black in coloration. *B. compacta* (p. 52) has a striking pattern of red-orange blotches or stripes on a jet black ground color.

Bolitoglossa robusta
Ring-tailed Webfoot Salamander

Least concern

Standard length to 5.28 in (134 mm), maximum known total length 10.24 in (260 mm). Females significantly larger than males. A large salamander with partly webbed hands and feet, a uniform dark brown to black coloration, and a unique light (cream, yellowish, or pale orange) ring encircling the base of the tail.

Ecoregions
1 2 3 4 5

Bolitoglossa robusta is the most widely distributed large black salamander in Central America, occurring from northwest Costa Rica (Cacao Volcano, in the Guanacaste Mountain Range), throughout the country's other mountain ranges, to western Panama, near La Fortuna (Chiriquí Province). Although one individual has been reported from an elevation as low as 1,640 ft (500 m), it is most commonly encountered at 4,920–6,725 ft (1,500–2,050 m).

Natural history

This is a nocturnal and terrestrial salamander that inhabits humid foothills and montane forests. Although it sometimes climbs up low vegetation, it is mostly seen in the leaf litter layer of pristine rainforests. Individuals have been found during the day hidden under logs, rocks or moss, inside decaying stumps, and occasionally in arboreal bromeliads.

Although it is infrequently seen, *B. robusta* is relatively widespread, and much of its preferred habitat is protected in reserves and parks.

B. robusta is the largest salamander in Costa Rica. In spite of its large size, and even though it generally lives on the forest floor, it is rarely seen.

Description

Eyes big and protruding; snout short and somewhat pointed, with weakly developed nasolabial protuberances.

Limbs robust and fairly short.

2–4 costal folds separate adpressed limbs.

Head large but not clearly differentiated from neck.

Ventral coloration black or slate, mottled with gray.

Has uniform, dark (brown to black) dorsal coloration and a ring (white, cream, yellow, orange, or reddish) around the base of the tail. In some individuals, the ring is suffused with black pigment and it is not easy to see. Juveniles with bright yellow to orange feet.

Tail moderately long, with a distinct constriction at its base.

Dark dorsal coloration sometimes marked with white or reddish flecking, especially on the sides and lower surfaces of the tail and limbs. The dorsal surfaces of the limbs may be marked with brown or copper blotches.

foot

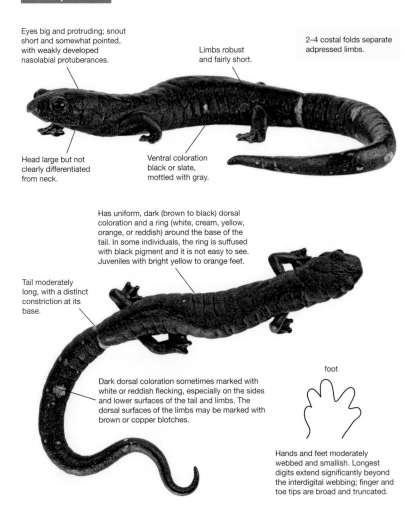

Hands and feet moderately webbed and smallish. Longest digits extend significantly beyond the interdigital webbing; finger and toe tips are broad and truncated.

Similar species

Other large black salamanders of the *Bolitoglossa nigrescens*-complex (*B anthracina* [p. 44], *B. nigrescens* [p. 70], *B. obscura* [p. 72], and *B. sombra* [p. 84]) are uniform black and never have a light ring at the base of the tail.

B. compacta (p. 52) displays a pattern of bright red to orange spots on its black background, but never a ring around the tail base.

Bolitoglossa schizodactyla
Wake's Webfoot Salamander

Least concern

Standard length for male and female to 2.44 in (62 mm); maximum known total length 5.79 in (147 mm). A moderate-sized salamander with completely webbed hands and feet. Light venter often marked with a bright yellow band, sometimes bisected by a midventral black stripe.

Ecoregions

Bolitoglossa schizodactyla is common in Panama, from El Valle de Antón (Cocle Province) and La Campana (Panama Province) on the Pacific slope; on the Atlantic slope, it occurs from central Panama to southeastern Costa Rica. On the Pacific slope, it is found from 985–2,790 ft (300–850 m); on the Atlantic slope, near sea level to 2,550 ft (780 m). Only one individual has been documented in Costa Rica, in 1984. However, as it occurs commonly in adjacent Panama, including on several islands in the Bocas del Toro region, it is likely to reappear in Costa Rica.

Natural history

This species occurs in undisturbed humid forests of lowlands and foothills. Like other webfoot salamanders, it is nocturnal and arboreal. *B. schizodactyla* is still regularly encountered in El Cope and other locations in Panama during nighttime surveys. Generally, individuals are found on large-leaved plants such as heliconias and understory palms. Although *B. schizodactyla* populations appear stable in appropriate habitat in protected areas, development, logging, and other alterations to habitat have taken a toll on the suitable available habitat in other places. In addition, this species was shown to be susceptible to the fungal pathogen *Bd*.

B. schizodactyla is quite variable in its coloration, but the fully webbed hands and feet are distinctive.

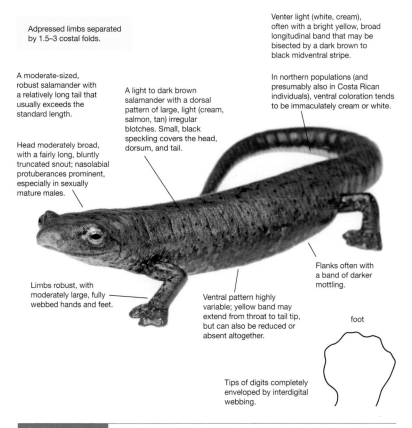

Adpressed limbs separated by 1.5–3 costal folds.

Venter light (white, cream), often with a bright yellow, broad longitudinal band that may be bisected by a dark brown to black midventral stripe.

A moderate-sized, robust salamander with a relatively long tail that usually exceeds the standard length.

A light to dark brown salamander with a dorsal pattern of large, light (cream, salmon, tan) irregular blotches. Small, black speckling covers the head, dorsum, and tail.

In northern populations (and presumably also in Costa Rican individuals), ventral coloration tends to be immaculately cream or white.

Head moderately broad, with a fairly long, bluntly truncated snout; nasolabial protuberances prominent, especially in sexually mature males.

Limbs robust, with moderately large, fully webbed hands and feet.

Ventral pattern highly variable; yellow band may extend from throat to tail tip, but can also be reduced or absent altogether.

Flanks often with a band of darker mottling.

foot

Tips of digits completely enveloped by interdigital webbing.

B. colonnea (p. 50) has fully webbed hands and feet but is distinguished by a fleshy, raised ridge across the head. *B. alvaradoi* (p. 42) has fully webbed hands and feet and a pattern of irregular light blotches on a dark background, but it has a much darker base color and a black venter. *B. striatula* (p. 89) is similar in size and has fully webbed hands and feet, but possesses a unique cream to yellow coloration streaked by many dark brown longitudinal lines on the head, body, and tail.

Bolitoglossa sombra
Shadowy Webfoot Salamander

Vulnerable

Standard length to 3.27 in (83 mm). Males considerably smaller than females. One of several large black salamanders that occurs in the uplands of the Talamanca Mountain Range, in Costa Rica and adjacent Panama. *Bolitoglossa sombra* may be difficult to identify in the field based on its morphology or coloration, but it is the only large, entirely black salamander found in its small range.

Bolitoglossa sombra occurs between Las Tablas and Cerro Frantzius, on the Pacific slope of the Talamanca Mountain Range, in southwestern Costa Rica and in adjacent western Panama; 4,925–7,550 ft (1,500–2,300 m).

Ecoregions
1 2 3 4 5

Natural history

This is a nocturnal species; individuals have been reported active at night on top of leaf litter, between buttressed roots of a tree, on moss-covered logs, and just above the ground on stumps and tree trunks. One individual was extracted from the leaf axil of a bromeliad during the day, presumably a preferred daytime retreat. *B. sombra* is an agile and active salamander; an individual was observed climbing rapidly up a tree trunk. It is known to use its prehensile tail to anchor itself as it performs its arboreal escapades.

This is essentially a dark salamander, although a few reports describe its ability to change color in response to differing light levels. A subadult *B. sombra* reportedly appeared light gray when captured, but turned black in captivity, revealing two white patches on the lower tail base. Other individuals have been observed to change variously from mottled charcoal, grayish-black, and pale brown-black patterns to a uniform dark, blackish-brown.

B. sombra occurs in relatively undisturbed rainforests, in foothills and montane habitats. It has also been observed in fairly close proximity to human settlements and on man-made structures. Due to its small distribution range and sensitivity to habitat alteration, this species is considered vulnerable. It occurs in protected areas in Costa Rica such as La Amistad International Park and the Las Tablas Protected Area, but even in pristine habitat it may still be at risk from chytrid fungal infections.

Recent morphological studies have suggested that *B. sombra* and *B. nigrescens* (p. 70) may be representatives of a single species.

B. sombra is one of several large black salamanders found in the highlands of Costa Rica.

Adpressed limbs separated by 1.5 costal folds in males and 2 in females.

A moderately large and relatively slender species; it is smaller than the other all-black salamanders.

Skin relatively rugose and glandular, most noticeably on top of the head.

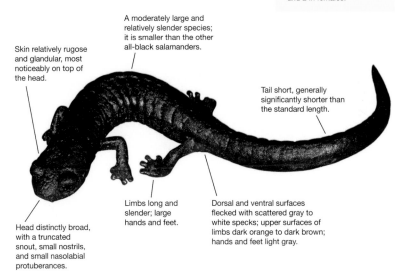

Tail short, generally significantly shorter than the standard length.

Limbs long and slender; large hands and feet.

Dorsal and ventral surfaces flecked with scattered gray to white specks; upper surfaces of limbs dark orange to dark brown; hands and feet light gray.

Head distinctly broad, with a truncated snout, small nostrils, and small nasolabial protuberances.

foot

Shows distinct sexual dimorphism; apart from size difference, males have more developed nasolabial protuberances, a shorter tail, and a wider head. Subadults and juveniles with small white patches on the flanks.

Fingers and toes bluntly truncated and significantly free of webbing, especially the longest digits.

Similar species

B. sombra may co-occur with *B. compacta* (p. 52); *B. compacta* is distinguished by bright red-orange blotches on its body and tail. *B. marmorea* (p. 66) is a mottled brown. *B. robusta* (p. 80) is a large black salamander typically distinguished by a light ring around its tail base.

The other large black salamanders of the *Bolitoglossa nigrescens*-complex in Costa Rica (*B. anthracina* [p. 44], *B. nigrescens* [p. 70], and *B. obscura* [p. 72]) are very similar in morphology and coloration, but their respective distribution ranges do not overlap with that of *B. sombra*.

Bolitoglossa sooyorum
Yellow-dotted Webfoot Salamander

Endangered

Standard length to 2.83 in (72 mm), maximum known total length 5.59 in (142 mm). Females slightly larger than males. A moderate-sized, slender montane salamander with long, skinny limbs; large, moderately webbed hands and feet; and a dark purplish-brown coloration marked with yellow flecks.

Ecoregions
1 2 3 4 5

Only known from the upper reaches of the northern Talamanca Mountain Range, on both the Atlantic and Pacific slopes; 7,725–9,850 ft (2,355–3,000 m). It may also occur in western Panama.

Natural history

This rare species favors wet forests on steep slopes and areas of talus (rock falls) located at high elevations. *B. sooyorum* seems to have decreased in numbers over the past years, though the reasons for this decline are unclear. Habitat loss, climate change, and disease are possible explanations. The latter two factors would explain why it has disappeared from seemingly pristine, suitable habitat.

Description

Head large and distinctly wider than neck; has large eyes and a bluntly rounded snout.

Displays a purplish-brown background color flecked to varying degrees with distinct yellow spots.

Adpressed limbs almost touch one another (separated by less than 1 costal fold).

Moderate-sized and slender, with a fairly long tail that measures about the same length as the standard length.

foot

Hands and feet are light yellowish-brown and contrast with dark dorsum and limbs.

Light markings most prominent on the limbs, hands, and feet.

Hands and feet large; fingers and toes have rounded tips and are significantly free from webbing.

Similar species

B. cerroensis (p. 48) has shorter limbs, smaller hands and feet, and a color pattern that consists of numerous yellowish-cream speckles that often fuse to form a pair of irregular dorsolateral lines; both species may co-occur in certain areas. *B. pesrubra* (p. 73) is stockier and usually has yellow, orange, or red limbs. *B. sooyorum* is morphologically similar to *B. marmorea* (p. 66), but apparently differs considerably in molecular features. The two species are separated geographically and can be most easily identified based on where they are found.

Bolitoglossa splendida
Splendid Webfoot Salamander

Not assessed

1.89 in (48 mm) in standard length. Unmistakable; the only Costa Rican black salamander with a striking orange-red band down the center of its back and bright yellow spots on its lower sides.

Ecoregions
1 2 3 4 5

A Costa Rican endemic, known only from a single female found in the upper reaches of the Río Coén region on the Atlantic slope of Cerro Arbolado, in the Talamanca Mountain Range; 5,985 ft (1,825 m).

Natural history

The only known individual of this species was discovered in 2007 and described in 2012. It was found at night, in mature cloud forest, and observed as it actively climbed about on a heliconia leaf, slightly higher than 3 ft (1 m) above the ground. At the time of discovery, the ambient air temperature was 55 °F (13 °C); the salamander's actvity may have been triggered by heavy rains that had occurred earlier that day.

Bolitoglossa splendida is considered a member of the *B. subpalmata*-species group (p. 91) and is most closely related to the other Costa Rican montane salamanders in this group (*B. bramei*, *B. gomezi*, *B. gracilis*, *B. kamuk*, *B. pesrubra*, *B. subpalmata*, and *B. tica*). Members of this speciose group of medium-sized salamanders occupy discrete distibution ranges at middle and upper elevations of the Talamanca Mountain Range, and different species rarely overlap in occurrence. Generally quite cryptic, none of the other members of this species group display the striking color pattern of *B. splendida*. The purpose of the bright coloration of *B. splendida* is not known and requires further study.

Description

Spectacular coloration; overall glossy black, with a broad, bright orange-red dorsal band that runs from the back of the head onto the tail stump (presumably further if the tail is complete).

Limbs short; adpressed limbs separated by 3.5 costal grooves.

Small to moderate-sized, slender.

Head broad, with rounded snout; eyes and nostrils small, nasolabial protuberances prominent.

Lower sides marked with irregularly shaped bright yellow spots on a black background.

The tail of this, the only known individual, is missing, and no estimate of total length can be made.

foot

Hands and feet moderate in size, with rounded digits that protrude significantly beyond the limited webbing.

Similar species

Bolitoglossa pesrubra (p. 73) often has a glossy black body with red limbs, but never an orange-red dorsal band and yellow spots.

Bolitoglossa striatula
Striated Webfoot Salamander

Least concern

Standard length to 2.56 in (65 mm); maximum known total length 5.12 in (130 mm). A smallish, slender salamander with completely webbed hands and feet. Displays a unique color pattern: yellowish-tan to reddish-brown with distinctive dark brown longitudinal bands on the sides, back, and belly, interspersed with irregular light lines.

Ecoregions
1 2 3 4 5

Bolitoglossa striatula inhabits humid lowlands and foothills. On the Atlantic slope of Central America, it occurs from northeastern Honduras to southeastern Costa Rica; on the Pacific slope, it occurs only in Costa Rica, in the Guanacaste Mountain Range and, locally, between Río Barranca and Río Barú; from near sea level to 4,525 ft (1,380 m).

Natural history

This relatively common salamander is most frequently encountered during warm, rainy nights, when it forages on vegetation for invertebrate prey. During the day, *B. striatula* retreats under logs or hides in leaf litter, where its cryptic coloration makes it almost invisible. It favors floodplains, edges of ponds, and other wetland types, and is most frequently encountered on leaves of low herbaceous and woody plants, tall grasses, reeds, and other kinds of plants that are found in wet areas.

Due to the dynamic character of its habitat, which regularly floods, *B. striatula* is relatively tolerant of habitat alteration, and individuals can be found in cultivated areas, plantations, and grassy vegetation. As a result, the creation of such open habitat areas due to the removal of native forest may actually be beneficial to this species.

Several species of *Bolitoglossa* are known to use their heavy tail as leverage during encounters with potential predators. They fling themselves off vegetation with a flick of the tail and drop into the relative safety of the leaf litter below. In an attempt to better camouflage itself, *B. striatula* will fold its limbs tightly against the body, flatten the padlike hands and feet against its ventral surfaces, and close its eyes to resemble a dead branch. When in this state, if they are touched with a finger they will simply roll back and forth in the leaf litter without extending their limbs. When camouflage fails, *B. striatula* uses other defensive strategies. If

restrained by the tail, the tail can break off easily (it will eventually regrow in its entirety). An unexpected case of predation on this species was recently reported when a Central American coral snake (*Micrurus nigrocinctus*) was observed regurgitating an adult *B. striatula*.

During the day, these salamanders retreat into the leaf litter and curl up into a tight coil, usually inside a rolled up dead leaf, with their hands and feet wrapped around the tail to hold the coils tightly in place. *B. striatula* is capable of significant color change in response to changing light levels between night and day. Its dark markings may fade at night, especially markings on the ventral surfaces, and the lateral brown bands may become bordered below by entirely cream undersurfaces.

During the day, these salamanders curl up into a tight coil.

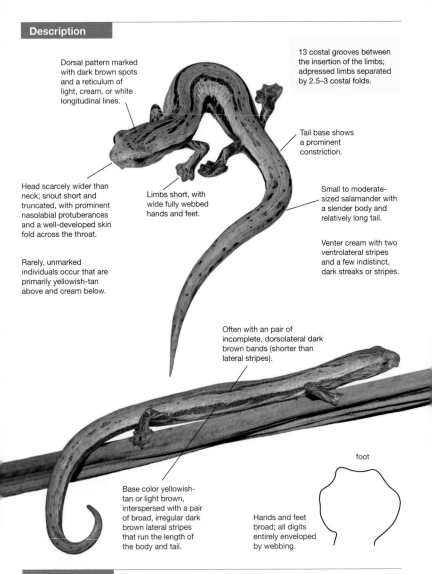

13 costal grooves between the insertion of the limbs; adpressed limbs separated by 2.5–3 costal folds.

Dorsal pattern marked with dark brown spots and a reticulum of light, cream, or white longitudinal lines.

Tail base shows a prominent constriction.

Head scarcely wider than neck; snout short and truncated, with prominent nasolabial protuberances and a well-developed skin fold across the throat.

Limbs short, with wide fully webbed hands and feet.

Small to moderate-sized salamander with a slender body and relatively long tail.

Venter cream with two ventrolateral stripes and a few indistinct, dark streaks or stripes.

Rarely, unmarked individuals occur that are primarily yellowish-tan above and cream below.

Often with an pair of incomplete, dorsolateral dark brown bands (shorter than lateral stripes).

foot

Base color yellowish-tan or light brown, interspersed with a pair of broad, irregular dark brown lateral stripes that run the length of the body and tail.

Hands and feet broad; all digits entirely enveloped by webbing.

Similar species

B. colonnea (p. 50) has a distinctive fleshy ridge across the top of its head. *B. lignicolor* (p. 64) only co-occurs with *B. striatula* in the southern Pacific lowlands (Barú area). It is similar in shape, size, and general appearance but lacks the brown longitudinal bands and light stripes. *B. alvaradoi* (p. 42) has a dark brown to black dorsal coloration, marked with a few, large, reddish-brown to tan blotches. *B. indio* (p. 60) has a dark brown dorsum with a pair of irregularly shaped, light dorsolateral bands.

Bolitoglossa subpalmata
Montane Webfoot Salamander

Endangered

Standard length to 2.36 in (60 mm), maximum known total length 5.16 in (131 mm). Males slightly smaller than females. This small to medium-sized highland salamander has moderately webbed hands and feet and prominently exposed finger and toe tips. Virtually impossible to distinguish from other salamanders in the *Bolitoglossa subpalmata*-species group by field characteristics alone, but all other known species occur in the Talamanca Mountain Range.

Ecoregions

Bolitoglossa subpalmata is endemic to Costa Rica and occurs in the Guanacaste, Tilarán, and Central Mountain Ranges of Costa Rica, on both Atlantic and Pacific slopes; 4,085–9,500 ft (1,245–2,900 m).

Natural history

B. supalmata is a nocturnal, secretive species. At night, it can be found actively climbing about on arboreal bromeliads or other vegetation; during the day, it hides under logs, rocks, and surface debris.

This species occurs at middle and high elevations, primarily in rainforest and cloud forest. It occasionally enters secondary growth, isolated trees in pastures, and other altered habitats, in areas where canopy cover is adequate to maintain sufficiently humid conditions. Forest fragmentation and agricultural encroachment leads to dry, open habitats that are uninhabitable for this species. Although its abundance is difficult to assess, *B. subpalmata* appears to have declined even in areas with seemingly undisturbed habitat, though no specific threat has been identified.

Several cryptic species may still be included in this variable, complex form, and geographic provenance seems to be the most reliable aid in identifying *B. subpalmata*. A number of highland salamanders in the *B. subpalmata*-species complex that occur in the Talamanca Mountain Range have been split off in recent years from *B. subpalmata* proper and are now considered unique species, several with a small distribution range. In the Talamanca Mountain Range, the most wide-ranging and common member of the *B. subpalmata*-species complex is *B. pesrubra*, which is morphologically similar to *B. subpalmata* but differs on a molecular level. As currently understood, all *subpalmata*-like highland salamanders found north of the Talamanca Mountain Range are

considered *B. subpalmata*, even though it is known that populations from the Tilarán and Guanacaste Mountain Ranges likely represent valid undescribed species. Further research is needed to formally describe these potentially new species.

B. subpalmata is known from several protected areas, including Cacao Volcano, Rincón de La Vieja National Park, Monteverde Cloud Forest Preserve, Poás Volcano, Irazú Volcano, Turrialba Volcano, and Braulio Carrillo National Park.

B. subpalmata is a mostly arboreal species.

Adpressed limbs separated by 1-3 costal folds.

Generally with some red in its coloration; many individuals have reddish limbs.

juvenile

Ventral surfaces vary from light (pinkish or tan) to dark (brown or black), with or without spots.

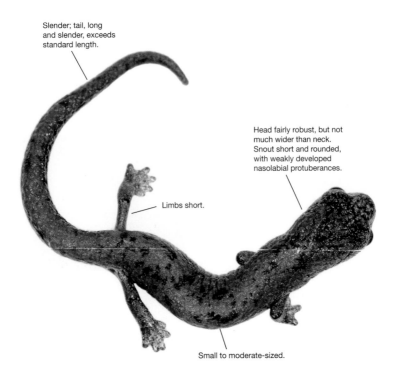

Slender; tail, long and slender, exceeds standard length.

Head fairly robust, but not much wider than neck. Snout short and rounded, with weakly developed nasolabial protuberances.

Limbs short.

Small to moderate-sized.

Dorsal coloration may be mottled or uniform reddish-brown, gray, or almost black; some individuals have a broad, middorsal light band.

Adult coloration extremely variable; several cryptic species are likely included in what is currently considered *B. subpalmata*.

Juveniles usually with proportionally larger eyes and nostrils and rugose skin on top of the head.

foot

Has moderately webbed hands and feet; all but the innermost fingers and toes are free (for a considerable distance) of webbing. Tips of digits bluntly rounded to truncated.

Similar species

B. pesrubra (p. 73) is similar in biology, habitat, and appearance but the two species do not co-occur. *B. robusta* (p. 80) is black, with a light-colored ring around the tail base. *B. nigrescens* (p. 70) is uniform black.

Bolitoglossa tica
Tico Web-footed Salamander

Endangered

Standard length to 2.36 in (60 mm). Largest known individual is a male, but females tend to be larger in other species of *Bolitoglossa*. An agile, dark-colored, arboreal salamander in the *Bolitoglossa subpalmata*-species group (p. 91). Slender, with long limbs. Broad hands and feet moderately webbed, with bluntly rounded digits. Its head is distinctly wider than its neck. A pattern of prominent white spots on its undersides, especially near the tail base, is diagnostic.

Ecoregions
1 2 3 4 5

Bolitoglossa tica is endemic to Costa Rica. It occurs on the northwestern slopes of the Talamanca Mountain Range, in an area along the Interamerican Highway south of El Palme; 7,225–8,200 ft (2,200–2,500 m). It also occurs in the Cerros de Escazú, on the southern boundary of the Central Valley, between the cities of Alajuela and Cartago, at elevations as low as 5,725 ft (1,745 m).

Natural history

This arboreal salamander is nocturnal. It is most frequently seen at night as it actively moves about in arboreal bromeliads, epiphytic clubmosses, ferns, and other vegetation, from 6 in (15 cm) to 10 ft (3 m) above the ground. During the day it retreats into bromeliads, under bark, or in leaf litter on the forest floor. These salamanders occur at high elevations, in humid cloud forest and in oak forests with mean annual temperatures that are relatively cool: 54–59 °F (12–15 °C). Individuals have been encountered in road cuts and disturbed oak forest, and this species may thus be somewhat tolerant of habitat alteration.

B. tica is known to occur in several protected areas, including the Cerros de Escazú protected zone, Tapantí National Park, and Cerro las Vueltas Biological Reserve. It was described in 2008, based primarily on a type series of several specimens collected in 1986. Multiple individuals were obtained on a single day, early in July that year, prior to the arrival of the chytrid fungal pathogen *Bd* in that area. These salamanders did not appear to persist in similar numbers for several years, but a 2015 survey of one site resulted in the discovery of four individuals in a short period of time, suggesting that this species is currently present in significant numbers. More research is needed to adequately assess the abundance and conservation status of this salamander.

B. tica is most closely related to *B. subpalmata* and *B. pesrubra* but differs significantly on a molecular level. In addition, *B. tica* is a more agile and active salamander that tends to be more arboreal than the other two species.

As it moves through vegetation at night, *B. tica* uses its prehensile tail as an anchor.

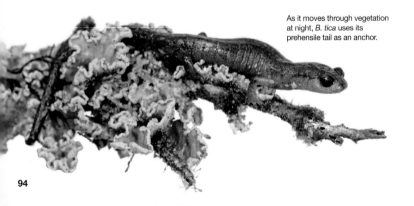

Description

A very dark (dark gray to nearly black) salamander during the day; at night, the dorsal coloration changes to reddish-brown or gray.

ventral view

Dorsum uniformly colored or patterned with irregular middorsal or dorsolateral stripes.

daytime coloration

Throat light; often a dark maroon or brown spot present on the chest, separating the light throat from the otherwise grayish ventral surfaces.

Well-defined head is distinct from the neck.

Lower sides and ventral surfaces are dark gray to black, contrasting sharply with any brown dorsal markings.

Prominent white spots in the area below the tail base may extend onto the underside of the body and tail.

nighttime coloration

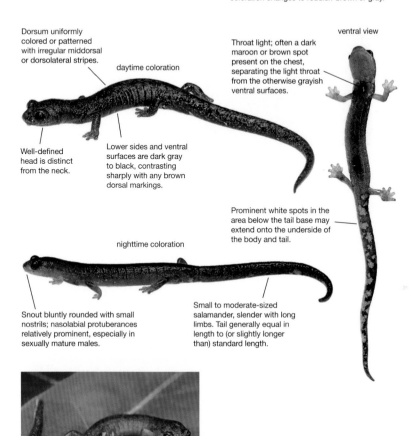

Snout bluntly rounded with small nostrils; nasolabial protuberances relatively prominent, especially in sexually mature males.

Small to moderate-sized salamander, slender with long limbs. Tail generally equal in length to (or slightly longer than) standard length.

juvenile

Juveniles with a dorsal pattern of bright reddish-brown or dark bronze-brown, marked with black streaks and cream-tipped nasolabial protuberances.

foot

Hands and feet are fairly broad, with clearly differentiated, bluntly rounded finger and toe tips that are exposed beyond the moderate webbing.

Similar species

B. pesrubra (p. 73) has smaller hands and feet, lacks the white spotting on the undersurfaces, and has a very different coloration; dark head, body, and tail contrast with red limbs. *B. gracilis* (p. 59) is smaller, with a shorter tail, a yellowish dorsal coloration, and black midventral stripe. *B. minutula* (p. 68) is much smaller and has fully webbed hands and feet.

Genus *Nototriton*
Moss Salamanders

Seventeen species of moss salamander are currently recognized; eight occur in Costa Rica, the other species are found in Nicaragua, Guatemala, and Honduras.

These are all small, stocky salamanders with 13–14 costal grooves, short limbs, and narrow hands and feet. Some species have syndactylous hands and feet, meaning that individual digits are (partly) fused into a pad-like structure encased in skin (interdigital webbing). However, in most moss salamanders, at least the tips of the digits extend visibly beyond the interdigital webbing. The condition of the hands and feet in *Nototriton* is somewhat intermediary between the broadly expanded and often extensively webbed hands and feet of *Bolitoglossa* (webfoot salamanders) and the extremely narrow, often syndactylous hands and feet of *Oedipina* (worm salamanders). Moss salamanders also lack the wormlike body and greatly elongated tail seen in salamanders of the genus *Oedipina*.

Costa Rican salamanders in the genus *Nototriton* are invariably mottled brown and cryptic. All have a distinct toxin-producing parotoid gland on each side of the neck; the color of this gland sometimes contrasts with the background color and is fairly visible.

Distinguishing moss salamander species in the field is very difficult because of their small size and the subtlety of the distinguishing features. In addition, most of the described species are known from very few specimens, and the full range of variation within and between species is poorly understood. *Nototriton* species are generally found in isolated populations on disjunct mountains. The distribution ranges of the eight Costa Rican species in this genus do not overlap and location may thus aid in identification.

Moss salamanders primarily inhabit cloud forest habitat at middle elevations. They owe their common name to the fact that many individuals have been found in moss banks on the ground or in clumps of moss attached to tree branches or trunks. However, *Nototriton* species found in countries north of Costa Rica tend to be mostly terrestrial and are generally found in or under rotting logs.

Moss salamanders undergo direct development and their egg clumps are deposited in the same moist microhabitats that adults retreat into during the day. Limited natural history observations suggest that adult *Nototriton* do not tend to their egg clutches once they have been deposited.

Costa Rican *Nototriton* have been divided into two distinct species groups. Members of the *Nototriton picadoi*-species group have clearly differentiated digits; their longest fingers and toes project significantly beyond the margins of the basal interdigital webbing and their finger and toe tips are rounded. This group consists of six Costa Rican species: *N. abscondens*, *N. gamezi*, *N. guanacaste*, *N. major*, *N. matama*, and *N. picadoi.*

Members of the *Nototriton richardi*-species group have syndactylous hands and feet and all digits are fused to form a single pad; their finger and toe tips are pointed. In addition, the two Costa Rican species included in this group, *N. richardi* and *N. tapanti*, have strikingly large nostrils.

Nototriton abscondens
Cordillera Central Moss Salamander

Least concern

Standard length to 1.30 in (33 mm); maximum known total length 2.76 in (70 mm). Compared to other moss salamanders, *Nototriton abscondens* has a relatively long and wide head, with tiny nostrils and long legs; it is smaller and more slender than most other species in the genus. The longest digits on its hands and feet project beyond the basal interdigital webbing.

Ecoregions
1 2 3 **4** 5

Endemic to Costa Rica. Found in humid regions of the Central Mountain Range; 3,300–8,200 ft (1,010–2,500 m) but most commonly above 4,925 ft (1,500 m).

Natural history

N. abscondens prefers cloud forest habitats. It is found in moss mats that grow on the banks of road cuts in disturbed habitats and on stumps and logs and tree branches in less disturbed areas. Individuals have also been found in arboreal bromeliads. This is a nocturnal species that may be terrestrial in secondary vegetation, but is generally arboreal in undisturbed habitat. Like all Costa Rican salamanders, it breeds through direct development. A clutch of 17 eggs was reportedly found under a mat of liverworts located on the vertical face of a road bank. A female salamander was found with eggs but did not appear to be brooding.

This tiny salamander can be locally common and is often observed in appropriate habitat. Historic records indicate that it formerly occurred within the Central Valley, but extensive habitat loss as a result of urban development undoubtedly reduced its ability to survive there. Fortunately, much of the known distribution range of *N. abscondens* is protected in Braulio Carrillo National Park.

Description

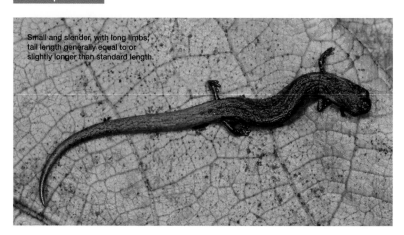

Small and slender, with long limbs; tail length generally equal to or slightly longer than standard length.

In some individuals, costal grooves are indicated by light brown to tan coloration.

Head with bluntly rounded, short snout; prominent parotoid glands in neck.

Mottled brown, occasionally with irregular, light dorsolateral or lateral stripes.

Venter brown, with irregular light spots.

Nostrils small compared with other Costa Rican *Nototriton*.

Fingers and toes not fused, with rounded tips that project significantly beyond basal webbing.

foot

Similar species

The only species of *Nototriton* that may co-occur is *N. richardi* (p. 107), which has syndactylous hands and feet (its digits are almost entirely enveloped in webbing). *N. gamezi* (p. 100) is very similar to *N. abscondens*, and the two were once considered representatives of the same species; it does not co-occur with *N. abscondens* but rather inhabits the Tilarán Mountain Range.

Nototriton gamezi
Tilarán Moss Salamander

Vulnerable

Standard length to 1.02 in (26 mm); maximum known total length is 2.13 in (54 mm). *Nototriton gamezi* has relatively large nostrils, prominent parotoid glands, and short legs. Venter black, with irregular light blotches.

Ecoregions
1 **2** 3 4 5

A Costa Rican endemic, only known from a small area near the continental divide in the Monteverde Cloud Forest Preserve, Alajuela Province, Costa Rica; 5,085–5,400 ft (1,550–1,650 m).

Natural history

Only a handful of *N. gamezi* individuals have been collected. All were found in dense moss mats, either growing on road banks or growing on trees, deep within undisturbed rainforest and cloud forest. Its distribution range is poorly known; as currently understood, its small range falls entirely within the Monteverde Cloud Forest Preserve. This species was previously grouped with *N. abscondens*, from the Central Mountain Range, but the two were split based on molecular characteristics and subtle morphological features. Their natural history is likely similar.

Description

Mottled brown overall but tail is paler brown. Dark brown blotches and white speckling scattered irregularly over dorsal surfaces.

Tiny to small salamander with a slender build.

Head fairly prominent and distinctly wider than neck. Parotoid glands conspicuous on back of head and neck; snout short and broadly truncated with medium-sized nostrils.

foot

Digits well-defined and free from basal webbing for much of their length. Finger and toe tips rounded.

Tail exceeds standard length; robust at base but tapering to a slender tip.

Ventral surface typically black, with irregular light blotches.

Limbs short, with moderate-sized hands and feet.

Similar species

This is the only species of *Nototriton* in the Tilarán Mountain Range; its close relative *N. abscondens* (p. 98) is restricted to the Central Mountain Range. *N. guanacaste* (p. 101) is limited to the Guanacaste Mountain Range.

Nototriton guanacaste
Guanacaste Moss Salamander

Vulnerable

Maximum known standard length 1.34 in (34 mm); reaches a total length of 2.87 in (73 mm). A robust, small moss salamander with a relatively large head and small nostrils. Its limbs are comparatively long; digits are well-defined, not fused.

Ecoregions
1 2 3 4 5

Nototriton guanacaste is endemic to northwestern Costa Rica. It is only known from Orosí Volcano and Cacao Volcano in Guanacaste Province; 4,600–5,200 ft (1,400–1,580 m).

Natural history

This uncommon salamander inhabits arboreal moss mats and bromeliads in wind swept pygmy forests on the northernmost volcanoes of the Guanacaste Mountain Range. Egg clutches found in these same microhabitats contained between 4 and 11 eggs. *N. guanacaste* reproduces through direct development and presumably does not tend to its egg clutches during incubation.

This species lives in an unlikely habitat. Prevailing weather patterns on the northernmost volcanoes of the Guanacaste Mountain Range allow rainforest habitat to form at relatively low

elevations in an otherwise arid region. Protection of this unique habitat is crucial to the survival of *N. guanacaste*; fortunately its entire range lies sheltered within Guanacaste National Park, and its population status appears stable, though vulnerable.

Like other Costa Rican salamanders, *N. guanacaste* has been observed deploying tail-flips and rapid coiling to escape predation. Using their heavy tail for leverage, these tiny creatures can launch themselves as far as 20 in (50 cm) with a single rapid tail flick.

Moss salamanders are small and indistinct, but no other *Nototriton* occur in the Guanacaste Mountain Range and location alone may help identify this species.

Quite variable in coloration; ranges from reddish-brown to bright orange, sometimes with dorsolateral rows of faint, darker spots.

Some individuals with small light markings on the dorsum and on the flanks.

Head proportionally long and broad; slightly wider than neck, with fairly prominent parotoid glands. Parotoid glands often lighter in color than dorsum.

Snout very short and broadly truncated, with small nostrils.

Ventral surfaces dark brown, with irregular light specks.

Small and rather stocky; tail moderate to long, generally longer than standard length. Tail usually not noticeably lighter in coloration than dorsum. Limbs long.

foot

On hands and feet, the longest digits clearly extend beyond basal webbing. Finger and toe tips rounded.

N. guanacaste is most closely related to *N. abscondens* (p. 98), *N. gamezi* (p. 100), *N. matama* (p. 104), and *N. picadoi* (p. 106), but none of those species occur in the Guanacaste Mountain Range.

Nototriton major
Large Moss Salamander
Critically endangered

Standard length of sole individual that has been recorded, an adult male, is 1.50 in (38 mm), total length 3.66 in (93 mm). Females of this species are likely somewhat larger than males. The largest of the Costa Rican moss salamanders, *Nototriton major* can be distinguished from other species of *Nototriton* in its range by its size, relatively short and narrow head, tiny nostrils, and relatively long legs.

Ecoregions
1 2 3 4 5

Nototriton major is endemic to the north-facing slope of the Talamanca Mountain Range in eastern Costa Rica. It was long known only from one individual found in Moravia de Chirripó, Cartago Province; at approximately 3,935 ft (1,200 m).

Natural history

N. major lives in premontane rainforest and is likely more widespread than is currently believed. It is probably nocturnal and may inhabit moss mats or leaf litter, as do its close relatives.

This species was described in 1993 based on a single individual. Many years passed without new sightings, but recently additional *N. major* have been observed.

Description

- Small and slender; moderately long tail is slightly longer than standard length.
- Limbs long and delicate.
- Head short and relatively narrow, with poorly defined parotoid glands and tiny nostrils.
- Coloration dark brown, with a faintly indicated pattern of light chevrons on dorsum.
- Parotoid glands slightly lighter than rest of dorsum.
- Ventral coloration grayish-brown with small light spots.

Hands and feet not syndactylous; longest fingers and toes, with rounded tips, project well beyond the margin of basal interdigital webbing.

foot

Similar species

The moss salamanders that range closest to this species are *N. matama*, *N. picadoi*, and *N. tapanti*. *N. matama* (p. 104) is much smaller, has a golden-tan tail, and bold white spots on its upper lip. *N. tapanti* (p. 109) has padlike, syndactylous hands and feet. *N. picadoi* (p. 106) does not have syndactylous feet, but it is smaller than *N. major* and has short limbs and distinctly large nostrils.

Nototriton matama
Matama Moss Salamander

Not assessed

Maximum known standard length 0.94 in (24 mm) in the only known adult female. The tail is incomplete and therefore an accurate tail length ratio cannot be ascertained for this species. A tiny *Nototriton* with very narrow hands and feet, well differentiated digits (not padlike and syndactylous), a narrow head, and enlarged, elongated nostrils.

Ecoregions
1 2 3 4 5

Nototriton matama is endemic to Costa Rica. It is only known from the type locality on the southeastern end of the Matama Ridge, in the Río Banano watershed, Limón Province; 4,265 ft (1,300 m).

Natural history

This very recently described species was discovered in 2007 in undisturbed, humid cloud forest habitat on the Matama Ridge. One adult female and three juveniles were found in or under clumps of moss attached to the vertical trunk of trees, or suspended from branches. Multiple individuals were found in the same tree, indicating that this species may occur in significant densities in appropriate habitat. However, additional surveys are needed to adequately assess the range of *N. matama* and its conservation status.

Description

Head small, barely wider than neck, with obvious parotoid glands on the back of the head; glands show as faint light spots.

Lower flanks and ventral surfaces dark brown with scattered light blotches.

Upper lip area below each eye is marked with a bold pattern of alternating dark and white spots.

Tiny and slender, with delicate, short limbs and very narrow hands and feet.

A relatively brightly colored moss salamander; dorsal surfaces copper to golden-brown, bordered on each side of the body by an irregular band consisting of short white and tan streaks.

Snout short and broadly truncated, with large, upward-pointing nostrils that are oblong in shape.

Tail predominantly golden-tan, lighter in color than the body and head.

foot

Digits are not fused and extend significantly beyond the basal interdigital webbing; tips of longest digits rounded, but shorter digits have blunt tips.

Similar species

N. major (p. 103), whose range is closest to this species, is much larger and has very small nostrils. *N. picadoi* (p. 106) is similar in appearance to *N. matama* but is larger and has broader hands and feet; these species do not co-occur.

Nototriton picadoi
Picado's Moss Salamander

Near threatened

Standard length to 1.10 in (28 mm); maximum known total length 2.60 in (66 mm). A small but robust, relatively short-legged species of *Nototriton* with a moderate-sized head and large nostrils. The hands and feet have basal webbing, and the tips of the longer digits are clearly differentiated.

Ecoregions
1 2 3 **4** 5

A Costa Rican endemic, known only from the northern end of the Talamanca Mountain Range, Costa Rica; 3,935–7,215 ft (1,200–2,200 m).

Natural history

Locally, this can be a fairly common salamander in its cloud forest habitat. Individuals have been collected from arboreal bromeliads and moss mats growing on tree trunks and branches. *N. picadoi* may be more arboreal than other moss salamanders, and individuals up to 26 ft (8 m) above the ground have been collected from suspended clumps of moss hanging from vines and aerial roots. These same microhabitats were found to contain egg clumps with 1–11 eggs. Larvae develop entirely inside the egg and hatch as tiny salamanders. The breeding season for this species seems to coincide with the wet months (May–December), but may be year-round in permanently wet areas.

Description

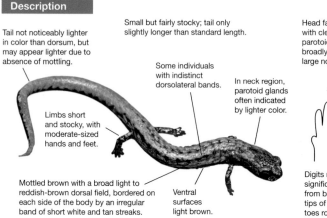

Small but fairly stocky; tail only slightly longer than standard length.

Tail not noticeably lighter in color than dorsum, but may appear lighter due to absence of mottling.

Some individuals with indistinct dorsolateral bands.

In neck region, parotoid glands often indicated by lighter color.

Head fairly robust, with clearly discernible parotoid glands. Snout broadly truncated, with large nostrils.

Limbs short and stocky, with moderate-sized hands and feet.

foot

Mottled brown with a broad light to reddish-brown dorsal field, bordered on each side of the body by an irregular band of short white and tan streaks.

Ventral surfaces light brown.

Digits not fused and significantly free from basal webbing; tips of fingers and toes rounded.

Similar species

N. tapanti (p. 109) occurs close to this species, in the northern Talamanca Mountain Range, but differs in having padlike, syndactylous hands and feet. *N. abscondens* (p. 98) is similar in appearance, albeit more slender, with smaller nostrils. It occurs farther north, in the Central Mountain Range.

Nototriton richardi
Richard's Moss Salamander

Near threatened

Standard length to 0.94 in (24 mm), total length to 1.97 in (50 mm). A tiny, slender moss salamander with short limbs, syndactylous hands and feet, pointed fingers and toes, and very large nostrils.

Ecoregions
1 2 3 4 5

Nototriton richardi is a Costa Rican endemic, known from a few localities along the Atlantic slopes of the Central Mountain Range; 4,500–5,900 ft (1,370–1,800 m).

Natural history

This uncommon species is known from only a few specimens but may be more common than originally thought. It is nocturnal and secretive. Unlike most other Costa Rican *Nototriton*, *N. richardi* is more commonly found on the ground and in leaf litter than in moss mats attached to trees. Most individuals have been found in leaf litter in undisturbed forest, although some occurred in moss clumps on logs and tree trunks and on the vertical banks of road cuts. Even though it is occasionally found in altered habitat, *N. richardi* does not appear very tolerant of habitat degradation. Habitat loss due to logging, agriculture, and urban development is likely the biggest threat to this species. Fortunately, suitable habitat in a large part of its range is protected in Braulio Carrillo National Park, and its population seems to be stable.

N. richardi differs subtly from other moss salamanders; it has very large nostrils and syndactylous hands and feet.

Tiny and delicate; tail fairly long, slightly exceeding standard length.

Coloration dark brown with irregular light spots on the dorsum, head, and tail.

Head moderate in size and only slightly wider than neck. Parotoid glands faintly indicated.

Concentrations of light spots low on the flanks, along the edge of the belly, may appear as a mottled ventrolateral light stripe. The venter is dark lavender-brown with lighter specks.

Some individuals with light-outlined costal grooves.

Limbs very short, with small padlike, syndactylous hands and feet.

Snout very short and bluntly rounded, with very large nostrils.

foot

Digits fused and only inner and outer toe indicated by a groove; tips of fingers and toes pointed.

Similar species

Other species of *Nototriton* that may co-occur have more clearly defined digits, with tips that extend beyond the margins of the interdigital webbing. Most similar to *N. tapanti* (p. 109) and difficult to distinguish; however, these species do not co-occur and geographic provenance may help in identification.

Nototriton tapanti
Tapantí Moss Salamander

Vulnerable

Standard length to 0.94 in (24 mm), total length to 2.01 in (51 mm). A tiny, slender moss salamander with short limbs, syndactylous hands and feet, pointed fingers and toes, and strikingly large nostrils.

Ecoregions
1 2 3 4 5

Endemic to Costa Rica. *Nototriton tapanti* is known from a single female found in the Orosí River Valley on the Atlantic slope of the northern Talamanca Mountain Range, at 4,265 ft (1,300 m).

Natural history

Presumably this is a nocturnal species. It was discovered in the leaf litter of a humid premontane forest and may be terrestrial. However, no further natural history information on this species exists. It was discovered in an area that has been cleared for development and agriculture. Seemingly suitable habitat exists within nearby Tapantí National Park, and this rare salamander may occur there, although repeated searches in the park have come up empty thus far.

Description

- Tiny and slender; tail slightly longer than standard length. Limbs short.
- Head moderate in size, with prominent parotoid glands in neck and with very large nostrils.
- Dorsum light brown, with slightly darker flanks.
- A pattern of faint light spots runs along the middle of the back; a pair of distinct reddish-brown spots appear on the back of the head, located between the eyes and parotoid glands.
- Parotoid glands well-developed and lighter in color than the dorsal coloration.
- Venter dark brown.

foot

Hands and feet tiny, with pointed digits fused to form padlike appendages.

Similar species

On other species of *Nototriton* in the same region, the tips of the longest digits project beyond the interdigital webbing. *Nototriton richardi* (p. 107) is similar in appearance but does not co-occur with *N. tapanti.*

Genus *Oedipina*
Worm Salamanders

Salamanders of the genus *Oedipina* are often referred to as worm salamanders because of their elongate and attenuate build, short limbs, and tiny hands and feet. Most species are small to medium-sized, although the largest species in the genus, *Oedipina collaris*, may nearly reach 10 in (250 mm) in total length.

All have a long to very long tail that is at least equal to the animal's standard length but may be almost three times as long. All worm salamanders are capable of caudal pseudoautotomy, which is the controlled release of sections of the tail when restrained by a predator. Autotomized parts regenerate quickly and completely, with all original tissues (bone, muscle, nerve, and the like) restored to their original condition; it is not uncommon to see individuals whose tail is not yet fully regenerated. Regenerated tail segments are sometimes lighter in color than the original tail.

Worm salamanders' fingers and toes are enveloped by varying amounts of webbing, and in many species the digits are syndactylous (fused). In some species, the digits are both fused and encased in webbing, making the hands and feet appear like fleshy pads.

Identifying individual species of *Oedipina* can be extremely difficult in the field. Members of the subgenus *Oedipinola* (a distinct species group within *Oedipina*) tend to be tan to brown, with distinctive light mottling or blotches. Most of the species assigned to the subgenus *Oedipina* are essentially uniform dark grey or black, without clearly distinguishable field marks. Diagnostic features include the number of transverse costal folds on the sides of the body that separate adpressed front and hind limbs; the shape and size of head, limbs, and feet; and the length of the tail relative to the length of the body. In some cases the number of maxillary teeth or molecular characteristics are the only conclusive identifiers, clearly not useful for field identification. Many *Oedipina* have a relatively restricted distribution range, and thus location is a helpful guide to identification.

Salamanders in this genus are nocturnal, and either terrestrial or fossorial. During the day they retreat into moist microhabitats in or under decaying logs, under cover objects, in moss mats, or in leaf litter. When uncovered, worm salamanders move surprisingly fast in spite of their puny limbs. Often they wriggle in serpentine fashion without using their limbs and rapidly disappear underground or in the leaf litter. If escape is impossible, they can roll into a ball, whip the tail around to distract predators, or flip themselves out of harm's way by using the long, heavy tail for leverage.

There is sexual dimorphism in *Oedipina*, and males are generally somewhat smaller than females. Adult males are also characterized by the presence of a pheromone-producing mental gland on the chin, enlarged teeth at the front of the upper jaw, and enlarged nasolabial protuberances. Little is known about the breeding biology of *Oedipina* specifically, but it is surmised that, as in other plethodontid salamanders, males use their enlarged teeth during courtship to scrape the skin of a receptive female. The minor abrasions created during this action allow the male to transfer pheromones from the mental gland directly into the bloodstream of the female, and this is thought to stimulate the female to mate. As far as is known, all *Oedipina* lay terrestrial eggs that undergo direct development. These salamanders are not known to display parental care.

The genus *Oedipina* is primarily found in southern Central America and is currently thought to contain at least 36 species. Several new species await formal description, and additional cryptic forms are undoubtedly hidden in already described species. Fifteen worm salamanders are known to occur in Costa Rica and a 16th as yet undescribed but distinct form is included in the species accounts.

Genus *Oedipina*

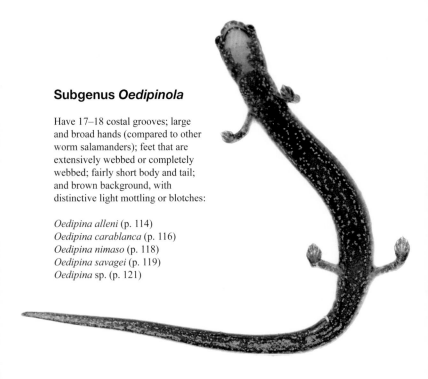

Subgenus *Oedipinola*

Have 17–18 costal grooves; large and broad hands (compared to other worm salamanders); feet that are extensively webbed or completely webbed; fairly short body and tail; and brown background, with distinctive light mottling or blotches:

Oedipina alleni (p. 114)
Oedipina carablanca (p. 116)
Oedipina nimaso (p. 118)
Oedipina savagei (p. 119)
Oedipina sp. (p. 121)

Subgenus *Oedipina*

Have 19–22 costal grooves; very small, syndactylous hands and feet that are barely wider than the limbs; extremely long body and tail; and coloration that is either uniform gray, dark brown, or black:

Oedipina alfaroi (p. 123)
Oedipina altura (p. 125)
Oedipina collaris (p. 127)
Oedipina cyclocauda (p. 129)
Oedipina gracilis (p. 131)
Oedipina grandis (p. 133)
Oedipina pacificensis (p. 134)
Oedipina paucidentata (p. 137)
Oedipina poelzi (p. 138)
Oedipina pseudouniformis (p. 140)
Oedipina uniformis (p. 142)

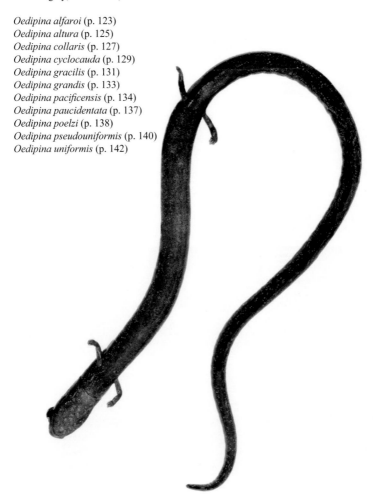

Oedipina alleni
Allen's Worm Salamander

Least concern

Standard length at least 2.28 in (58 mm), total length to 5.91 in (150 mm); females tend to grow somewhat larger than males. A moderate-sized, short-tailed worm salamander with brown to tan dorsal coloration and distinct light (white or cream) markings on the head and body; widespread in the southern Pacific lowlands.

Ecoregions
1 2 **3** 4 5

Oedipina alleni occurs in the lowlands and foothills of southwestern Costa Rica and adjacent western Panama, up to 2,900 ft (880 m).

Natural history

O. alleni generally occurs in moist and wet pristine forest habitat, but is sometimes found in altered habitat and areas with secondary growth (old overgrown plantations, for example). It is found in leaf litter accumulations on the forest floor or between buttressed tree roots. Individuals are sometimes uncovered during the day when flipping logs or surface debris. One *O. alleni* individual was seen active at night as it walked up a branch of a fallen tree at a height of approximately 4 ft (125 cm) above the ground. Upon discovery it launched

itself off its perch (by whipping its heavy tail around) and rapidly wriggled away, disappearing into the leaf litter.

Worm salamanders are generally secretive and rarely encountered. Although uncommon, *O. alleni* is among the more regularly observed *Oedipina* because it occurs in several protected areas that receive significant numbers of visitors, including Carara National Park, Manuel Antonio National Park, Hacienda Barú, Las Esquinas, and Corcovado National Park.

O. alleni is one of the more commonly observed salamanders in the Pacific lowlands. Within its range, the light blotch on its head is diagnostic.

A cream-colored blotch (size and shape varies) usually present on the top of the head; in some individuals, blotch may be indistinct and incorporated in a light-colored, broad middorsal band.

Costal grooves 17; adpressed limbs separated by 6.5–8.5 costal grooves.

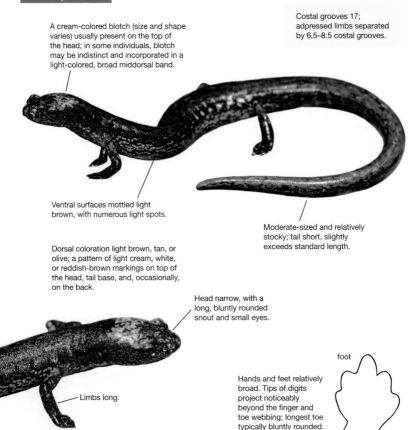

Ventral surfaces mottled light brown, with numerous light spots.

Moderate-sized and relatively stocky; tail short, slightly exceeds standard length.

Dorsal coloration light brown, tan, or olive; a pattern of light cream, white, or reddish-brown markings on top of the head, tail base, and, occasionally, on the back.

Head narrow, with a long, bluntly rounded snout and small eyes.

foot

Limbs long.

Hands and feet relatively broad. Tips of digits project noticeably beyond the finger and toe webbing; longest toe typically bluntly rounded.

Similar species

O. pacificensis (p. 134) is uniform dark and has an extremely long tail, short limbs, syndactylous hands and feet, and 19–20 costal grooves.

O. savagei (p. 119) is quite similar in overall appearance to *O. alleni* but is restricted to higher elevations (3,935–4,600 ft [1,200–1,400 m]) in the Costa Rica-Panama border region. It differs in having a broader head and rounded snout. In addition, the longest toe in this species extends beyond the foot webbing and is distinctly pointed (bluntly rounded in *O. alleni*).

Bolitoglossa lignicolor (p. 64) and *Bolitoglossa colonnea* (p. 50) are much more robust, with broad, padlike hands and feet, and have fewer than 15 costal grooves.

Oedipina carablanca
White-headed Worm Salamander

Endangered

Standard length to at least 2.13 in (54 mm), maximum known total length 5.04 in (128 mm); generally females are somewhat larger than males. The dark brown to black dorsum, marked with a pattern of bold white spots and blotches, mostly on the head and tail, is unique among Costa Rican salamanders.

Ecoregions
1 2 3 4 5

Oedipina carablanca is endemic to the Atlantic lowlands of Costa Rica. It is known from a few scattered areas in the vicinity of Guápiles, Siquirres, and Limón; 195–2,460 ft (60–750 m).

Natural history

Very few individuals of this species have ever been observed, and it is poorly understood. *O. carablanca* is currently known from only five localities: Pococí and Finca Los Diamantes (in Guápiles), Pocora, Brisas de Alegría Reserve, and Alto Guayacán. It likely occurs in other areas.

A recently observed individual was spotted about one meter above the ground in a branch of a rotten log that was filled with termites. Other individuals have been found inside rotten logs or under bark of fallen trees in banana plantations. Several of the known *O. carablanca* were found in secondary vegetation or in disturbed agricultural areas, indicating that this species is somewhat tolerant of habitat modification. Undoubtedly, removal of the primary lowland moist forest that originally covered this species' range poses a major threat, but insufficient information is currently available to adequately assess its conservation status.

O. carablanca is unmistakable; its black-and-white bold coloration is unique among Costa Rican salamanders.

Description

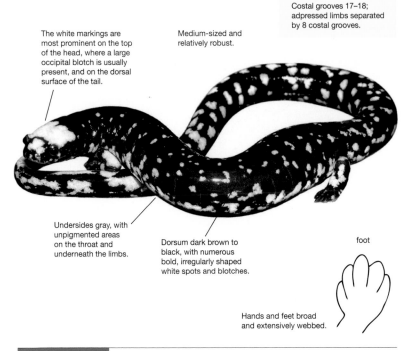

The white markings are most prominent on the top of the head, where a large occipital blotch is usually present, and on the dorsal surface of the tail.

Medium-sized and relatively robust.

Costal grooves 17–18; adpressed limbs separated by 8 costal grooves.

Undersides gray, with unpigmented areas on the throat and underneath the limbs.

Dorsum dark brown to black, with numerous bold, irregularly shaped white spots and blotches.

foot

Hands and feet broad and extensively webbed.

Similar species

Other worm salamanders that co-occur are generally uniform dark brown, gray, or black, have an extremely long tail, minute extremities, and more costal grooves (19–20).

The undescribed *Oedipina* species (p. 121) is the only other member of the *Oedipinola* subgenus in northeastern Costa Rica, but it is smaller and mottled brown rather than black and white.

O. carablanca somewhat resembles webfoot salamanders of the genus *Bolitoglossa*, and several of those salamanders occur in its range. However, *Bolitoglossa* species invariably have fewer than 15 costal grooves and none are black and white.

Oedipina nimaso
Nimaso Worm Salamander

Not assessed

Standard length in only known male 1.22 in (31 mm), total length 2.87 in (73 mm). A very small, short-tailed worm salamander, with dark dorsal coloration, distinct light (white or cream) markings on the head, but with relatively few contrasting light markings on the remaining dorsal surfaces; only known from Cerro Nimaso in southeastern Costa Rica.

Ecoregions
1 2 3 4 5

Endemic to Costa Rica. Known from a single male individual collected in 1984 on Cerro Nimaso, Limón Province, at 3,585 ft (1,093 m). No additional individuals have been found since. It was eventually described in 2012.

Natural history

Oedipina nimaso is only known from mature forest habitat in the humid tropical forest zone of the type locality. It is not known whether this secretive species ranges widely outside of the currently known locale.

Description

Adpressed limbs separated by 9.5 costal grooves.

- Very small and slender, with long limbs.
- Head narrow, with a long snout that is rounded in dorsal profile.
- Coloration of live individual was never recorded. Preserved specimen is brown, with white markings on the head, primarily between the eyes; undersides lighter than dorsum, with white speckling.
- Presumably this species is mostly dark brown to black, with light markings on the top of the head.

Hands and feet very narrow. The longest digit of the syndactylous, padlike hands and feet extends beyond the adjacent digits, lending the extremities a very narrow and somewhat pointed appearance.

foot

Similar species

O. carablanca (p. 116) is larger, with broad, webbed hands and feet, and a striking pattern of bold white spots on the head, body, and tail. *Oedipina* sp. (p. 121) appears to be similar to *O. nimaso* in size, and possibly coloration, but is only known from northeastern Costa Rica. Webfoot salamanders in the genus *Bolitoglossa* are much more robust, with broad, padlike hands and feet; they invariably have fewer than 15 costal grooves.

Oedipina savagei
Savage's Worm Salamander   **Data deficient**

Standard length to 1.54 in (39 mm), total length to 3.86 in (98 mm). A smallish, short-tailed worm salamander with dark brown dorsal coloration and distinct light (white or cream) markings on the head; only known from middle elevations in the Talamanca Mountain Range.

Ecoregions
1 2 3 4 5

Oedipina savagei has only been recorded from a small area on the Pacific slope of the Talamanca Mountain Range that straddles the Costa Rica-Panama border; 3,935–4,600 ft (1,200–1,400 m).

Natural history

This species was described relatively recently, and its biology and exact distribution range are still incompletely known. Much of its known range, as currently understood, is severely fragmented and has been mostly converted to coffee plantations. It is unknown whether *O. savagei* is tolerant of habitat alteration, but such disturbance may pose a serious threat to its survival. Like most worm salamanders, this species appears to be primarily terrestrial, and individuals have been collected from leaf litter in premontane wet forest habitat. It has been observed in the forested preserve surrounding Las Cruces Biological Station.

O. savagei is superficially similar to *O. alleni*, but it occurs at higher elevations.

Description

Usually with irregularly shaped light tan or cream-colored blotch on the top of the head.

Dorsum brown, generally with a lighter brown middorsal band and darker brown flanks and tail.

Costal grooves 17 (rarely 18); adpressed limbs separated by 6–7 costal grooves.

Small to moderate-sized, with a fairly short tail that only slightly exceeds the standard length.

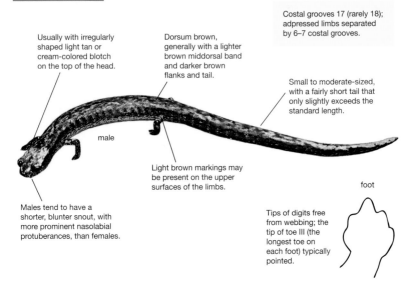

male

Light brown markings may be present on the upper surfaces of the limbs.

foot

Males tend to have a shorter, blunter snout, with more prominent nasolabial protuberances, than females.

Tips of digits free from webbing; the tip of toe III (the longest toe on each foot) typically pointed.

Similar species

O. pacificensis (p. 134) is uniform dark in coloration and has an extremely long tail, short limbs, syndactylous hands and feet, and 19–20 costal grooves; it generally occurs at lower elevations.

O. alleni (p. 114) is similar in overall appearance to *O. savagei* but is restricted to lower elevations (below 2,900 ft [880 m]). It has a narrower head and a long, bluntly rounded snout. In addition, the longest toe in this species extends beyond the foot webbing and is rounded (pointed in *O. savagei*).

All webfoot salamanders in the genus *Bolitoglossa* found in the area are much more robust, with broad, padlike hands and feet; they invariably have fewer than 15 costal grooves.

Oedipina sp.
Sarapiquí Worm Salamander

Not assessed

The only known individual has a standard length of 1.38 in (35 mm) and measures 2.72 in (69 mm) in total length. A small, short-tailed worm salamander with mottled brown coloration; only known from the Sarapiquí region, in northeastern Costa Rica.

Ecoregions
1 2 3 4 5

An undescribed, endemic species of worm salamander that is only known from a single individual found in 2011 in the Rara Avis Rainforest Preserve, Sarapiquí region, Heredia Province; 2,035 ft (620 m).

Natural history

This species has been tentatively assigned to the subgenus *Oedipinola* based on the low number of costal grooves (18), its relatively short tail, and fairly broad hands and feet, which are not syndactylous.

The single known individual, seemingly an adult male, was discovered within a large tank bromeliad that had fallen from the canopy a considerable time before the salamander was sighted. It is not known whether the salamander fell from the canopy with the plant, but given that Costa Rican worm salamanders are mostly terrestrial it seems most likely that this animal simply used the dense rosette of leaves of the bromeliad as a daytime retreat.

The actual range of this species is not known, but it likely favors the lowland and middle elevation wet pristine forest habitat in which it was found. The Rara Avis Rainforest Preserve is located in the buffer zone of Braulio Carrillo National Park, and this species likely occurs within the protected forests of that park.

Description

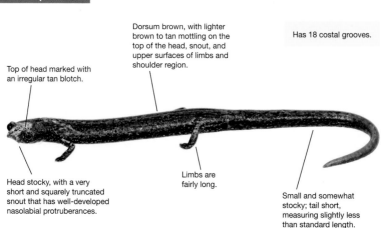

Dorsum brown, with lighter brown to tan mottling on the top of the head, snout, and upper surfaces of limbs and shoulder region.

Has 18 costal grooves.

Top of head marked with an irregular tan blotch.

Head stocky, with a very short and squarely truncated snout that has well-developed nasolabial protruberances.

Limbs are fairly long.

Small and somewhat stocky; tail short, measuring slightly less than standard length.

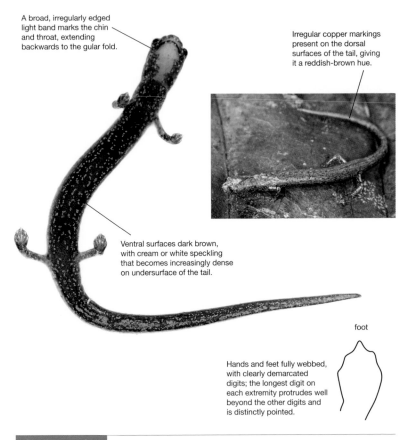

A broad, irregularly edged light band marks the chin and throat, extending backwards to the gular fold.

Irregular copper markings present on the dorsal surfaces of the tail, giving it a reddish-brown hue.

Ventral surfaces dark brown, with cream or white speckling that becomes increasingly dense on undersurface of the tail.

foot

Hands and feet fully webbed, with clearly demarcated digits; the longest digit on each extremity protrudes well beyond the other digits and is distinctly pointed.

Similar species

Other *Oedipina* species in the area are mostly uniform dark in coloration and have a proportionally longer tail, shorter limbs, syndactylous hands and feet, and 19–20 costal grooves.

O. carablanca (p. 116) is the only worm salamander in the subgenus *Oedipinola* in northeastern Costa Rica. It differs in having a striking dorsal pattern of numerous white spots on a dark brown to black background and broad, fully webbed hands and feet.

O. nimaso (p. 118) appears to be a close relative of this species; it is also small and has the same overall appearance. It is currently only known from Cerro Nimaso.

Bolitoglosa alvaradoi (p. 42), *B. striatula* (p. 89), *B. colonnea* (p. 50), and any other web-footed salamanders that might occur in the area are much more robust, with broad, padlike hands and feet and fewer than 15 costal grooves.

Oedipina alfaroi
Alfaro's Worm Salamander

Vulnerable

Maximum known standard length 2.36 in (60 mm) in females, males likely somewhat smaller; maximum known total length 6.10 in (155 mm). A poorly defined, moderate-sized salamander that is morphologically similar to several other species in the genus. Distinguished in the field by its small eyes and elongated, pointed snout.

Ecoregions
1 2 3 4 5

Inhabits humid lowland forest habitats along the Atlantic slope of southern Costa Rica and adjacent Panama; sea level to 2,800 ft (850 m).

Natural history

With only few individuals known, this is apparently a rare species. All were collected from fallen logs in humid lowland and premontane forest. As *O. alfaroi* does not appear to tolerate habitat degradation, it is threatened by the conversion of its habitat to agricultural use. The phylogenetic status of this salamander is poorly understood, but it appears most closely related to *O. gracilis*, *O. pacificensis*, and *O. uniformis*. Identifying these species in the field based on variable and subtle morphological characteristics is a challenge. In some instances, location is an aid to identification.

Slender, with a fairly long tail that exceeds the standard length.

20 costal grooves between the insertion of the front and hind limbs; 12–14 costal grooves separate adpressed limbs.

Has a relatively pointed, elongated snout and small eyes.

Limbs tiny.

Uniform dark brown; head and venter a lighter brown.

foot

Hands and feet are syndactylous.

Similar species

O. gracilis (p. 131) has a bluntly rounded snout and proportionally longer tail. *O. cyclocauda* (p. 129) has a more rounded snout profile. *O. pseudouniformis* (p. 140) has a more rounded snout and larger eyes.

Oedipina altura
Cartago Worm Salamander

Critically endangered

Standard length to 2.28 in (58 mm), total length to least 5.16 in (131 mm). A uniformly dark worm salamander that may not be reliably distinguished from similar looking species when direct comparison with individuals of other species is not possible.

Ecoregions
1 2 3 4 5

Oedipina altura is endemic to Costa Rica and is only known from two areas along the Pan-American Highway, at the extreme northern end of the Talamanca Mountain Range, near El Empalme, San José Province; 7,500–7,600 ft (2,286–2,320 m).

Natural history

An extremely rare species. Only three male specimens (two adults and one juvenile) were collected during the early 1960s, and no individuals have been observed since. Presumably this species' range is restricted to a very small area within the northern Talamanca Mountain Range. Originally found in montane rainforest; urbanization, agriculture, and cattle farming in the area have since led to significant loss of habitat in its range. In recent decades, repeated searches for this species have come up empty.

Very little is known about the natural history of *O. altura*. This is a terrestrial species, and one individual was found under moss. Presumably it reproduces through direct development, like other neotropical salamanders.

O. altura is similar in overall apearance to several other relatively dark colored worm salamanders with 19–20 costal grooves and tiny, syndactylous hands and feet (*O. alfaroi, O. cyclocauda, O. gracilis, O. pacificensis, O. paucidentata, O. pseudouniformis,* and *O. uniformis*). Of those, only the exceedingly rare *O. paucidentata* (p. 137) is known from the same area as *O. altura*.

Description

Costal grooves 19–20; adpressed limbs separated by 11–13 costal grooves.

Medium-sized and slender, with a long tail.

Dark gray to black, with small white spots on the upper surfaces of the head, neck, and limbs.

Head moderately wide, with a truncated snout.

adult

Limbs relatively short.

In adults, ventral surfaces and tail solid black. Only known juvenile has a black venter; dorsal surfaces bronze, with a thin silver line separating the dorsal and ventral coloration.

foot

Narrow, syndactylous hands and feet.

125

O. paucidentata (p. 137) has a narrower head with a rounded snout, shorter limbs, and narrower hands and feet. *O. uniformis* (p. 142) commonly occurs in locations close to *O. altura*, but has smaller hands and feet, shorter limbs, and a rounded snout.

Oedipina collaris
Snouted Worm Salamander

Data deficient

Maximum known standard length 3.03 in (77 mm), total length to at least 10.04 in (255 mm). Males generally slightly smaller than females. A large worm salamander with a distinctively elongated, pointed snout and an extremely long, slender tail. Adult size alone distinguishes *Oedipina collaris* from most other Costa Rican worm salamanders.

Ecoregions
1 2 3 4 5

Oedipina collaris is known from scattered localities on the Caribbean lowlands, from central Nicaragua to central Panama; there is a single known locality in Costa Rica, near Turrialba, Cartago Province. Throughout its range, individuals have been observed from near sea level to 1,975 ft (600 m).

Natural history

O. collaris is a rare species known from isolated sightings at widely scattered localities. If all animals assigned to this species truly belong to one species, then the potential distribution range of *O. collaris* is quite extensive. However, it is possible that multiple species are represented by the limited number of individuals available for study. More research is needed to clarify this situation.

This is a terrestrial, nocturnal species. One individual was collected under a log at the edge of a stream bank. All known sightings were in humid lowland forests; *O. collaris* probably suffers from the extensive conversion of its habitat to logging, agriculture, and human settlements.

Size alone distinguishes *O. collaris* from other Costa Rica worm salamanders.

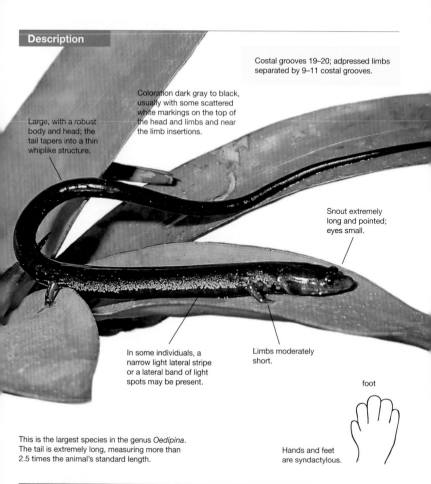

Costal grooves 19–20; adpressed limbs separated by 9–11 costal grooves.

Coloration dark gray to black, usually with some scattered white markings on the top of the head and limbs and near the limb insertions.

Large, with a robust body and head; the tail tapers into a thin whiplike structure.

Snout extremely long and pointed; eyes small.

In some individuals, a narrow light lateral stripe or a lateral band of light spots may be present.

Limbs moderately short.

foot

This is the largest species in the genus *Oedipina*. The tail is extremely long, measuring more than 2.5 times the animal's standard length.

Hands and feet are syndactylous.

Similar species

O. alfaroi (p. 123) is similar in coloration and body shape to *O. collaris*, but *O. alfaroi* is much smaller and its snout is far less elongated and pointed. *O. carablanca* (p. 116) is more robust, with a striking pattern of white spots on a black base color. *O. cyclocauda* (p. 129), *O. gracilis* (p. 131), and *O. pseudouniformis* (p. 140) are much smaller than *O. collaris* and have a shorter, more rounded snout. *O. poelzi* (p. 138) is much smaller, has a short, rounded snout, and a brown dorsal coloration separated from its black venter by a narrow light lateral stripe.

Oedipina cyclocauda
Round-tailed Worm Salamander

Least concern

Standard length to 1.81 in (46 mm), maximum known total length to 5.51 in (140 mm). Females slightly larger than males. One of the smaller *Oedipina* in the country. Uniform grayish-black. Very long-tailed, with a relatively broad head and short, rounded snout. This species is currently poorly defined; it may not be possible to identify it conclusively in the field.

Ecoregions
1 2 3 4 5

As currently understood, *Oedipina cyclocauda* ranges from southeastern Nicaragua to western Panama; from near sea level to 1,975 ft (600 m).

Natural history

The taxonomic status of this species and several other uniform dark *Oedipina* is still unresolved. There may be several cryptic species involved in this complex. Molecular data show differences between individuals from the northern part of the range and those from the south. Until recently, *O. cyclocauda* was thought to range from Honduras to extreme northeastern Panama. In recent years, however, several Honduran and Nicaraguan populations have been split off as separate species. Populations from Panama and extreme southeastern Costa Rica may also turn out to be distinct from those in northeastern Costa Rica and adjacent Nicaragua.

This species is a terrestrial or fossorial occupant of wet habitats in humid tropical forests. *O. cyclocauda* can be fairly common locally. It is primarily found near streams, where it retreats under or into rotting logs, piles of vegetable matter, or leaf litter around buttressed trees. Desiccation of habitat due to removal of tree cover for agriculture or lumber extraction is perhaps the biggest threat to its survival.

O. cyclocauda is a small worm salamander with short limbs and narrow, paddle-like hands and feet.

129

Moderately robust, with short limbs and a long tail.

Costal grooves 19–20; adpressed limbs separated by 11–12.5 costal grooves.

Morphologically similar to other uniformly dark gray to black worm salamanders.

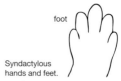

foot

Syndactylous hands and feet.

Similar species

O. alfaroi (p. 123) has an elongated, bluntly pointed snout. All species in the *Oedipina uniformis*-complex (*O. gracilis*, *O. pacificensis*, and *O. uniformis*) have a narrower head, shorter legs, smaller hands and feet, and an overall stockier build (*O. cyclocauda* only overlaps in distribution range with *O. gracilis*, p. 131). *Oedipina pseudouniformis* (p. 140) has a somewhat broader head and broader feet, but it is difficult to distinguish the two species.

Oedipina gracilis
Long-tailed Worm Salamander

Endangered

Standard length to 1.85 in (47 mm), maximum known total length 5.91 in (150 mm). This is a moderate-sized, black to dark brown salamander, with a very long, slender tail, short limbs, and tiny hands and feet. The head is moderately broad, with a bluntly rounded snout. Morphologically, *Oedipina gracilis* is very similar to other co-occurring *Oedipina* and very difficult to identify in the field.

Ecoregions
1 2 3 4 5

Oedipina gracilis occurs in the Atlantic lowlands of Costa Rica and extreme northwestern Panama; from near sea level to 2,330 ft (710 m).

Natural history

O. gracilis is morphologically indistinguishable from *O. pacificensis* and *O. uniformis*; these three species make up a cryptic-species complex whose individual species are only reliably distinguished based on molecular differences. Their distribution ranges never overlap, however, so location is a good key to identification.

This species was once fairly common in low- and middle-elevation rainforest, but it appears to have declined in numbers. Like other *Oedipina*, this terrestrial and fossorial species is rarely encountered and therefore difficult to survey. Most have been found under cover objects (logs and rocks, for example) or, at night, active in leaf litter. Occasionally, individuals have been discovered low on tree trunks, in moss cushions.

O. gracilis generally does not tolerate habitat alteration and is found only in relatively pristine forests. The drying effects of openings in the canopy result in less suitable habitats; the conversion of forested land to agricultural use is likely the biggest threat to this species. *O. gracilis* is known to occur in a few protected areas, including La Selva Biological Station and Braulio Carrillo National Park.

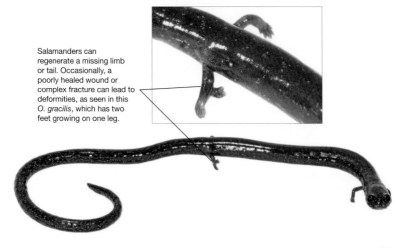

Salamanders can regenerate a missing limb or tail. Occasionally, a poorly healed wound or complex fracture can lead to deformities, as seen in this *O. gracilis*, which has two feet growing on one leg.

Costal grooves 19–20; adpressed limbs separated by 12.5–14.5 grooves.

Uniformly dark brown, black, or gray; small white spots sometimes visible on the head and body (noticeable only up close).

Exceedingly slender and elongate; extremely long tail is sometimes nearly three times the standard length, if intact.

Upper surfaces of limbs may show an irregular rust-brown to brown blotch.

Head moderately broad, with a bluntly rounded snout; adult males with well-developed nasolabial protuberances.

Head sometimes with a brownish hue or a pair of irregular brown blotches on either side of the nape. The salamander shown here has a unique color pattern, with rusty-orange nape patches and a reddish-brown dorsal field. Normally the brown patches are far more subtle.

foot

Hands and feet tiny, syndactylous.

O. cyclocauda (p. 129) is smaller. *O. alfaroi* (p. 123) has a more elongated, pointed snout. *O. pseudouni-formis* (p. 140) has a broader head and broader feet.

Oedipina grandis
Brown-backed Worm Salamander

Endangered

Standard length to 2.83 in (72 mm), maximum known total length 8.31 in (211 mm). The large size and mottled brown dorsal coloration on an otherwise black salamander (black head, venter, and tail) readily identify this species.

Ecoregions
1 2 3 4 5

Oedipina grandis is found only in the southern Talamanca Mountain Range, in the border region with Panama; 5,940–6,400 ft (1,810–1,950 m).

Natural history

O. grandis is a nocturnal and fossorial species, occasionally found in and under rotting logs or in thick leaf litter around tree buttresses. It does not appear to be tolerant of habitat degradation and has only been encountered in relatively undisturbed humid lower montane forests.

This species was described from adjacent Panama in 1970 and only detected in extreme southern Costa Rica in 1990. Although it was reportedly once common in the Las Tablas region, a dead individual with a chytrid fungal infection was discovered there in 1991, and its numbers have since declined drastically. The current conservation status of this species in Costa Rica is unclear, but it is still observed, infrequently, in western Panama.

Description

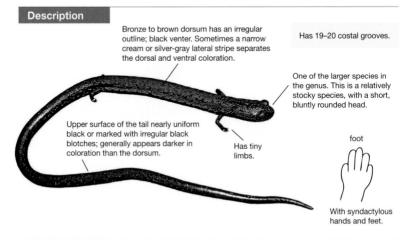

Bronze to brown dorsum has an irregular outline; black venter. Sometimes a narrow cream or silver-gray lateral stripe separates the dorsal and ventral coloration.

Has 19–20 costal grooves.

One of the larger species in the genus. This is a relatively stocky species, with a short, bluntly rounded head.

Upper surface of the tail nearly uniform black or marked with irregular black blotches; generally appears darker in coloration than the dorsum.

Has tiny limbs.

foot

With syndactylous hands and feet.

Similar species

Other long-tailed, slender species of *Oedipina* in its range are uniform dark brown, gray, or black. *Oedipina poelzi* (p. 138) is the only other Costa Rican worm salamander with a similar, brown-backed coloration. It is smaller, with distinctly wider hands and feet. No range overlap.

133

Oedipina pacificensis
Pacific Worm Salamander   **Least concern**

Standard length to 2.01 in (51 mm), maximum known total length 6.89 in (175 mm). Males generally slightly smaller than females. Within its Pacific lowland distribution range, this is the only uniform black or dark brown salamander with an extremely long tail, short limbs, syndactylous hands and feet, and 19–20 costal grooves.

Ecoregions
1 2 **3** 4 5

Oedipina pacificensis inhabits the humid forests of the Pacific lowlands of southwestern Costa Rica and neighboring Panama; from near sea level to 2,400 ft (730 m).

Natural history

This cryptic member of the *O. uniformis*-species complex cannot reliably be distinguished morphologically from the other forms in the complex, *O. gracilis* (p. 131) and *O. uniformis* (p. 142). These three forms are primarily distinguished on the basis of genetic differences, though their ranges do not overlap.

O. pacificensis is a nocturnal, fossorial salamander, and possibly one of the more commonly encountered worm salamander species in the country. It is most frequently found in leaf litter

accumulations between buttressed roots of large trees or under decaying logs or other cover objects. Individuals can be seen actively moving through the leaf litter at night. *O. pacificensis* is fairly tolerant of habitat alteration and can occasionally be found in densely vegetated plantations and other agricultural areas. It is regularly observed in several protected areas along the southern Pacific lowlands, including Corcovado National Park and Manual Antonio National Park, but also can be found in many other forested areas.

On first glance, *O. pacificensis* appears uniformly dark, though a more thorough inspection reveals subtle markings over the surface of the body.

Seemingly uniform dark brown, gray, or black salamander, although close examination in bright light will show suffusion of brown on the dorsum or on the head, and a pattern of irregular bronze blotches on the body and tail or a mottled pattern of brown and slate-gray.

Costal grooves 19–20; 12.5–13.5 costal grooves separate adpressed limbs.

Generally with numerous minute white spots on the upper surfaces of the body and tail.

Head moderately broad, with a bluntly rounded snout; adult males have pronounced nasolabial protuberances.

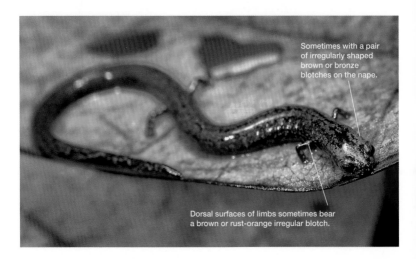

Sometimes with a pair of irregularly shaped brown or bronze blotches on the nape.

Dorsal surfaces of limbs sometimes bear a brown or rust-orange irregular blotch.

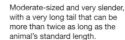

Moderate-sized and very slender, with a very long tail that can be more than twice as long as the animal's standard length.

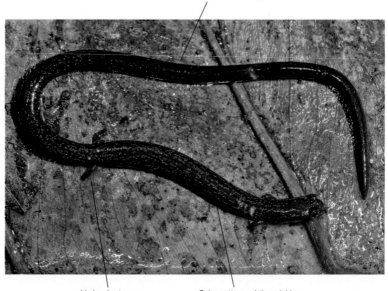

Limbs short.

Color pattern subtly variable on close examination.

foot

Tiny syndactylous hands and feet.

Similar species

O. alleni (p. 114) is brown, with light markings on the top of the head and a much shorter tail. It has 17 costal grooves and relatively wide hands and feet.

Bolitoglosa lignicolor (p. 64) and *Bolitoglossa colonnea* (p. 50) are much more robust, with broad, padlike hands and feet, and have fewer than 15 costal grooves; neither species has a nearly uniform dark brown to black color pattern.

Oedipina paucidentata
Highland Worm Salamander

Critically endangered

Standard length to 2.44 in (62 mm), total length to 5.43 in (139 mm). Males are slightly smaller than females. This uniform dark worm salamander is difficult to distinguish from similar looking species when direct comparison with individuals of other species is not possible.

Ecoregions
1 2 3 4 5

Oedipina paucidentata is endemic to Costa Rica and only known from the type locality at the northern end of the Talamanca Mountain Range, near El Empalme, in the vicinity of Cartago, at 7,500 ft (2,286 m), which is the same general area as the type locality of *O. altura* (p. 125).

Natural history

No individuals of this rare species have been observed since 1952; it is known from a very small area that has since been transformed by urbanization, agriculture, and cattle ranching. It is possible that *O. paucidentata* is now extinct. A co-occurring species, *O. altura*, is likely suffering from similar environmental pressures and has not been seen since the 1960s.

O. paucidentata is similar in appearance to all other uniformly dark worm salamanders in the country. However, most of those species are restricted to lower elevations and only two (*O. altura* and *O. uniformis*) occur at comparable altitudes in the highlands of the Talamanca and Central Mountain Ranges, respectively.

Description

- Moderate-sized and very slender, with a long tail, short limbs.
- Head very short and narrow, with a rounded snout.
- Dorsal coloration uniform slate to black; venter gray.
- *O. paucidentata* differs from closely related forms primarily in having very few maxillary teeth (*paucidentata* means few teeth), a feature not visible in the field.

Costal grooves 19–20; adpressed limbs separated by 12.5–15 costal grooves.

foot

Tiny syndactylous hands and feet.

Similar species

O. altura (p. 125) has longer limbs, relatively wider hands and feet, and a broader head with a truncate snout. *O. uniformis* (p. 142) has a broader head and a more robust build than does *O. paucidentata.*

Oedipina poelzi
Poelz's Worm Salamander

Endangered

Standard length to 2.52 in (64 mm); up to 7.13 in (181 mm) in total length. There appears to be no distinct size difference between the sexes. A worm salamander with mottled brown dorsal coloration and a black venter, separated by a cream to silver-gray lateral line. This species is also characterized by a relatively broad head, blunt snout, and fairly long limbs.

Ecoregions
1 2 3 **4** 5

A Costa Rican endemic. *Oedipina poelzi* is known from middle elevations in the Tilarán and Central Mountain Ranges; it also occurs in isolated valleys on the Atlantic slope of the northern Talamanca Mountain Range; 2,550–6,725 ft (775–2,050 m).

Natural history

This species prefers very wet rainforests of lower and middle elevations. It favors stream corridors or areas of seepage where water flows along the surface. During the day, it retreats below cover objects such as logs and rocks; some individuals have been found inside moss mats in wet microhabitats. Although it has been encountered in altered habitats (e.g., wet road cuts and quarries), removal of trees leads to a drying of the habitat, which may render an area unsuitable for this species.

Habitat alteration, destruction, and fragmentation appear to be the biggest threats to *O. poelzi*, although chytrid fungal infection may also play a role. It was formerly quite common in appropriate habitat, but appears to have significantly declined in numbers. Individuals are still encountered infrequently in suitable habitat. It is known to occur in several protected areas, including the Monteverde Biological Reserve and Braulio Carrillo National Park.

The genetic composition of populations of *O. poelzi* that occur near Moravia de Chirripo in the northern Talamanca Mountain Range appears to differ significantly from that of other populations of this species. These animals may represent a separate species, but too few individuals from that area exist in museum collections to carry out the morphological analyis needed to adequately separate these populations.

Description

Costal grooves 19–20; when adpressed, limbs are separated by 9.5–11.5 costal grooves.

Relatively broad-headed, with a blunt snout.

Adult males with distinct nasolabial protuberances on the snout; this feature is far less pronounced in females, as is shown here.

Limbs fairly long.

In some individuals, head and tail marked with yellowish or brown bands.

Brown dorsal color can vary from light orange-brown to dark brown.

Moderate-sized and slender, with a long tail.

Color pattern distinctive; typically with a bronze to brown dorsum and black venter; a cream or silver-gray lateral stripe usually separates the dorsal and ventral coloration.

Dorsal surfaces of tail and limbs black, variously marked with brown or bronze.

foot

Hands and feet small and syndactylous.

Similar species

Other co-occurring long-tailed, slender *Oedipina* display a nearly uniform coloration (dark brown, gray, or black). *O. poelzi* is similar in overall appearance to *O. grandis* (p. 133), but the latter species is larger and has a narrower head and relatively shorter limbs. No range overlap.

Oedipina pseudouniformis
Cryptic Worm Salamander   **Endangered**

Standard length to 2.09 in (53 mm), total length to 6.50 in (165 mm). There appears to be no sexual size dimorphism. A small to moderate-sized, uniform dark gray to almost black salamander. Has a long, robust tail, a relatively broad head, bluntly rounded snout, and large eyes and feet. Positive field identification of this species may be impossible.

The distribution range of *Oedipina pseudouniformis* is not well understood. Occurs on the Atlantic and Pacific slopes of mountains in northern Costa Rica; also occurs on the Atlantic slope, in the lowlands and foothills in the center of Costa Rica's Caribbean coast; near sea level to 4,100 ft (1,250 m). Nicaraguan populations previously assigned to this species have recently been split off as a separate, cryptic species: *Oedipina koehleri*.

Ecoregions
1 2 3 **4** 5

Natural history

This fossorial species is predominantly found in lowland and middle elevation wet forests, where it is found in leaf litter and under or inside decaying logs and clumps of moss. It can survive minor

O. pseudouniformis is a poorly defined species without distinctive diagnostic features; it may be impossible to identify in the field.

habitat alteration but disappears from degraded areas. Even though it was previously quite common locally, its numbers have been greatly reduced in recent years due to conversion of its forested habitat for agriculture or real estate development. No confirmed sightings of this species have been reported in almost two decades, though it likely persists in appropriate habitat. The difficulty of correctly identifying this species makes it challenging to assess its conservation status.

O. pseudouniformis is most closely related to *O. cyclocauda*. Future research may reveal that these two forms—morphologically almost indistinguishable—are representative of a single species.

O. pseudouniformis is generally smaller and stockier than the members of the *O. uniformis*-species complex (*O. gracilis*, *O. pacificensis*, and *O. uniformis*). Due to the superficial similarity in morphology, the population status and distribution ranges of all of these long-tailed black salamanders are poorly known.

Costal grooves 19–20; adpressed limbs separated by 9–12 costal grooves.

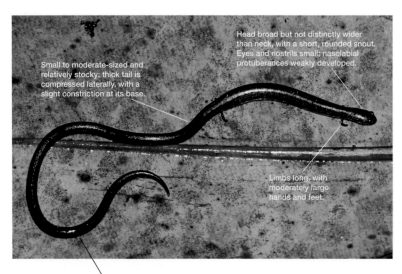

Small to moderate-sized and relatively stocky; thick tail is compressed laterally, with a slight constriction at its base.

Head broad but not distinctly wider than neck, with a short, rounded snout. Eyes and nostrils small; nasolabial protuberances weakly developed.

Limbs long, with moderately large hands and feet.

Dorsum dark gray to deep black; on close examination, ventral surfaces and limbs sometimes slightly lighter than the dorsal surfaces, which are often a deep orange-brown to dark grayish-brown.

Light markings sometimes present on the throat and upper lip, but these markings are not diagnostic.

foot

Fingers and toes syndactylous, with partial webbing.

Members of the *O. uniformis* complex (*O. gracilis* [p. 131], *O. pacificensis* [p. 134], and *O. uniformis* [p. 142]) tend to have a narrower head, truncated snout, and shorter, more slender limbs.

O. poelzi (p. 138) has a brown dorsal field and black venter, separated on each side of the body by a thin silvery-white stripe.

O. cyclocauda (p. 129) has a slightly narrower head and is smaller than *O. pseudouniformis*, but it may not be possible to distinguish between them in the field.

Oedipina uniformis
Common Worm Salamander

Near threatened

Standard length to 2.24 in (57 mm), maximum recorded total length 8.46 in (215 mm). Females generally slightly larger than males. This is a cryptic species of *Oedipina*, morphologically indistinguishable from *O. gracilis* and *O. pacificensis*, and separated from those species primarily by genetic differences. Only tentatively identified under field conditions, based on geographic occurrence.

Ecoregions
1 2 3 4 5

Oedipina uniformis ranges widely throughout Costa Rica's humid premontane and montane areas, including the Central Valley. This species can be found as far north as Tenorio Volcano in the Guanacaste Mountain Range, throughout the Tilarán and Central Mountain Ranges, and south, from the Talamanca Mountain Range to near the Panamanian border; 2,460–7,050 ft (750–2,150 m). Although its distribution range likely continues into adjacent Panama, this has yet to be confirmed.

Natural history

This species is relatively common and often encountered; it is likely the most commonly observed worm salamander in the Central Valley. However, its numbers have apparently decreased significantly in some areas.

These salamanders are nocturnal, fossorial, and secretive. They spend much of their time hiding from view in leaf litter, insect burrows, and moss banks, and under cover objects such as decaying logs. As long as sufficient shelter and humid microclimates are available, *O. uniformis* can withstand significant levels of habitat alteration, and this species persists in pastures, gardens, and even urban environments. It also occurs in several protected areas (e.g., Irazú Volcano, Turrialba Volcano, Tapantí National Park, Poás Volcano, Monteverde, and Tenorio Volcano), but has declined for unknown reasons in some areas where habitat alteration would not be considered a threat.

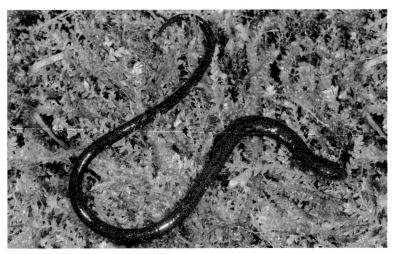

O. uniformis is still relatively common in the Central Valley, even in urban gardens.

Description

Moderate–sized and extremely slender. Very long tail is fragile and breaks easily; when intact, it can be more than twice the standard length.

Costal grooves 19–20; adpressed limbs separated by 12.5–14.5 costal grooves.

Coloration uniform dark gray, brown, or black. On close examination, the head may show slightly lighter markings.

Ventral surfaces usually slightly lighter in hue than the dorsum, but still mostly uniform dark.

Limbs very short.

Head small and only barely wider than the neck, with a bluntly rounded snout. Adult males with pronounced nasolabial protuberances.

Careful examination in good light reveals presence of numerous, minute brown and white specks on the body and tail.

foot

Hands and feet miniscule, with padlike syndactylous digits.

Similar species

O. uniformis co-occurs locally with the similarly colored species *O. pseudouniformis* (p. 140). The latter species has a broader head with larger eyes and more robust and longer limbs and feet. However, these differences are subtle and difficult to assess without being able to compare specimens in the field.

    O. uniformis is geographically separated from the other two, nearly identical, species in the *O. uniformis* complex (*O. gracilis* [p. 131] and *O. pacificensis* [p. 134]), and is best identified based on geographic provenance.

Frogs and Toads
(Order Anura)

The order anura includes more than 6,600 species worldwide, with new species identified every year. By far the largest and most successful group of living amphibians, the tailless amphibians, or frogs and toads as they are more commonly called, occur throughout most of the world and are absent only from some oceanic islands and the two polar regions.

Frogs and toads, like all amphibians, have thin, moist, permeable skin that facilitates the exchange of fluids and gasses. Thus, any anuran that ventures outside its hiding place during the heat of the day faces the threat of dehydration, especially in hot tropical environments. As a result, most species have adopted behaviors to minimize that risk. Some species are mainly active at night, when it tends to be cooler and more humid. Frogs that are active during the day either live in (or near) water or inhabit the relatively cool, shaded microclimates found in the leaf litter layer of the forest interior. In addition, many tropical amphibians minimize their surface activity during the dry season and in some cases retreat underground or seek shelter in some other moist environment. They return to activity (and to the surface) once the first heavy rains signal the start of the wet season.

The arrival of the rainy season not only triggers increased surface activity in many frogs but also often initiates the start of the breeding season. Frogs that breed in temporary rain pools or permanent ponds tend to synchronize their peak breeding activity to coincide with the wet season. Species that breed in streams display two distinct breeding strategies. Stream-breeding species such as glass frogs, which rely on rainfall to help move their tadpoles toward streams, breed during the wet season; other species—including certain toads, tree frogs, and poison dart frogs—prefer to avoid the risks associated with frequent flooding and breed during the dry season, when water levels tend to remain low.

In many species, the male approaches a breeding pool or stream and emits an advertisement call to attract a female. A pair is formed when a male climbs on top of a female and clasps her with his front limbs (a behavior called amplexus). While thus linked, the pair searches for a suitable place to lay the eggs. In all Costa Rican anuran, fertilization of the eggs takes place outside the body (external fertilization); the female lays the eggs, and the male fertilizes them immediately after they leave her body.

The sense of hearing in male and female frogs is excellent, and in most species a conspicuous eardrum is visible on each side of the head. In species without visible eardrums, it is presumed that the frogs are still capable of detecting auditory cues; however, communication using other means, such as through visual cues, may be more important. Some species, including the harlequin frogs (genus *Atelopus*), inhabit noisy streamside habitats; they rely more on visual communication than on auditory cues. The brightly colored markings seen in several Costa Rican frogs are not there simply for beauty, but rather serve important biological roles, to signal toxicity or fitness, for example. Sometimes, brightly colored markings are located on parts of the body that are usually hidden from view; these so-called flash colors can be suddenly revealed when the frog moves, thus confusing a potential predator and giving the frog a split-second chance to escape.

Throughout their developmental stages, frogs and tadpoles form an important part of the food web. Predators range widely and include fish, other amphibians, reptiles, birds, mammals, and a variety of predatory invertebrates, including waterbugs, spiders, scorpions, katydids, and mantids. Prey is nearly equally diverse; although the tadpoles of most anura eat algae and other plant matter, adult anurans are exclusively carnivorous, consuming a wide variety of insects, spiders, and other invertebrates. Larger frogs and toads may also occasionally devour a small vertebrate, and Costa Rican frogs and toads are known to opportunistically prey on other frogs, lizards, snakes, and bats and other mammals. Frogs and toads locate their prey primarily by sight; their vision is particularly acute when recording rapid motion. A long, projectile tongue attached in the front of the mouth is used to capture prey.

Costa Rica is home to 147 species of frogs and toads, divided into 12 families.

Costa Rican Frogs and Toads at a Glance

The 12 families of frogs and toads in Costa Rica are distinguished by the following traits.

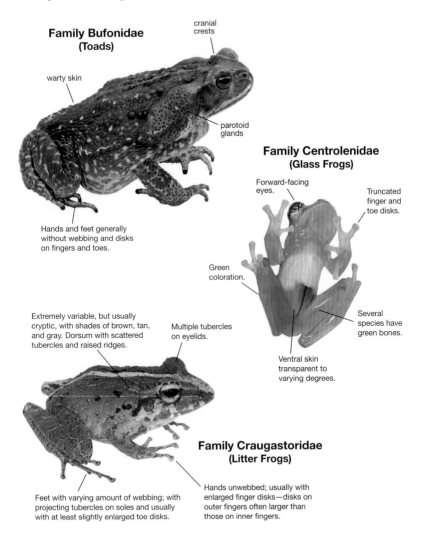

**Family Bufonidae
(Toads)**

cranial crests

warty skin

parotoid glands

Hands and feet generally without webbing and disks on fingers and toes.

**Family Centrolenidae
(Glass Frogs)**

Forward-facing eyes.

Truncated finger and toe disks.

Green coloration.

Several species have green bones.

Ventral skin transparent to varying degrees.

Extremely variable, but usually cryptic, with shades of brown, tan, and gray. Dorsum with scattered tubercles and raised ridges.

Multiple tubercles on eyelids.

**Family Craugastoridae
(Litter Frogs)**

Feet with varying amount of webbing; with projecting tubercles on soles and usually with at least slightly enlarged toe disks.

Hands unwebbed; usually with enlarged finger disks—disks on outer fingers often larger than those on inner fingers.

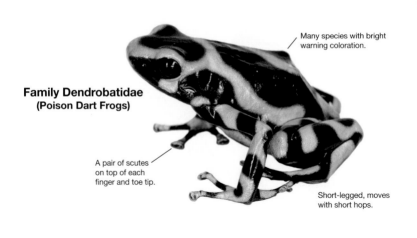

**Family Dendrobatidae
(Poison Dart Frogs)**

Many species with bright warning coloration.

A pair of scutes on top of each finger and toe tip.

Short-legged, moves with short hops.

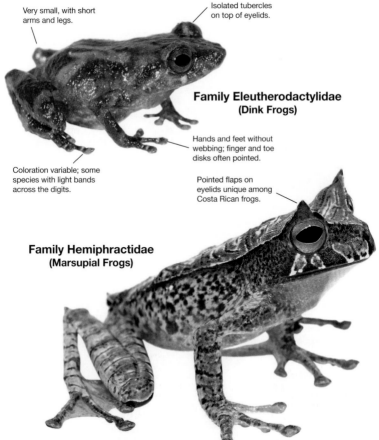

Very small, with short arms and legs.

Isolated tubercles on top of eyelids.

**Family Eleutherodactylidae
(Dink Frogs)**

Hands and feet without webbing; finger and toe disks often pointed.

Coloration variable; some species with light bands across the digits.

Pointed flaps on eyelids unique among Costa Rican frogs.

**Family Hemiphractidae
(Marsupial Frogs)**

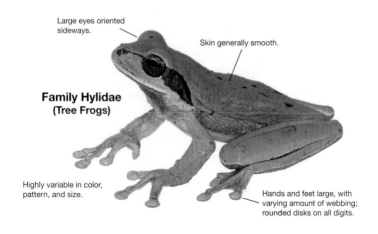

Large eyes oriented
sideways.

Skin generally smooth.

**Family Hylidae
(Tree Frogs)**

Highly variable in color,
pattern, and size.

Hands and feet large, with
varying amount of webbing;
rounded disks on all digits.

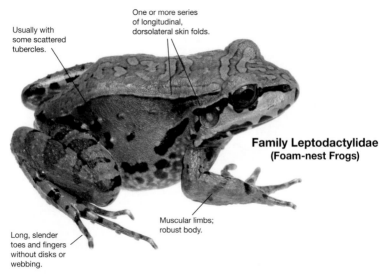

Usually with
some scattered
tubercles.

One or more series
of longitudinal,
dorsolateral skin folds.

**Family Leptodactylidae
(Foam-nest Frogs)**

Muscular limbs;
robust body.

Long, slender
toes and fingers
without disks or
webbing.

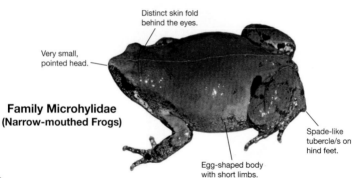

Distinct skin fold
behind the eyes.

Very small,
pointed head.

**Family Microhylidae
(Narrow-mouthed Frogs)**

Spade-like
tubercle/s on
hind feet.

Egg-shaped body
with short limbs.

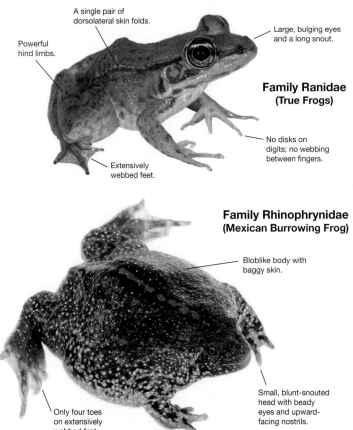

A single pair of dorsolateral skin folds.

Powerful hind limbs.

Large, bulging eyes and a long snout.

Family Ranidae
(True Frogs)

No disks on digits; no webbing between fingers.

Extensively webbed feet.

Family Rhinophrynidae
(Mexican Burrowing Frog)

Bloblike body with baggy skin.

Only four toes on extensively webbed feet.

Small, blunt-snouted head with beady eyes and upward-facing nostrils.

Family Strabomantidae
(Rain Frogs)

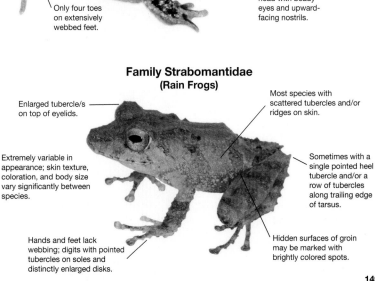

Enlarged tubercle/s on top of eyelids.

Most species with scattered tubercles and/or ridges on skin.

Extremely variable in appearance; skin texture, coloration, and body size vary significantly between species.

Sometimes with a single pointed heel tubercle and/or a row of tubercles along trailing edge of tarsus.

Hands and feet lack webbing; digits with pointed tubercles on soles and distinctly enlarged disks.

Hidden surfaces of groin may be marked with brightly colored spots.

Family Bufonidae
(Toads)

The amphibians colloquially referred to as toads comprise a group of almost 600 species found nearly worldwide. They are simply a distinct group of frogs that share a combination of several morphological features (some difficult to observe, such as the absence of teeth). While most people are familiar with the overall appearance of the characteristic "true" toad—a robust, plump body, relatively short limbs, and conspicuously warty skin—some toads don't look like that at all. Costa Rican species can be divided into two distinct subgroups: the true toads and the harlequin toads (sometimes called stubfoot toads). The country has 14 species of true toad (12 species of *Incilius*, 1 species each of *Rhaebo* and *Rhinella*) and 4 species of harlequin toad (genus *Atelopus*). The species currently included in *Incilius*, *Rhaebo*, and *Rhinella* were formerly assigned to the genus *Bufo*, which is now thought to be restricted to the Old World. The three obscure spiny toadlets that previously comprised the endemic genus *Crepidophryne* are now placed in *Incilius*.

Genus *Atelopus*

The genus *Atelopus* currently includes 97 described species of toads that occur in Central and South America. It represents one of the most endangered taxa of amphibians in the world: several species are considered extinct, a few only survive in carefully managed captive maintenance facilities, and more than 90% of the remaining species are critically endangered.

Costa Rican members of the genus *Atelopus* are small froglike toads; relatively slender-bodied and smooth-skinned, they have long limbs and fleshy, padlike hands and feet. They tend to move slowly and deliberately, choosing to walk (not jump) except when escaping predators.

Typically, these animals are diurnal and terrestrial, inhabiting moist microhabitats along cascading mountain streams in forested areas. Strong skin toxins protect them from potential predators, and their often bright colors and bold markings undoubtedly serve as a clearly visible warning signal to potential predators. The bright coloration also plays a role in communication among individual frogs. The *Atelopus* in Costa Rica lack an externally visible eardrum; even though many *Atelopus* species are known to call, vision may play a more important role than hearing as these frogs defend territories and attract mates in their typically noisy streamside habitat.

Four species of *Atelopus* are known from Costa Rica. Three of these are currently presumed extinct, while a fourth (*Atelopus varius*) has re-appeared in recent years in a few isolated areas—after it disappeared from all previously known sites for more than 13 years.

Atelopus chiriquiensis
Chiriquí Harlequin Toad

Critically endangered

Has distinctly raised parotoid, dorsal, and limb glands and a clearly defined glandular area on the tip of its snout. Males to 1.34 in (34 mm) in standard length; females to 1.93 in (49 mm).

Ecoregions
1 2 3 4 5

Atelopus chiriquiensis is known from the cool, humid cloud forests of the Talamanca Mountain Range in Costa Rica and the Chiriquí region of adjacent western Panama. It ranges from 4,600 to 8,200 ft (1,400 to 2,500 m), although all Costa Rican populations occur above 5,900 ft (1,800 m).

Natural history

A. chiriquiensis is a diurnal and terrestrial species that inhabits humid, relatively cool microclimates along high-gradient mountain streams. Males are very territorial; they are usually found on a preferred perch alongside a stream. During the breeding season, males emit a buzzy call to defend their territory and attract females; both sexes rely on bright coloration to attract mates. Males resort to physical combat with intruding males if all else fails. Adult females appear to spend considerable time away from streams—in surrounding forest—and only venture into the stream habitat during the breeding season.

The sex ratio in this species is skewed, with as many as seven males for each female in some locations. Reproductive activity mainly takes place in the early wet season (May–July); pairs in amplexus sometimes remain attached for several days or weeks prior to oviposition. During this period, single males will try to physically dislodge an amplecting male. Eggs are typically laid in strings and are attached to submerged rocks; eggs and larvae develop in the stream. Tadpoles of this species

have a large ventral sucker that enables them to attach to rocks in strong currents.

Unlike other species in its genus, *A. chiriquiensis* has relatively inconspicuous poison glands, which are scattered over its head, dorsum, and limbs. These glands are known to contain alkaloids, steroids, and other highly toxic components, including tetrodotoxin. Most of these chemical compounds do not readily break down when ingested and therefore provide an effective defense against predators.

This species was once commonly observed in appropriate habitat, but all known populations disappeared in the early 1990s. In spite of intensive searches, no *A. chiriquiensis* have been seen in the past two decades and it is now feared extinct. Studies have shown that chytridiomycosis caused disastrous amphibian declines in the distribution range of this species, but introduction of trout (a predatory fish) into mountain streams, pollution, and extensive conversion of habitat for agriculture are other probable factors in the decline of *A. chiriquiensis.*

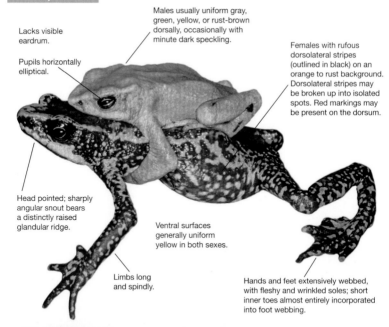

Lacks visible eardrum.

Pupils horizontally elliptical.

Males usually uniform gray, green, yellow, or rust-brown dorsally, occasionally with minute dark speckling.

Females with rufous dorsolateral stripes (outlined in black) on an orange to rust background. Dorsolateral stripes may be broken up into isolated spots. Red markings may be present on the dorsum.

Head pointed; sharply angular snout bears a distinctly raised glandular ridge.

Ventral surfaces generally uniform yellow in both sexes.

Limbs long and spindly.

Hands and feet extensively webbed, with fleshy and wrinkled soles; short inner toes almost entirely incorporated into foot webbing.

Moderate-sized, with distinct sexual size dimorphism; females are considerably larger than males.

A. chirripoensis (p. 154)

NO PHOTO

Stocky, short-limbed, and all black, with distinct wartlike glandular areas on its flanks and limbs. Only known from Cerro Chirripó.

A. varius (p. 157)

Smooth-skinned, without cleary differentiated glandular areas on the head, back, and limbs. It generally occurs at lower elevations.

153

Atelopus chirripoensis
Chirripó Harlequin Toad

Not assessed

A stocky, short-limbed uniformly black *Atelopus*. Toadlike, with a smooth dorsum and characteristic concentrations of warts on the flanks and upper surfaces of its limbs. The only specimen available in museum collections, an adult female, measures 1.69 in (43 mm) in standard length; males are presumably smaller.

Ecoregions
1 2 3 4 5

Only known from an area near the summit of Cerro Chirripó Grande, at approximately 11,320 ft (3,450 m).

Natural history

Noted Costa Rican biologist Luis Diego Gómez discovered a large group of these frogs on the Fila Norte, Cerro Chirripó, in March, 1980. In this highland paramó habitat, the frogs were found gathered around boggy ponds, where they presumably bred. Dr. Gómez assumed this observation represented a new location for an already known species, *A. chiriquiensis*, and so collected a single individual to document his observation. *A. chirripoensis* was eventually recognized and described as a new species in 2009, based on the single frog collected by Dr. Gómez. It is more similar in appearance to some South American *Atelopus* found in the Andean highlands of Ecuador than it is to any of the other three species known from

Costa Rica. Repeated subsequent visits to the area over a span of 28 years have come up empty and the species is now feared extinct.

Other Costa Rican *Atelopus* are slender, long-limbed, and typically breed in rocky mountain streams. The appearance, preferred habitat, and breeding biology of *A. chirripoensis* is similar to that of a group of stubfoot toads found in isolated locations in the South American Andes, the nearest known population of which is more than 625 mi (1,000 km) from Cerro Chirripó. These Andean highland species have all undergone dramatic population declines in recent decades and most, like *A. chirripoensis*, have not been seen since the 1980s.

Description

- An unusual *Atelopus*, with uncharacteristically short limbs and a stocky build.
- Snout short and rounded; lacks external tympanum.
- Skin on the dorsal area mostly smooth.
- Distinctive clumps of spine-tipped warts located on the flanks, upper arms, and thighs.
- Hands and feet padlike, with short, stubby digits encased in fleshy webbing.
- Dorsal coloration uniform dark brown or black.
- Undersurfaces pale orange-cream.

Similar species

A. chiriquiensis (p. 152)
Lacks spine-tipped warts; it is uniform reddish-brown, green, or yellow with red stripes or spots.

Atelopus senex
Glandular Harlequin Toad   **Critically endangered**

Atelopus senex is the only *Atelopus* in Costa Rica that has strongly developed, raised skin glands on the head, back, and limbs; these glands give the frog an angular appearance. Males to 1.26 in (32 mm) in standard length, females to 1.69 in (43 mm).

Ecoregions
1 2 3 4 5

This species has been documented in three discrete areas: the Barva Volcano region in the Central Mountain Range, 6,430 to 6,690 ft (1,960 to 2,040 m), and two regions on the extreme northern slopes of the Talamanca Mountain Range: the Macizo de Cedral south of San José, 7,050 ft (2,150 m), and the upper reaches of the Reventazón River south of Cartago, 4,200–4,330 ft (1,280–1,320 m).

Natural history

Little is known about the biology of this species. Like other Costa Rican *Atelopus*, it is diurnal and breeds in streams, with peak reproduction during the wet season (July–August). Sexually active males can be identified by the presence of brown nuptial pads on the base of their thumbs. Presumably, they lay their eggs directly in the stream, in strings that are attached to rocks. There is no information about the tadpoles, though they are likely small, with a short tail and a ventral sucker on the body for clinging to rocks in the swift current.

Although there is sexual dichromatism in some populations (in particular, animals from the Macizo de Cedral), males and females usually display a similar color pattern.

No vocalizations have ever been recorded for this species. Due to its relatively cryptic coloration, it is not known whether extensive visual communication takes place in this species. Although *A. senex* is less aposematically colored than other species in the genus, the extreme development of its skin glands indicates that it may be quite toxic.

A. senex was once very common locally along streams in humid montane rainforest habitat. But its populations declined dramatically in the late 1980s, and the species has not been observed since. Additional surveys are needed in its historic range but it is possible that this species is now extinct. Overall habitat degradation through pollution, logging, agricultural development, and the release of non-native trout in its stream habitat are undoubtedly key threats. Nonetheless, it has also disappeared from Braulio Carrillo National Park and other areas protected from such threats, indicating that a chytrid fungal infection is another likely cause of the disappearance of *A. senex*.

The skin glands lining the canthus rostralis and adorning the dorsum are so pronounced that they sometimes form a depression on the top of the head and back.

A short bony ridge extends backward behind each eye.

adult male

Pupils horizontally elliptical. There are no visible eardrums.

Males usually a shade of gray, bluish-green, olive, or black; the raised skin glands, slighly paler than the background color, show cream, tan, salmon, or pink.

Limbs long and spindly; head and body slender and angular.

adult female

In some populations, females have a cream, yellowish, or reddish-brown dorsum bordered on each side of the head and body by a broad, dark lateral band.

Females are considerably larger than males.

Digits short and stout, with fleshy basal webbing between fingers and extensive webbing between the toes. Undersides of the hands and feet fleshy and wrinkled.

The undersurfaces in both sexes tend to be light gray, with yellow, orange, or reddish suffusion, especially along the edges of the venter.

Similar species

A. varius (p. 157)

Smooth-skinned, without cleary differentiated glandular areas on the head, back, and limbs.

Atelopus varius
Variable Harlequin Toad

Critically endangered

A smooth-skinned *Atelopus* with an angular body, long limbs, and a pointed snout. Lacks clearly differentiated, raised glandular areas on the head, back, and limbs. Males to 1.38 in (35 mm) in standard length, females to 1.77 in (45 mm).

Atelopus varius has been found throughout premontane and lower montane zones on the Atlantic and Pacific slopes of Costa Rica and western Panama. Occurs from near sea level to 6,550 ft (2,000 m); most populations are known from elevations above 2,790 ft (850 m).

Ecoregions
1 2 3 4 5

Natural history

A. varius is a terrestrial and diurnal species. It was formerly common along high-gradient mountain streams, where it perches on rocky ledges, boulders, log jams, or in crevices. It prefers spots that are fully exposed to direct sunlight but that also receive mist from nearby rapids or waterfalls. Moisture absorbed from the surfaces on which they sit protects them from the dehydrating effect of the sun. At night, individuals sleep on leaves of low streamside vegetation or in rock crevices in the banks.

Male *A. varius* are highly territorial during the breeding season and vigorously defend their perch. They emit short, explosive *buzz* calls that are barely audible in their turbulent, noisy environment. Occasionally, a territorial male will wave its front limbs (first one limb then the other) over his head and signal to its neighbors, a form of visual communication called semaphoring.

The reproductive season of *A. varius* coincides roughly with the wet season (August–November). During this period sexually mature males set up territories along a suitable stream corridor. Mature males can be recognized by the rough, brown nuptial pads on the base of their thumbs. Females are far fewer in number than males and, when a gravid female enters male territories along a breeding stream, they will pounce on her and grapple with one another for a chance to engage in amplexus. Mated pairs, who can remain in amplexus for a long period of time (days or even weeks), have been seen between July and early December. Their small, white eggs are laid in long strings attached to rocks, logs, or tree roots; the eggs hatch after four to seven days.

Tadpoles are black with minute light dots. They initially hide in loose gravel or debris on the bottom of streams; during early tadpole development, the mouth parts develop into a strong suction device that extends onto the ventral surface and is used to cling to rocks and logs in the strong current. These tiny tadpoles (to 0.47 in [12 mm] total length) have a very short, poorly developed tail.

On *A. varius*, the poison glands are scattered evenly over the surface of the skin (unlike other Costa Rican *Atelopus*, which have concentrations of glandular tissue in clearly defined areas of the head, back, or limbs). The skin contains significant amounts of toxins, including tetrodotoxin, a potent nerve toxin capable of immobilizing or even killing potential predators. The bold color pattern seen in most *A. varius* may serve as warning coloration. Known predators include many invertebrates such as spiders, crabs, and a species of fly, *Notochaeta bufonivora*. Females of the latter parasite deposit their eggs on the frog; when the maggots hatch they eat their way into the victim's body and consume the frog's intestines, causing it to die within days.

Like other *Atelopus*, this species has experienced a dramatic crash in numbers in recent decades. Previously, *A. varius* was quite common in many of the more than one hundred sites in which it was known to occur in Costa Rica. In 1982–83, survey data of a population near Monteverde reported an average of 751 adults along a 650 ft (200 m) transect. Similarly high densities were anecdotally reported from other areas, until all populations disappeared within a short period of time in the late 1980s; the last individual *A. varius* was seen in 1996. In 2003, however, a remnant population was found in the Fila Chonta, near Quepos; and in 2008, a second population was found near Las Tablas, in the San Vito area. A few additional sightings have recently been reported from other Costa Rican locations and at least one population persists in western Panama. The Fila Chonta population is very small (31 individuals have been documented there), while the surviving population near Las Tablas is more robust, with more than 200 individuals documented thus far.

Nonetheless, the future of *A. varius* is far from secure, and more research is needed to better understand the circumstances that caused its decline. Most likely chytrid fungi played a critical role, as museum specimens collected in the 1980s and 1990s have tested positive for chytrid infection. However, the impact of climate disturbances, habitat alteration, and the introduction of predatory trout in breeding streams have probably also played a role.

Description

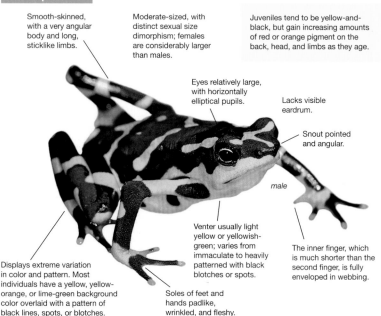

Smooth-skinned, with a very angular body and long, sticklike limbs.

Moderate-sized, with distinct sexual size dimorphism; females are considerably larger than males.

Juveniles tend to be yellow-and-black, but gain increasing amounts of red or orange pigment on the back, head, and limbs as they age.

Eyes relatively large, with horizontally elliptical pupils.

Lacks visible eardrum.

Snout pointed and angular.

male

Venter usually light yellow or yellowish-green; varies from immaculate to heavily patterned with black blotches or spots.

The inner finger, which is much shorter than the second finger, is fully enveloped in webbing.

Displays extreme variation in color and pattern. Most individuals have a yellow, yellow-orange, or lime-green background color overlaid with a pattern of black lines, spots, or blotches.

Soles of feet and hands padlike, wrinkled, and fleshy.

Similar species

A. chiriquiensis (p. 152)

Has weakly developed skin glands on the head, back, and limbs that are rounded and difficult to discern; generally occurs at higher elevations.

Dendrobates auratus (p. 297)

Has a bold pattern of black and green markings, as do some populations of *A. varius*; it has a less angular body and head and its limbs are not as long and spindly. The two species co-occur at low elevations on the Pacific slope.

Incilius periglenes (p. 180)

NO PHOTO

Female *Incilius periglenes* share a pattern of yellow, orange, and black; they differ in having cranial crests and parotoid glands.

Genus *Incilius*

This genus currently includes 40 species. Most of these were previously included in the genus *Bufo*, but that genus is now thought to contain Old World species only. Three unusually spiny toadlets that until recently formed the genus *Crepidophryne* are now also placed in *Incilius*.

These animals are readily identified by a combination of several easily observed traits. Members of the genus *Incilius* are stocky, short-limbed amphibians, often with relatively dry and warty skin, distinct cranial crests on the top of the head, and a pair of parotoid glands in the neck region. These parotoid glands, as well as additional glands scattered over the dorsal skin, produce powerful defensive toxins. Note that recently metamorphosed juveniles can be difficult to identify, as they lack cranial crests and parotoid glands.

Costa Rica is home to 12 toads in the genus *Incilius.* Though they are morphologically similar, their individual habitat requirements are varied; some species inhabit only seasonally dry lowland areas while others are restricted to cool climates in isolated high elevation mountains. Several Costa Rican toads are common and widespread in the region, but a few of the more specialized montane species have undergone dramatic population declines and may be extinct.

Incilius aucoinae
Pacific Forest Toad

Least concern

The only toad in southwestern Costa Rica with well-developed cranial crests, a prominent concave depression between the parietal crests, and small, triangular parotoid glands. *Incilius aucoinae* is often orange-brown in coloration. Males to 2.64 in (67 mm) in standard length, females to 4.09 in (104 mm).

Incilius aucoinae inhabits the Pacific region of Costa Rica, from the center of the country (south of Puntarenas) toward the southern border; from near sea level to 1,975 ft (600 m). Recently, this species was also observed in the Chiriquí region of adjacent Panama.

Ecoregions
1 2 3 4 5

Natural history

This widespread, common species inhabits moist and wet forests in lowlands and foothills. It prefers the interior of relatively undisturbed forests, where it is most often found near large rivers and streams. *I. aucoinae* is nocturnal and terrestrial. During the daytime, its cryptic coloration allows it to blend in well with leaf litter that covers the forest floor. Adult males have a single internal vocal sac and can produce a soft trill call to attract mates to suitable breeding habitat. In the dry season (January–February), when water levels are low, females lay their eggs in long strings in quiet sections of rivers and streams.

This species is closely related to *I. melanochlorus*, and both species are very similar in appearance, but their distribution ranges do not overlap. *I. aucoinae* differs most notably in lacking transverse skin folds between the parietal crests, and in having a relatively light and unmarked venter, chest, and throat.

The cryptic color pattern of *I. aucoinae* allows it to blend in well with leaf litter that covers the forest floor.

Cranial crests well-developed but thin. Area between parietal crests with a large concave depression; skin on top of head smooth and fused to underlying bone.

Skin on body and limbs covered with scattered small tubercles. Males have low, rounded tubercles.

Color pattern variable, but dorsum generally brown, gray, or (most often) reddish-brown; a pattern of paired dark spots and a middorsal light stripe usually prominent.

adult male

Sexually active males with brown nuptial patches on inner two fingers of each hand.

Tympanum clearly visible.

Hind limbs often with alternating dark and light transverse bars.

Vocal sac unpigmented; not visible as a dark spot on the throat of adult males.

Limbs fairly slender. Hands lack webbing; feet with basal webbing between toes.

Often with an irregular light stripe that follows the linear series of enlarged lateral tubercles; stripe is bordered below by a dark chocolate-brown field.

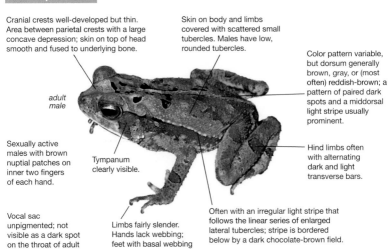

A series of prominently enlarged tubercles forms a lateral row on each side of the body.

Dorsal tubercles large and pointed in females.

adult female

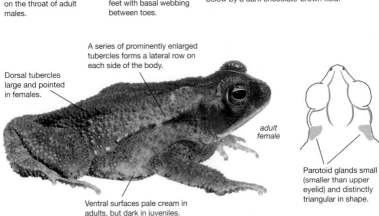

Parotoid glands small (smaller than upper eyelid) and distinctly triangular in shape.

Ventral surfaces pale cream in adults, but dark in juveniles.

I. coniferus (p. 166)

Has distinctly spiny tubercles on its body and a prominent series of linearly arranged, large spiny warts along each side.

Rhaebo haematiticus (p. 185)

Much larger, lacks cranial crests; also note smooth skin and large parotoid glands.

Rhinella marina (p. 188)

Much larger and has enormous parotoid glands.

Engystomops pustulosus (p. 451)

Has a pair of parotoid glands, but has extremely warty skin on the top of its head instead of bony crests.

161

Incilius chompipe
Chompipe Spiny Toadlet

Vulnerable

A small, extremely spiny, globular toadlet with a sharply pointed snout, prominent cranial crests, small parotoid glands, and fleshy padlike hands and feet that have extensive webbing. *I. chompipe* lacks an externally visible eardrum. Males to 1.06 in (27 mm) in standard length, females to 1.30 in (33 mm).

Ecoregions
1 2 3 4 5

Endemic to Costa Rica; only known from a few locations in the Central Mountain Range near Cerro Chompipe and Cerro Dantas; 4,925 to 5,250 ft (1,500 to 1,600 m).

Natural history

I. chompipe is one of three species previously assigned to the genus *Crepidophryne* (also see *I. epioticus* [p. 168] and *I. guanacaste* [p. 172]). They are all small and have extremely rugose skin, distinct cranial crests, a sharply pointed snout, small parotoid glands, and fleshy, padlike hands and feet with very extensive webbing. These toadlets all lack an externally visible eardrum. The three species can be distinguished by subtle differences in skin texture, though location may be a better clue to identification as their limited ranges do not overlap.

This diurnal toadlet is rarely seen, at least in part because it is secretive (thus it might be more common than the number of sightings suggests). It is found only in undisturbed primary cloud forest with a deep layer of leaf litter. *I. chompipe* is fossorial, and likely escapes detection by hiding in leaf litter, beneath moss cushions, or possibly in underground burrows, for long periods of time. It is usually seen early in the morning, walking slowly and deliberately or moving in short single hops. During its peak activity times, several individuals may be observed in close proximity to one another.

Since these forest denizens often occur far from standing water, it was long thought that they might reproduce through direct development, depositing their eggs in a moist terrestrial location. This reproductive strategy was very recently confirmed for *I. chompipe* and constitutes the first record of direct development in any Central American toad species. Likely, the two other spiny toadlets in Costa Rica (*I. epioticus* and *I. guanacaste*) breed in a similar fashion.

Observations of a captive pair of *I. chompipe* showed that, after engaging in axial amplexus, they laid a clutch of approximately 60 large, whitish eggs. The eggs were hidden beneath a sheath of moss and were guarded by the female throughout the incubation period, which lasted about seven weeks. During this period the female would position herself on top of the eggs and flatten her body, possibly to hide the clearly visible light-colored eggs from potential predators. Upon hatching the juveniles measured approximately 0.18 in (4.5 mm), had a short tail stub remaining, and were able to walk immediately.

I. chompipe can produce skin secretions that are capable of causing noticeable irritation to human eyes. In addition, this species feigns death when handled. This defensive strategy—remaining motionless and faking death—can help avoid predation in some cases.

The unusual body shape, padlike hands and feet, and pointed head of *I. chompipe* are quite distinctive.

Young juveniles are more colorful; they are marked with orange to brick-red rounded spots and irregular white lines. A dark brown lateral line may be present on each side of the body and head. Small juveniles lack crests.

A row of enlarged, pointed tubercles runs from the parotoid gland to the groin on each side of the body.

Coloration dull grayish-brown to brown, with irregular dark brown markings.

Rough, tuberculate skin and pronounced cranial crests are distinctive.

On each side of the body, a light, irregularly shaped, dorsolateral band may be present between the parotoid gland and the groin.

Very small and stocky, with a distinctly pointed snout. Lacks visible eardrums.

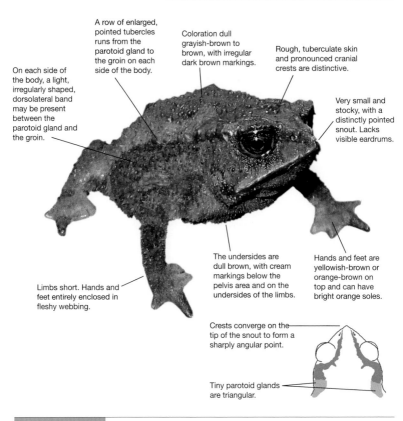

The undersides are dull brown, with cream markings below the pelvis area and on the undersides of the limbs.

Hands and feet are yellowish-brown or orange-brown on top and can have bright orange soles.

Limbs short. Hands and feet entirely enclosed in fleshy webbing.

Crests converge on the tip of the snout to form a sharply angular point.

Tiny parotoid glands are triangular.

Few other toads occur in the general range of *I. chompipe*. All but *I. holdridgei* have visible eardrums and clearly differentiated fingers and toes (not fleshy, padlike hands and feet).

I. holdridgei (p. 174)

Larger and lacks the sharply pointed snout and overall spiny appearance of *I. chompipe*.

I. epioticus (p. 168)

Very similar in overall appearance, but its parotoid glands are round (not triangular) and the inner finger extends beyond the edge of webbing.

163

Incilius coccifer
Dry Forest Toad

A medium-sized, rough-skinned toad. Tubercles covering dorsal surfaces are of relatively uniform size; lacks a prominent series of enlarged tubercles on its sides. Cranial crests robust; parotoid glands oval, about 1–1.5x the size of the upper eyelid. Almost invariably with a thin light middorsal stripe and a light mark on top of its head, between the eyes. Males to 2.44 in (62 mm) in standard length, females to 3.23 in (82 mm).

Ecoregions
1 2 3 **4 5**

Incilius coccifer inhabits the Pacific slope of Guatemala, El Salvador, Honduras, Nicaragua, and northwestern Costa Rica, and is also found in the Mosquitia region of Atlantic Honduras and Nicaragua; from near sea level to 4,700 ft (1,435 m).

Natural history

I. coccifer is abundant throughout its range, in a variety of habitats, including dry lowland forest, humid montane forests, relatively open areas, pastures, roadside ditches, gardens, and vacant lots in urban areas. It is adaptable, common, and widespread and currently does not appear to face any significant threats.

This species spends much of the drier months in hiding, likely in underground burrows. Breeding takes place during the early rainy season (mid-May through mid-June). Males congregate around temporary rain pools, ponds, and flooded pastures. Their advertisement call is a high-pitched buzzing trill, usually made from the edge of the water but occasionally while in the water. Females tend to arrive at the breeding pools a few weeks later than the males. Amplexus in this species is axillary. Eggs are typically produced in long strings that often break into shorter segments during oviposition. The larvae rapidly complete metamorphosis in about five weeks. Tadpoles of *I. coccifer* do not form large aggregations as do the tadpoles of other toad species.

Description

Usually a pale, irregular broad stripe extends on each side of the body, from the parotoid gland to the groin.

adult male

Sexually active males with brown nuptial patches on inner two fingers. Males with extendable single vocal sac, indicated by darker throat color when not in use.

Ventral surfaces white, cream, or pale yellow, sometimes with diffuse gray or brown mottling or spots.

Limbs short and stocky; toes with basal webbing.

Most individuals with a light interorbital bar bordered by dark markings anteriorly and posteriorly.

Skin on the dorsum and flanks covered with densely spaced, low, rounded tubercles that gradually become spine-tipped on the lower lateral surfaces. Ventral skin granular (with or without small spiny tips).

Tympanum prominent.

There is no distinct series of enlarged lateral warts in *I. coccifer.*

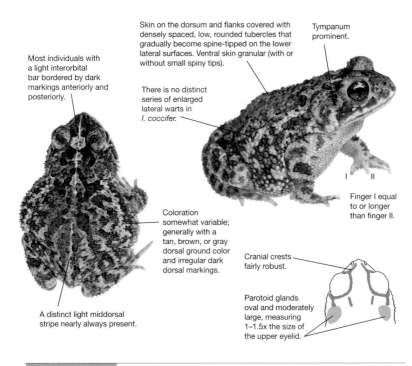

Finger I equal to or longer than finger II.

Coloration somewhat variable; generally with a tan, brown, or gray dorsal ground color and irregular dark dorsal markings.

A distinct light middorsal stripe nearly always present.

Cranial crests fairly robust.

Parotoid glands oval and moderately large, measuring 1–1.5x the size of the upper eyelid.

Similar species

female *I. luetkenii* (p. 176)

Quite similar in appearance but differs in having black-tipped cranial crests, very small parotoid glands, and no pale interorbital bar.

I. coniferus (p. 166)

Shows a distinctly different color pattern; it also has a prominent series of linearly arranged, enlarged, spiny warts on each side of the body.

I. melanochlorus (p. 178)

Has small, triangular parotoid glands and more prominent cranial crests.

Rhinella marina (p. 188)

Significantly larger and has enormous parotoid glands.

Rhaebo haematiticus (p. 185)

Much larger, lacks cranial crests, and has relatively smooth skin and large parotoid glands.

Engystomops pustulosus (p. 451)

Has a pair of parotoid glands, but note extremely warty skin (rather than bony crests) on top of its head.

165

Incilius coniferus
Green Climbing Toad
Least concern

A moderate-sized toad with well-developed, black-tipped cranial crests, small parotoid glands, a prominent linear series of enlarged, pointed tubercles on each flank, and black-tipped spiny tubercles on the body and limbs. Males to 2.83 in (72 mm) in standard length, females to 3.70 in (94 mm).

Ecoregions
1 2 3 4 5

Incilius coniferus is relatively widespread in Atlantic lowlands and foothills, from north-central Nicaragua to central Panama; it also occurs in the Pacific lowlands of central and southern Costa Rica and, discontinuously, in Colombia and northern Ecuador; sea level to 5,075 ft (1,550 m).

Natural history

This is a fairly common species in humid lowland and foothill forests. *I. coniferus* tolerates minor habitat disturbance; it is most commonly seen in relatively open settings along forest edges, in streambeds, and in road cuts passing through forested areas.

I. coniferus reproduces in temporary rain pools throughout wetter months of the year; breeding activity may be limited during the dry season, when there is greater risk that breeding pools will dry up. The male's mating call is a long, high-pitched melodious trill. Mating pairs engage in axillary amplexus; eggs (black with a gray pole) are laid in long strands, directly in the water. Larval development is very fast, an adaptation that helps diminish the risk of dying in ephemeral breeding pools (note that tadpoles of this species are capable of surviving a period of time buried in mud if their breeding pool temporarily dries up). Eggs hatch within five days, and tadpoles can complete metamorphosis and leave the water as soon as 33 days later. They have been observed to fall prey to predatory tadpoles of *Leptodactylus savagei* (family Leptodactylidae), and to the snakes *Erythrolamprus epinephelus* and *Rhadinaea decorata*. Although *I. coniferus* is mainly nocturnal, juveniles are sometimes active on rainy, overcast days.

In spite of their chunky build, these toads are skilled climbers; individuals hold their fingers in a clawlike fashion and use long arms and legs to ascend tree trunks or dense tangles of vegetation. During the day, individuals have been observed several meters high in epiphytic vegetation, on top of a tree fern, or in a hole in the trunk of a dead tree. Notwithstanding such climbing skills, during the daytime these toads also retreat into burrows and underneath dense tangles of vegetation on the forest floor. At night *I. coniferus* emerges to forage on small prey, mostly ants and mites.

Although it is still regularly encountered in suitable habitat throughout its range, local populations of this species declined or even disappeared in some areas in the early 1990s. Its habitat is protected in several reserves and parks, but pathogens such as chytrid fungi may have negatively impacted some populations, even in protected areas.

Adult males call at night from breeding pools to attract mates.

Short head, with a sharply truncated snout and prominent cranial crests, has an angular appearance.

Coloration variable; adults range from shades of gray, green, brown, olive, and yellow to brick-red, marked with darker spots and blotches.

Prominent series of large, pointed tubercles forms a line connecting the parotoid gland to the groin on each side of the body.

Cranial crests tipped with black keratin.

Skin has scattered, enlarged tubercles that are tipped with black spines.

Eardrum distinct.

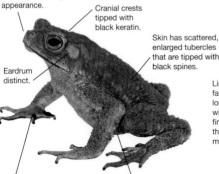

Limbs long and fairly slender, with long digits. Hands without webbing, finger I shorter than finger II. Feet moderately webbed.

Sexually mature males with dark brown nuptial patches on inner fingers and a single, internal vocal sac on throat, indicated by darker pigmentation.

Ventral surfaces usually dirty white to cream, uniform or mottled in the flanks.

Parotoid glands small, variably shaped, and indistinct.

juvenile

Juveniles are more colorful than adults, often with a bright green coloration punctuated by reddish-brown to orange tubercles; they have orange hands and feet. Younger individuals sometimes display a light mark below each eye.

During the breeding season the dorsal coloration intensifies: males may be bright yellow to orange-red, while females turn bright mossy-green.

I. luetkenii
(p. 176)

Shares the black-edged cranial crests but lacks spine-tipped tubercles on its body and limbs.

I. melanochlorus
(p. 178)

I. melanochlorus, *I. aucoinae*, and *I. valliceps* lack the black-edged cranial crests and prominent, spine-tipped tubercles.

Rhinella marina
(p. 188)

Much larger, with very large parotoid glands.

Rhaebo haematicus
(p. 185)

Larger, lacks cranial crests, and has smooth skin.

167

Incilius epioticus
Talamanca Spiny Toadlet   **Least concern**

A small, distinctly spiny, globular toadlet. Note sharply pointed snout, prominent cranial crests, small parotoid glands, and fleshy, padlike hands and feet with very extensive webbing. *I. epioticus* lacks an externally visible eardrum. Geographic provenance may be the easiest way to distinguish this species from related species. Males to 1.30 in (33 mm) in standard length, females to 1.57 in (40 mm).

Ecoregions
1 2 3 4 5

Incilius epioticus is more widespread than the two other spiny toadlets that occur in Costa Rica. It is known from isolated localities along the Atlantic slope of the Talamanca Mountain Range, both in southeastern Costa Rica and adjacent Panama; 3,450–6,700 ft (1,050–2,040 m).

Natural history

This is one of three distinctly spiny toadlets previously assigned to the genus *Crepidophryne*. All three species have extremely rugose skin, distinct cranial crests, a sharply pointed snout, small parotoid glands, and fleshy, padlike hands and feet with very extensive webbing. These toadlets all lack an externally visible eardrum. Subtle differences in skin texture distinguish each species. Because there is no known overlap in range, location may be the surest key to identification.

Though rarely seen, this toadlet is perhaps more widespread than is currently thought. It may occur in appropriate habitats that lie between the few known localities. All reports indicate that this is a terrestrial species that inhabits the deep leaf litter of pristine rainforest, cloud forest, highland oak forest, and mature secondary forest with an intact canopy. Likely fossorial, it escapes detection by hiding in leaf litter, or possibly in burrows underground. When uncovered in the leaf litter, these toadlets tend

to remain still at first, relying on their camouflage. When handled, they may remain motionless for an extended period of time and feign death, a defensive strategy that can discourage predation.

These toads are primarily observed early in the morning, as they walk slowly and deliberately or move in short single hops across trails or the forest floor. Increased surface activity seems to take place in April, but the reason for this is unknown. Various observers have reported aggregations of 2–8 of these toads moving as a group, in relatively close proximity to each other. It is unclear whether these aggregations are related to reproductive behavior. The breeding biology of *I. epioticus* is unknown, but direct development as a breeding strategy was very recently confirmed in the closely related species *I. chompipe*. It is likely that *I. epioticus* reproduces in the same fashion, and that its young hatch from terrestrial eggs as fully formed miniature toadlets.

Description

Adults are uniform dull brown with lighter (yellowish-brown or orange-brown) hands and feet. Some individuals have irregular black markings on the dorsum.

adult

Limbs short. Hands and feet nearly entirely enclosed in fleshy webbing, except for inner finger, which is free from webbing for almost half its length.

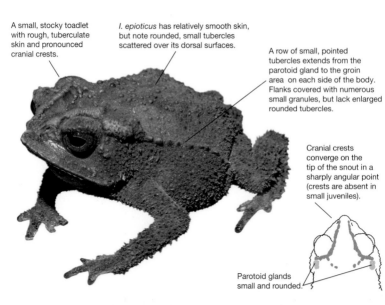

A small, stocky toadlet with rough, tuberculate skin and pronounced cranial crests.

I. epioticus has relatively smooth skin, but note rounded, small tubercles scattered over its dorsal surfaces.

A row of small, pointed tubercles extends from the parotoid gland to the groin area on each side of the body. Flanks covered with numerous small granules, but lack enlarged rounded tubercles.

Cranial crests converge on the tip of the snout in a sharply angular point (crests are absent in small juveniles).

Parotoid glands small and rounded.

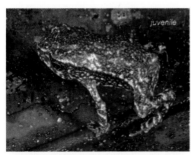
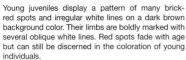

Young juveniles display a pattern of many brick-red spots and irregular white lines on a dark brown background color. Their limbs are boldly marked with several oblique white lines. Red spots fade with age but can still be discerned in the coloration of young individuals.

Colorful individuals of this species display a beautifull cryptic pattern, composed of subtle hues of brown, tan, and orange that blend in perfectly with the leaf litter on the forest floor.

Similar species

Few other toads occur in the general range of *I. epioticus*; those that do all have visible eardrums and clearly differentiated fingers and toes (not fleshy, padlike hands and feet), except for *I. fastidiosus* (p. 170), a larger toad that lacks a sharply pointed snout and has expanded cranial crests that cover much of the top of its head.

I. chompipe (p. 162)

Very similar in overall appearance but only occurs in the Central Mountain Range. It has triangular parotoid glands and its inner fingers are completely enveloped in webbing.

169

Incilius fastidiosus
Talamanca Montane Toad

Critically endangered

This is a small montane toad with very strongly developed cranial crests and unique coloration: dark brown to black dorsum and cream to orange cranial crests and dorsal tubercles. Lacks visible eardrums. Males to 2.05 in (52 mm) in standard length, females to 2.36 in (60 mm).

Incilius fastidiosus occurs in Costa Rica in isolated localities on both slopes of the southern Talamanca Mountain Range; it also occurs in immediately adjacent Panama, on the Atlantic slope; 2,490 to 6,890 ft (760 to 2,100 m).

Ecoregions
1 2 3 4 5

Natural history

This secretive species was formerly abundant, though even then rarely observed. Individuals have been sighted between February and June; adults are very rarely seen when not reproducing. Throughout the year, however, they have been excavated from deep leaf litter. There have been scattered sightings of juveniles almost year-round; they are generally seen among the rocky margins of streams.

I. fastidiosus is a diurnal species that spends most of its life burrowed in leaf litter or underground retreats. This is an explosive breeder; large numbers gather simultaneously around temporary rain pools, triggered by the heavy rains that kick off the wet season (April–May). Males apparently do not produce a mating call, but do make a trilling release call if inadvertently clasped by another male. Amplexus is inguinal, with males clasping females around the waist. Males appear to significantly outnumber females at breeding pools, and single females have been observed at the center of toad balls that consisted of up to 10 amplectant males. Occasionally females drown under the weight of these males. *I. fastidiosus* lays its eggs in a single, long strand consisting of 80–90 yellow-and-black eggs.

I. fastidiosus inhabits foothill and middle elevation forests within its limited range. It is similar in biology and morphology to the two other endemic,

earless highland toads in Costa Rica, *I. holdridgei*, p. 174, found locally in the Cental Mountain Range, and *I. periglenes*, p. 180, only known from the Tilarán Mountain Range. Like these two species, *I. fastidiosus* has undergone a dramatic decline since the late 1980s, and no recent sightings have been recorded in spite of repeated searches. This species may be extinct. Most likely, chytrid fungal infections and significant habitat loss and habitat alteration underly this decline.

I. fastidiosus is an usual little toad, with a hardened, helmetlike head. The function of this structure may never be known, as this species has not been seen in decades.

Color pattern distinctive. Dark brown to black dorsal ground color, often with a light middorsal stripe or broad band; cranial crests and dorsal tubercles are cream, tan, orange, or pinkish.

Sexually active males have swollen front limbs and light nuptial patches on inner fingers.

Small, with a robust, armored head.

Dorsum and limbs mostly with small rounded tubercles, interspersed with massive pointed warts; has a prominent linear series of large smooth tubercles on its flanks, between parotoids and groin.

Ventral surfaces are mottled brown to black, and also covered with tubercles.

Some individuals with extensive brick-red markings on the dorsum and/or limb surfaces.

Limbs short and stocky. Hands and feet lack tubercles on their palms and soles; they are fleshy and padlike, with significant webbing between digits.

Skin on top of the head co-ossified and fused with the skull; thick cranial crests expand into a globular structure located between each eye and parotoid gland.

Parotoid glands oblong, smaller than upper eyelid.

I. coniferus
(p. 166)

Has clearly visible eardrums and a very different color pattern.

I. epioticus
(p. 168)

Smaller, with a pointed head, uniformly spiny skin, and padlike hands and feet.

Rhinella marinus
(p. 188)

Much larger, with massive parotoid glands.

Rhaebo haematicus
(p. 185)

Much larger, has smooth skin, and lacks cranial crests.

171

Incilius guanacaste
Guanacaste Spiny Toadlet

Data deficient

A small, globular toadlet with rugose skin, a sharply pointed snout, prominent cranial crests, small parotoid glands, and fleshy, padlike hands and feet with very extensive webbing. *I. guanacaste* lacks an externally visible eardrum. Geographic provenance may be the easiest way to distinguish this species from related species. Males to 0.87 in (22 mm) in standard length, females to 1.26 in (32 mm).

Ecoregions
1 2 3 4 5

Incilius guanacaste is endemic to northwestern Costa Rica. It is known from just a handful of individuals encountered in the isolated wet forests found near the summit of two volcanoes (Miravalles and Rincón de la Vieja) in the Guanacaste Mountain Range; 6,235–6,560 ft (1,900–2,000 m).

Natural history

Hardly anything is known about the biology of this rare toadlet. It has been observed in deep leaf litter in the wet and windy cloud forests and elfin forests that can be found at high elevations on the summits of Miravalles and Rincón de la Vieja Volcanoes; these volcanoes are surrounded by the dry forests that characterize Guanacaste Province. Like the other two spiny toadlets that occur in Costa Rica (*I. chompipe* [p. 162] and *I. epioticus* [p. 168]), *I. guanacaste* is thought to be primarily a diurnal leaf litter inhabitant.

This toad is most often encountered as it ambles through its forest habitat with a slow, deliberate gait. When uncovered within the leaf litter, it tends to remain motionless and rely on its camouflage. However, when found in an exposed location, it invariably attempts to escape with short hops. When caught, it may feign death, remaining motionless, with limbs held limp and eyes open, in an attempt to thwart predation.

No tadpoles or eggs have ever been observed for this species; it has been speculated that it lays its eggs in a moist microclimate within the leaf litter layer and that the larvae develop directly into toadlets within the eggs, skipping a free-swimming larval stage, a reproductive strategy seen in ecologically similar South American toadlets. This particular reproductive breeding strategy has very recently been confirmed in the closely related species

I. chompipe. Given that *I. guanacaste* generally inhabits leaf litter environments far removed from any standing water, it is likely that it uses the same breeding strategy.

This is one of three small, distinctly spiny toadlets formerly assigned to the genus *Crepidophryne*. Subtle differences in skin texture distinguish each species; *I. guanacaste* lacks the pointed tubercles that typically cover the skin of *I. chompipe* and *I. epioticus*.

I. guanacaste is an exceedingly rare species, endemic to the Guanacaste Mountain Range of northwestern Costa Rica. Only a handful of individuals have ever been observed.

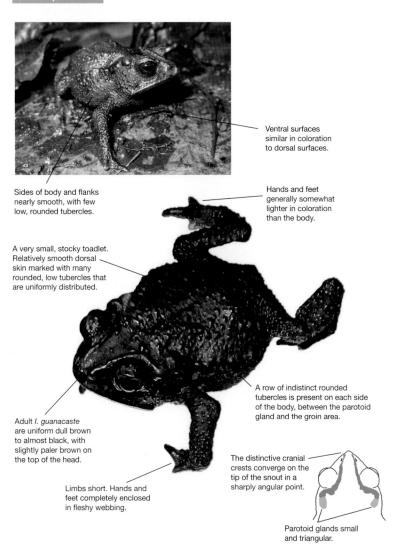

Ventral surfaces similar in coloration to dorsal surfaces.

Sides of body and flanks nearly smooth, with few low, rounded tubercles.

Hands and feet generally somewhat lighter in coloration than the body.

A very small, stocky toadlet. Relatively smooth dorsal skin marked with many rounded, low tubercles that are uniformly distributed.

Adult *I. guanacaste* are uniform dull brown to almost black, with slightly paler brown on the top of the head.

A row of indistinct rounded tubercles is present on each side of the body, between the parotoid gland and the groin area.

Limbs short. Hands and feet completely enclosed in fleshy webbing.

The distinctive cranial crests converge on the tip of the snout in a sharply angular point.

Parotoid glands small and triangular.

Similar species

Other toads that occur in the general vicinity of *I. guanacaste* (*I. coccifer* [p. 164], *I. luetkenni* [p. 176], *I. melanochlorus* [p. 178], *Rhaebo haematiticus* [p. 185], and *Rhinella marina* [p. 188]) all have clearly visible eardrums and well-differentiated fingers and toes (not fleshy, padlike hands and feet).

Incilius holdridgei
Holdridge's Montane Toad

Critically endangered

A small montane toad that lacks visible eardrums. The soles of its hands and feet lack tubercles; has fleshy, padlike hands and feet. Cranial crests are low except for a prominently expanded section behind each eye that connects with a parotoid gland. Males to 1.81 in (46 mm) in standard length, females to 2.09 in (53 mm).

Ecoregions
1 2 3 4 5

Incilius holdridgei is endemic to Costa Rica. It is known only from the Barva Volcano region, in the Central Mountain Range; 6,560–7,220 ft (2,000–2,200 m). Not having been seen since 1986, it was feared extinct until a surviving population was discovered in 2008.

Natural history

I. holdridgei occurs in middle elevation rainforests. This secretive species spends most of its time underground; individuals have been found under cover objects during rainy spells, but apparently they hide in the banks of mossy streams and roads during the dry season.

At the onset of the wet season (April–May), this explosive breeder congregates around rainwater-filled depressions on the forest floor. Males lack a vocal sac and perhaps do not produce a mating call; they generally outnumber females at the breeding pools and several males sometimes attempt to mate with a single female (or even with other species of frog), creating mating balls that can lead to the drowning death of the animal at its center. Amplexus is inguinal, with males clasping the female in the waist area. Eggs are

I. holdridgei was thought to be extinct for more than 20 years, until a surviving population was discovered in 2008.

deposited directly in the water and larvae develop rapidly, completing metamorphosis in about one month.

I. holdridgei was formerly common in its limited range (although it was rarely observed). In a study in the 1970s, 2,765 males were recorded visiting two breeding pools over an 8 day period! In the mid-1980s, its populations declined precipitously (as did those of many other amphibians in Costa Rica). In 2008, after 25 years of repeated site visits and intensive searches had failed to turn up any individuals, this species was formally declared extinct. But in that same year, several juvenile toads were observed in its original habitat, and additional monitoring led to confirmation in 2010 that *I. holdridgei* still persists in low numbers in at least one location on Cerro Dantas.

This relict population is currently under study, providing an opportunity to assess what might have caused its earlier decline and current recovery. Other amphibian species that were formerly common in the area have also declined dramatically (e.g., *Atelopus senex, Isthmohyla angustilineata, I. rivularis*, and *Rana vibicaria*). Interestingly, with the exception of *Atelopus senex*, all of these species have recently been detected there again in low numbers, leaving biologists hopeful but with many unanswered questions. Although the exact reasons for the demise of these species are still unclear, the main cause is likely chytridiomycosis (the fungal pathogen *Bd* has been confirmed in the area). Habitat alteration, climate change, and airborne pollution may also pose threats to this species.

The surviving *I. holdridgei* currently occur in an area of regenerating oak forest habitat that once had been cut and used as pasture. Although this area—and the rest of this species' historic distribution range—fall within Braulio Carrillo National Park, their habitat is nonetheless significantly impacted by road infrastructure, intensive recreational use of the site by visitors (e.g., mountain biking and off-road driving), and other factors.

Description

Sexually mature males have light nuptial patches on inner fingers and significantly swollen arms.

Smooth, rounded tubercles on dorsal surfaces and limbs; note irregular series of enlarged tubercles on each side of the body, between parotoids and groin.

No visible eardrums.

Cranial crests also black or brown; in some individuals, light blotches may adorn the lips and head.

Small with a robust head; skin on top of the head fused to skull (co-ossified).

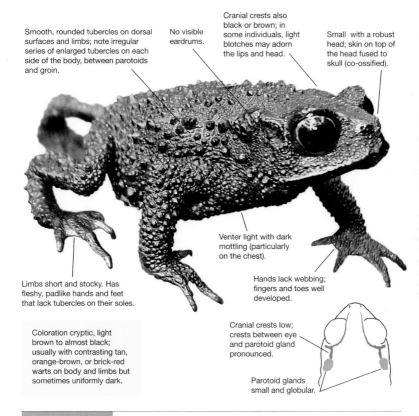

Venter light with dark mottling (particularly on the chest).

Limbs short and stocky. Has fleshy, padlike hands and feet that lack tubercles on their soles.

Hands lack webbing; fingers and toes well developed.

Coloration cryptic, light brown to almost black; usually with contrasting tan, orange-brown, or brick-red warts on body and limbs but sometimes uniformly dark.

Cranial crests low; crests between eye and parotoid gland pronounced.

Parotoid glands small and globular.

Similar species

I. chompipe (p. 162)
The only other toad that occurs in the same area; it is smaller, has a pointed head, and its hands and feet are almost completely enveloped in webbing.

Incilius luetkenii
Yellow Toad

Least concern

A large toad with small parotoid glands and black-tipped cranial crests. Male's uniform bright yellow, tan, or lime-green coloration unmistakable; female with more cryptic coloration, showing shades of olive or brown. Males to 3.78 in (96 mm) in standard length, females to 4.21 in (107 mm).

Ecoregions
1 2 3 **4 5**

On Pacific slope, ranges from Chiapas, Mexico, to northwestern Costa Rica; on Atlantic slope, occurs in Guatemala, Honduras, and Nicaragua. In Costa Rica, found from near sea level to about 1,475 ft (450 m).

Natural history

I. luetkenii occurs mostly at low elevations, generally in dry forest habitat, but it also ventures into more humid forests at slightly higher elevations. It is common throughout much of its range. It is usually seen at night, in a variety of open areas, including streambeds, disturbed pasturelands, and even man-made bodies of water in towns and cities.

Like most other dry forest amphibians, *I. luetkenii* spends a lot of time in hiding during the drier months (December–May), but resumes surface activity after the first rains of the wet season. It hides under rocks and logs during the day and hops about while foraging at night. The start of the rainy season also signals the onset of the breeding season. *I. luetkenii* breeds in temporary ponds, roadside ditches, and quiet pools in rivers and streams. As males sit at the water's edge, they produce a whistling trill call, lasting approximately 4 seconds and repeated at short intervals. Amplexus is axillary; eggs are laid in long strings, typical of toads. *I. luetkenii* eggs and tadpoles develop rapidly; after slightly more than 30 days, larvae begin metamorphosis.

Description

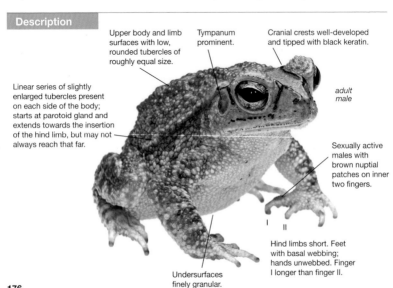

Upper body and limb surfaces with low, rounded tubercles of roughly equal size.

Tympanum prominent.

Cranial crests well-developed and tipped with black keratin.

Linear series of slightly enlarged tubercles present on each side of the body; starts at parotoid gland and extends towards the insertion of the hind limb, but may not always reach that far.

adult male

Sexually active males with brown nuptial patches on inner two fingers.

I   II

Hind limbs short. Feet with basal webbing; hands unwebbed. Finger I longer than finger II.

Undersurfaces finely granular.

Adult males easily recognized, with yellow, tan, or lime-green coloration; cream undersurfaces usually mostly free of markings.

Adult males have a large, extendable vocal sac on the throat that inflates during calls.

adult male

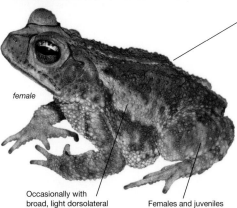

female

Females and juveniles variably colored, generally more or less uniform dark chocolate-brown, rust-brown, or olive-green, with a broad cream to yellow middorsal stripe. Paired dark spots or chevrons may be present along either side of the middorsal stripe.

Occasionally with broad, light dorsolateral stripes that may be bordered above and below by dark pigment.

Females and juveniles have alternating dark and light transverse bands on the hind limbs.

Parotoid glands small, about half the size of the upper eyelid.

Similar species

I. coniferus
(p. 166)

Has black-tipped cranial crests, but differs in having a prominent row of pointed warts on each side of the body; it usually shows green in its coloration.

I. coccifer
(p. 164)

May co-occur; lacks black-tipped cranial crests, has larger parotoid glands, and generally has a light mark on top of the head, between its eyes.

I. melanochlorus
(p. 178)

Has thin, raised cranial crests without black keratin; it also has triangular parotoid glands.

I. valliceps
(p. 182)

Has larger parotoid glands; its cranial crests are more prominent and lack the black keratin.

Incilius melanochlorus
Atlantic Forest Toad   **Least concern**

A fairly large toad of forest interiors. Has prominent cranial crests, small triangular parotoid glands, a row of distinctly enlarged tubercles on each side of the body, and a dark throat and chest. Males to 2.87 in (73 mm) in standard length, females to 4.21 in (107 mm).

Ecoregions
1 2 3 4 5

Incilius melanochlorus inhabits the Atlantic lowlands and the foothills of the Guanacaste, Tilarán, Central, and Talamanca Mountain Ranges; sea level to 3,540 ft (1,080 m).

Natural history

This is a widespread but secretive species that is rarely encountered except during the breeding season. In the dry season (January–February), individuals congregate to breed in large forest streams when the water levels are low. Males call from pools along rocky streams or from ponds to attract a mate; the call is a short trill lasting a few seconds and repeated regularly at short intervals. Amplexus is axillary. Females lay eggs in streams, in long strings. After eggs are deposited, adults leave the stream and disperse into the surrounding forest.

Infrequently, this species is found on the forest floor of dense rainforest, or along edges and trails of relatively undisturbed forest habitats. It is more commonly found at middle elevations than at low elevations.

The closely related species *I. aucoinae*, which occurs in the southern and central Pacific lowlands of Costa Rica, was once thought to represent a geographically isolated population of *I. melanochlorus*. The latter species differs most notably in

having its venter, chest, and throat distinctly marked with black pigment and in having small, transverse skin folds in the concave depression between the parietal crests on its head.

I. melanochlorus inhabits the dimly lit forest floor, where it is hard to see as it sits still in leaf litter.

178

Description

Adults generally brown or gray, with a pattern of irregular dark brown or black blotches. Most individuals have a light middorsal stripe; some also have pale gray dorsolateral stripes.

Body and limbs covered with small scattered tubercles.

Sexually active males with brown nuptial patches on inner two fingers of each hand.

On each side of the body, a series of prominently enlarged tubercles forms a row extending from parotoid gland to groin.

Tympanum clearly visible.

adult

Lateral rows of enlarged tubercles often cream or white, (generally a lighter color than the dorsum) and bordered below by a dark brown or dark gray band.

Juveniles have a more contrasting color pattern than do adults. Often with bold dark triangular markings aligned on either side of the middorsal strip; this pattern fades with age.

Limbs fairly slender.

Undersurfaces mottled dark, nearly black on the throat and chest, making it impossible to discern the single internal vocal sac in males.

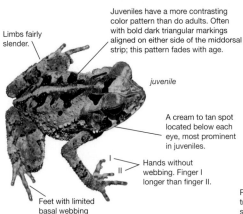

juvenile

A cream to tan spot located below each eye, most prominent in juveniles.

Well-developed cranial crests high and thin. Area between parietal crests is a large concave depression marked with several transverse folds.

I
II
Hands without webbing. Finger I longer than finger II.

Feet with limited basal webbing between toes.

Parotoid glands triangular and small (smaller than upper eyelid).

Similar species

I. coniferus
(p. 166)

Has black-tipped cranial crests, usually shows green in its coloration, and is covered with spine-tipped tubercles.

I. valliceps
(p. 182)

Has cranial crests that are even more prominent and has larger parotoid glands.

I. coccifer
(p. 164)

Has a thin, light middorsal line and a light mark on top of its head (between the eyes).

I. luetkenii
(p. 176)

Has black-tipped cranial crests and rounded parotoid glands.

Rhinella marina
(p. 188)

Much larger and has enormous parotoid glands.

Incilius periglenes
Golden Toad

Extinct

The combination of parotoid glands, cranial crests, and brilliant coloration (males uniform orange; females black or greenish with a pattern of yellow-outlined red spots) makes this species unmistakable. Males to 1.89 in (48 mm) in standard length, females to 2.13 in (54 mm).

Ecoregions
1 2 3 4 5

Incilius periglenes is an endemic species known only from an elfin forest less than 4 mi² (10 km²) in size, located on a ridge top within the Monteverde Cloud Forest Preserve, in the Tilarán Mountain Range; 4,850 to 5,250 ft (1,480 to 1,600 m).

Natural history

Although these extraordinary toads were discovered in 1964, knowledge of their existence remained limited until their description in 1966. The creation of the Monteverde Cloud Forest Preserve in 1972 protected their minute distribution range. Until 1987, the status of *I. periglenes* seemed secure, and hundreds of individuals congregated annually around several known breeding pools. But at the start of the breeding season in 1988, only 8 males and 2 females were observed at the two most important breeding ponds instead of the more than 1,500 seen there in the previous year. In 1989, only a single male was observed and none have been seen since, despite extensive searches.

I. periglenes inhabits undisturbed cloud and elfin forest. It is a very secretive, fossorial species that spends most of its time underground and only emerges briefly during the breeding season, which lasts from March to June. Males greatly outnumbered females at the breeding sites and reports indicate that 8 to 20 times more males than females would be present. Amplexus is axillary and may last more than 24 hours, during which time the amplectant male is attacked from all sides by other males attempting to dislodge him from the female's back, occasionally forming writhing toad balls of up to 10 toads. During the breeding season, *I. periglenes* is mainly active during the day, although egg-laying appears to occur predominantly at night. The average clutch size of 200 to 250 eggs is small compared to the several thousands of eggs that most toads produce, but *I. periglenes* eggs are larger than those of most other toads, with an average diameter of 0.1 in (3 mm). Eggs are deposited on the forest floor in small to very small temporary rain pools such as shallow water-filled depressions between tree roots, or even human footprints. Tadpoles require about five weeks to metamorphose.

The reasons for the demise of *I. periglenes* are still under debate but chytridiomycosis, climate change, and possibly airborne pollution, combined with this species' small range, likely contributed to its extinction. The precarious status of *I. periglenes* was recognized internationally when the International Union for the Conservation of Nature (IUCN) listed it as Endangered in 1979. In 1996, the status of the golden toad was revised to Critically Endangered, and in 2001 the IUCN declared the golden toad formally Extinct. It is one of only a few amphibian species in the world to officially receive this diagnosis.

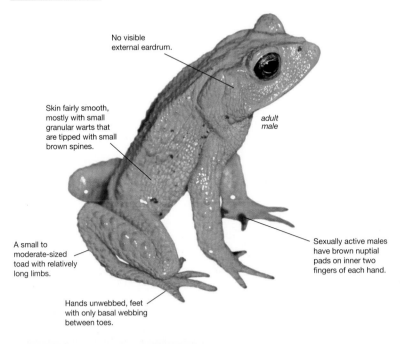

No visible external eardrum.

Skin fairly smooth, mostly with small granular warts that are tipped with small brown spines.

adult male

A small to moderate-sized toad with relatively long limbs.

Sexually active males have brown nuptial pads on inner two fingers of each hand.

Hands unwebbed, feet with only basal webbing between toes.

Adults display striking sexual dimorphism. Males are yellow to orange, females are black or dark olive-green with a pattern of yellow-outlined red spots. Juveniles display a faded version of the female color pattern.

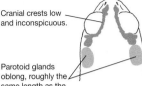

Cranial crests low and inconspicuous.

Parotoid glands oblong, roughly the same length as the upper eyelid.

Similar species

No other toads occur in the range of *I. periglenes*, although *Atelopus varius* (p. 157) formerly inhabited streams in the area. The latter species sometimes shows a color pattern of black, yellow, and red markings, like female *I. periglenes*, but has a distinctly different body shape and lacks cranial crests and parotoid glands.

Incilius valliceps
Gulf Coast Toad   **Least concern**

A moderate-sized toad with prominent cranial crests that surround a deep concave depression on the center of the head. A distinct row of pointed tubercles runs along each side of its body. It has medium-sized elliptical parotoid glands that are similar in size or larger than the upper eyelid. Males to 2.99 in (76 mm) in standard length, females to 3.31 in (84 mm).

Ecoregions
1 2 3 4 5

Incilius valliceps occurs from central Veracruz, Mexico, to extreme northeastern Costa Rica on the Atlantic slope and from the Isthmus of Tehuantepec, Mexico, to southeastern Honduras on the Pacific slope. In Costa Rica, this species is only found in the upper San Juan River watershed; sea level to 1,640 ft (500 m).

Natural history

Like other toads, *I. valliceps* is primarily nocturnal and terrestrial, although individuals of this species have been reported climbing up trees toward a daytime retreat in a hollow 10–15 ft (3–5 m) above the ground. This is an adaptable species that inhabits lowland areas and foothills, generally preferring open, altered habitats. Costa Rican populations appear to be associated with streams and breed in temporary pools formed when water levels drop in the dry season. Males produce a short trilling call that is repeated at brief intervals; females produce long strands of black-and-white eggs that are deposited directly in the water and sink to the bottom. The eggs of this species are toxic to some predators. Tadpole development is rapid and metamorphosis occurs after about one month.

Northern populations of this species (from Veracruz, Mexico, north to the southern US), formerly included in this species, are now considered a separate species, *I. nebulifer*.

The sharp-edged, raised crests of *I. valliceps* give its head a distinctive, angular shape.

Females and juveniles with a pattern of bold, dark triangles along each side of the vertebral stripe; triangles may fuse to form dark blotches.

During breeding season, sexually mature males have dark brown nuptial pads on inner fingers.

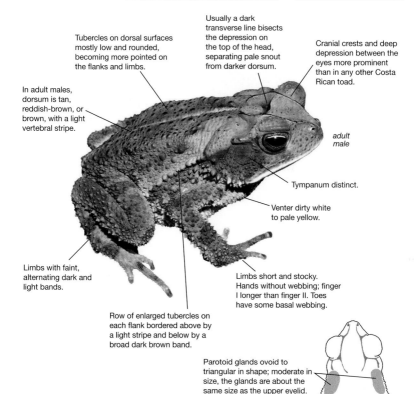

Tubercles on dorsal surfaces mostly low and rounded, becoming more pointed on the flanks and limbs.

Usually a dark transverse line bisects the depression on the top of the head, separating pale snout from darker dorsum.

Cranial crests and deep depression between the eyes more prominent than in any other Costa Rican toad.

In adult males, dorsum is tan, reddish-brown, or brown, with a light vertebral stripe.

adult male

Tympanum distinct.

Venter dirty white to pale yellow.

Limbs with faint, alternating dark and light bands.

Limbs short and stocky. Hands without webbing; finger I longer than finger II. Toes have some basal webbing.

Row of enlarged tubercles on each flank bordered above by a light stripe and below by a broad dark brown band.

Parotoid glands ovoid to triangular in shape; moderate in size, the glands are about the same size as the upper eyelid.

I. melanochlorus (p. 178)

May co-occur in the Atlantic lowlands; it is distinguished by a black throat and chest and by parotoid glands that are smaller and triangular.

I. luetkenii (p. 176)

May co-occur locally on the Atlantic slopes of the Guanacaste Mountain Range; it has black-tipped cranial crests and smaller parotoid glands.

I. coccifer (p. 164)

Lacks a distinct row of enlarged spiny tubercles on the flanks; it has a light middorsal pinstripe and a light bar on top of the head, between its eyes.

Genus *Rhaebo*

This genus includes 13 species of toads that mostly occur in South America. All lack cranial crests and have prominently visible eardrums and large parotoid glands. Superficially, they appear more smooth-skinned than other toads. This group was previously recognized as a sub-group (the *Bufo guttatus*-species group) within the large, cosmopolitan genus *Bufo*. Recent revisions have elevated it to its own genus level. Only one species of *Rhaebo* occurs in Central America.

Rhaebo haematiticus
Smooth-skinned Toad

Least concern

This fairly large species is easily recognized by its oversized parotoid glands and its relatively smooth dorsal skin; it also lacks any trace of cranial crests. Males to 2.44 in (62 mm) in standard length, females to 3.15 in (80 mm).

On the Pacific slope, *Rhaebo haematiticus* ranges from Costa Rica to northern Colombia and northwestern Ecuador; on the Atlantic slope, its range extends farther north, to Honduras; sea level to 4,265 ft (1,300 m).

Ecoregions
1 2 3 4 5

Natural history

Adults of this species are nocturnal toads that inhabit the leaf litter of relatively undisturbed humid forests in the lowlands and foothills. This common species is easily overlooked as its cryptic color pattern beautifully creates the impression of a dead leaf, rendering it nearly invisible when it sits motionless in leaf litter.

Reproduction takes place from the end of the dry season (March) to the middle of the wet season (July). Males call from exposed boulders along the edge of forest streams. Their advertisement call is very unlike the trills produced by other toads; at regular intervals, they repeat a series of stuttering, birdlike whistles. Black-and-cream eggs are deposited in long strands, directly in the water of rocky pools, where they develop into free-swimming tadpoles. Between April and August, metamorphs and small juveniles can be

abundant among the rocks that line the water's edge; they are often seen during the day. In spite of its large size, *R. haematiticus* consumes very small prey; ants, mites, beetles, and other small arthropods ranging from 0.02 to 0.19 in (0.4 to 4.8 mm) in length appear to make up the bulk of its diet.

This species is generally quite common in Costa Rica, but locally appears to undergo dramatic population fluctuations; it has reportedly all but disappeared from some areas but remains abundant in others. The exact mechanisms and threats that underlie these patterns are not known, but they likely include pollution, habitat alteration, and the effects of pathogen infestations. *R. haematiticus* is somewhat tolerant of habitat alteration, but only survives in areas that are in close proximity to large tracts of undisturbed forest.

Description

Usually with a distinct light blotch on the upper lip, located slightly in front of each eye.

Dorsal coloration quite variable, ranging from yellowish-tan to purplish gray. A broad reddish-brown to deep chocolate-brown lateral band covers the sides of the head and runs along the side of the body to the groin; it is bordered above by a light (cream, tan, orange) thin line that separates the dorsal coloration from the lateral dark band.

185

Has large eyes and a short snout.

Note clearly visible eardrums.

adult male

Adult males always have a single, internal vocal sac on the throat.

Dorsal skin relatively smooth, but close examination shows scattered, low tubercles; limbs more tuberculate than dorsum and flanks.

On hands, fingers lack webbing; finger I longer than finger II. Feet have basal webbing.

Sexually mature males have indistinct, brown nuptial pads on their inner fingers.

Fairly large and long-legged, with slender arms and legs and long digits.

juvenile

Dorsal surfaces may be heavily marked with light-outlined dark brown spots and poorly defined brick-red to burnt-orange blotches, particularly in juveniles.

Red and orange markings often present on the hidden surfaces of the limbs.

Ventral coloration is yellowish-orange to brown; the posterior half of the belly and the undersides of the hind limbs are usually marked with cream to white spots.

Entirely lacks cranial crests.

Parotoid glands very large and elongated.

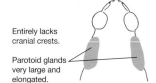

Similar species

R. haematiticus is the only toad in Costa Rica that does not have cranial crests; its skin is less warty than that of other toads. The co-occurring but unrelated *Leptodactylus savagei* (p. 462) can be similar in size and coloration, but does not have parotoid glands.

Genus *Rhinella*

This genus currently contains 89 species, most of which occur in South America. The sole Costa Rican member is the cane toad (*Rhinella marina*).

This is a confusing genus. There is some molecular data that supports the notion that these toads may be more closely related to Old World toads than to the Central American toads now assigned to the genus *Incilius*. Taxonomic research on these groups is ongoing.

Due to these taxonomic uncertainties, *Rhinella marina* has also been included in recent literature under the names *Bufo marinus* and *Chaunus marinus*.

Rhinella marina
Cane Toad, Marine Toad

Least concern

A distinctively large toad with enormous parotoid glands and prominent cranial crests, commonly seen throughout the country in close proximity to human settlements. Males to 5.71 in (145 mm) in standard length, females to 6.89 in (175 mm).

Ecoregions
1 2 3 4 5

Rhinella marina is native to the Americas, but has been introduced to Australia and many other countries, where it is now an invasive species. Its range has expanded significantly in recent times, in part because of continued destruction of closed-canopy forests (it thrives in open areas). It now ranges from southern Texas and southern Florida, US, through tropical Mexico and Central America, to northern South America. In Costa Rica it is abundant from sea level to 5,250 ft (1,600 m).

Natural history

This is one of the largest frogs in the world; South American and Australian individuals grow to 9.45 in (240 mm) in standard length and can weigh more than 3 lbs (1.5 kg). Like most other toads, this species is nocturnal and terrestrial. During the day, it hides under cover objects, debris, rocks, and logs; at dusk it emerges to hunt. A voracious predator, *R. marina* consumes a wide variety of invertebrate prey and also vertebrate prey small enough to swallow whole. These toads are usually seen when they gather at night around buildings and in streets where lights attract insect prey; in a pinch, they have been observed eating cat and dog food, as well as plant matter.

Wherever *R. marina* has been introduced, native amphibians and other small animals have fallen prey to its indiscriminate eating habits. And, domestic dogs and other predators with no prior exposure to this toad have experienced the deleterious effects of its toxins. The parotoid and scattered skin glands of *R. marina* excrete a milky white secretion that contains a mixture of bufotenins, bufogenins, and other highly toxic compounds capable of debilitating or even killing a predator. The amount of toxin present in the parotoid glands of an adult *R. marina* is sufficiently powerful to kill large mammals, and reportedly even adult humans have died from eating this species. Some snake species that specialize in eating toads have developed an increased immunity to this species' toxins, while other predators (notably mammals and some birds) will flip over and eviscerate a

R. marina and consume its innards, carefully avoiding contact with the skin glands.

Males produce a mating call that has been described as a low pitched, drawn-out, machine-gun trill. Breeding takes place year-round. Female *R. marina* are extremely prolific and deposit between 5,000 and 25,000 eggs in the shallow margins of rivers and permanent or temporary pools, which are typically located in open areas. Like adults, eggs and tadpoles are toxic and have few predators; they displace co-occurring tadpoles and exacerbate the negative impact *R. marina* can have on an ecosystem.

Under duress, the parotoid and scattered skin glands of *R. marina* excrete a milky white secretion that contains highly toxic compounds capable of debilitating or even killing a predator.

Head large, with prominent cranial crests.

adult female

Females show blotches of varying colors.

Limbs short. Fingers without webbing; finger I longer than finger II. Feet with basal webbing.

In both sexes, venter dirty white to cream, with irregularly spaced brown blotches.

adult male

Skin covered with irregularly scattered tubercles. In mature males, each tubercle is distinct and raised, bearing one or more small keratinized spines; females have large, indistinct smooth tubercles on the dorsum and limbs.

Eardrums clearly visible.

Adult males with a single, internal vocal sac on the throat.

Males generally uniform gray, brown, or olive.

Sexually mature males have brown or black nuptial pads on the inner fingers.

metamorph

juvenile

Metamorphs and very small juveniles can be more difficult to identify because they lack the diagnostic cranial crests and parotoid glands.

Juveniles are more boldly colored than adults, with blotchy brown dorsal coloration, marked with contrastingly colored (cream to orange) tubercles. Hind limbs often with alternating dark and light bands.

189

adult female

adult male

juvenile

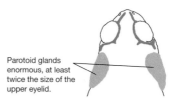

Parotoid glands enormous, at least twice the size of the upper eyelid.

Similar species

Rhaebo haematiticus (p. 185)

Has large parotoid glands but lacks cranial crests

Family Centrolenidae
(Glass Frogs)

These small, arboreal frogs occur in Central and South America. Biologists currently recognize 154 species, 14 of which occur in Costa Rica. The common name refers to their transparent undersides; in many species it is possible to see the frog's intestines, bones, and, in a few cases, a beating heart through the pigmentless skin on the belly. From above, glass frogs usually show a translucent shade of green, and most species sport yellow, white, blue, or black markings. The dorsal coloration allows ambient colors to be visible through the body of the frog, which means that glass frogs are able to blend in perfectly with their surroundings. Indeed, they can be very difficult to spot as they perch on the top or bottom surfaces of leaves.

Glass frogs are strictly arboreal; the tips of their fingers and toes are slightly expanded and bear adhesive disks, an adaptation to life in the trees. Species in this family are sometimes confused with certain small species of tree frog. In glass frogs, however, the eyes face forward; in tree frogs, the eyes are directed sideways. In addition, glass frogs have truncated disks on their fingers and toes, while tree frogs have round disks.

All glass frogs lay their eggs on vegetation over rapidly flowing streams, sometimes several meters above the surface of the water. Some species deposit their eggs on the upper surface of leaves, and others place them on the underside; a few species may occasionally lay their eggs on wet boulders or moss-covered rock faces in the spray zone of waterfalls. Typically, species that lay eggs on the top of leaves or rocks produce dark clutches—usually dark brown or black—while species that deposit eggs on the underside of leaves produce eggs that are white, cream, or light green. Presumably this color difference reduces the visibility of the clutches.

The number of eggs produced per clutch is relatively small, usually between 20 and 40. The tadpoles undergo their initial development within the egg and typically hatch when it is raining; they wriggle themselves free from the egg and drop into the stream below. Timing the moment of hatching so as to coincide with a period of rain increases the chance that tadpoles that get washed off the leaf and land on the banks of the stream below will ultimately reach the water. Also, the turbidity of the stream during a downpour makes the larvae less visible to predators. After the tadpoles enter the stream, their body coloration becomes bright red, and they bury themselves in the leaf litter and debris that collects in quiet eddies. Sometimes tadpoles are found in the small puddles that are left behind after a stream's water level goes down. The tadpoles complete their development into small frogs several months after hatching.

Certain species, especially those in the genus *Hyalinobatrachium*, display elaborate parental care behavior. During the larval development period, males guard and tend one or more egg clutches at night. They keep the eggs moist, protect the clutches from parasitic flies, and wrestle other males that intrude into their territory.

The taxonomic arrangement of Central American glass frogs has undergone a lot of change in the past few decades. Several species have been moved in and out of different genera repeatedly since the early 1990s, when all Costa Rican glass frogs were still placed in a single genus. Currently, the 14 species known to occur in the country are divided over 5 genera. Half of these species are in the genus *Hyalinobatrachium*, which forms a relatively recognizable group of small, flattened glass frogs that call from the undersurfaces of leaves, on which they also lay their eggs. The remaining 7 species of Costa Rican glass frog vary in different aspects of behavior, morphology, and call. Some can be recognized by their unique dorsal color pattern, the degree to which their undersides are transparent, or the color of their bones.

Genus *Cochranella*

This genus has undergone several taxonomic revisions in the past two decades. As currently understood, *Cochranella* contains 19 species of glass frog, 2 of which occur in Costa Rica.

These 2 Costa Rican species are quite different from one another in general appearance, but like all other species of *Cochranella* share the following traits: green bones; no humeral spine on each upper arm; ventral parietal peritoneum (a membrane that covers most of the internal organs) is white but becomes transparent towards the posterior half of the belly; in addition, the digestive tract of these frogs is covered in a separate white membrane, and it is therefore difficult to see any of their organs.

Males of the 2 Costa Rican *Cochranella* call while perched on the upper surfaces of leaves. Females deposit black-and-cream eggs on the upper surface of leaves; these frogs are not known to show any form of parental care. In the days following the moment when eggs are laid, the jelly surrounding the clutch swells to enormous proportions, leaving the eggs encased in a gelatinous, drooping mass that hangs from the drip-tip of a large leaf near, or overhanging, a stream.

Cochranella euknemos
Slope-snouted Glass Frog

Least concern

A bluish-green glass frog marked with small white to yellow spots. Has a long sloping snout and fairly small eyes. The combination of a distinct white lip stripe and fleshy, white fringes along the trailing edges of its limbs is unique. Males to 0.98 in (25 mm) in standard length, females to 1.26 in (32 mm).

Ecoregions

1 2 3 4 5

In Costa Rica, *Cochranella euknemos* is only known from a few locations, located on the Atlantic slope of the Central Mountain Range; 3,275–4,925 ft (1,000–1,500 m). It occurs more widely in Panama (300 to 4,150 ft [90 to 1,270 m]) and in western Colombia.

Natural history

This is a relatively rare species in Costa Rica, known only from a few scattered localities on the Atlantic slope of the Central Mountain Range. Nocturnal, this species is most easily located by its call, a loud shrill *creep* repeated one to three times (similar to the call of the closely related *C. granulosa*, p. 196). Frequency of calling is highest during the rainy season, when most reproductive activity takes place.

Like most other glass frogs, *C. euknemos* inhabits vegetation that lines relatively fast-flowing streams; but unlike other glass frogs, observations suggest that this species prefers drier locations at certain times. Egg masses have been reported from vegetation overhanging dry streambeds and very small trickles during the rainy season. Presumably these eggs hatch during periods of heavy rain and the tadpoles wash into a nearby stream. This unusual breeding strategy may be an adaptation to avoid competition with other glass frogs over scarce oviposition sites during their peak breeding season.

Females lay a clutch of 40–70 black-and-white eggs on top of a leaf; the jelly surrounding the eggs swells up to form a large gelatinous mass that contains the eggs. This gelatinous mass turns cloudy white, making it very difficult to see the enclosed eggs. Generally the egg mass is suspended from the tip of a leaf, causing the tip to droop down and ensure a regular flow of water over the eggs when it rains. On hatching, larvae drop into the stream below and complete their development within the sediment at the bottom of the stream. There

appears to be no parental care in this species, and adults do not guard their egg masses.

This species is rarely encountered in Costa Rica but still seems to occur in low numbers in appropriate habitat. Due to its preference for relatively undisturbed rainforest habitat, the larger threat to its survival may be habitat loss and alteration, but no information exists on how disease might be affecting *C. euknemos*. Most of the known locations for this species are currently included in national parks or other protected areas.

The taxonomic composition of this species is uncertain and more than one species may be included in what is now considered to be *C. euknemos*.

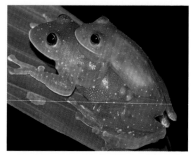

Distinct sexual size dimorphism; males are considerably smaller than females. Even though most of the internal organs are obscured by opaque membranes, the eggs in this gravid female are clearly visible through her skin.

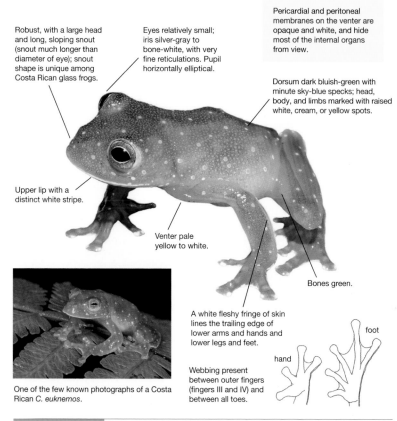

Robust, with a large head and long, sloping snout (snout much longer than diameter of eye); snout shape is unique among Costa Rican glass frogs.

Eyes relatively small; iris silver-gray to bone-white, with very fine reticulations. Pupil horizontally elliptical.

Pericardial and peritoneal membranes on the venter are opaque and white, and hide most of the internal organs from view.

Dorsum dark bluish-green with minute sky-blue specks; head, body, and limbs marked with raised white, cream, or yellow spots.

Upper lip with a distinct white stripe.

Venter pale yellow to white.

Bones green.

One of the few known photographs of a Costa Rican *C. euknemos*.

A white fleshy fringe of skin lines the trailing edge of lower arms and hands and lower legs and feet.

Webbing present between outer fingers (fingers III and IV) and between all toes.

hand

foot

C. granulosa
(p. 196)

Has a shorter snout. It lacks white or yellow spots and fleshy fringes along its limb margins.

Sachatamia albomaculata
(p. 217)

Lacks the long, sloping snout and fleshy fringes on the limbs.

Teratohyla pulverata
(p. 222)

Has a short sloping snout, large eyes, and webbing between fingers II and III.

Cochranella granulosa
Granular Glass Frog

Least concern

A glass frog with characteristically granular skin; yellowish-green to dark green with minute turquoise specks. Often with a few scattered dark blue to black spots on its dorsum and a white stripe on the upper lip. Males to 1.14 in (29 mm) in standard length, females to 1.26 in (32 mm).

Ecoregions
1 2 3 4 5

On the Atlantic slope, *Cochranella granulosa* ranges from southeastern Honduras to central Panama; on the Pacific slope, it occurs from central Costa Rica to southwestern Panama; near sea level to 4,925 ft (1,500 m).

This is a common inhabitant of low and middle elevation forests, where it is found on the leaves of streamside trees. Males call almost year-round; they are not easy to see, however, since they generally perch on the upper surfaces of leaves high overhead. Generally, individuals are located 12–30 ft (4–10 m) above the ground. In areas with few tall trees, these frogs can occasionally be found closer to the ground.

C. granulosa makes a loud, raspy series of high-pitched trills: *creep-creep-creep*; their call usually consists of three notes, with several minutes between each series. The male's vocalizations attract mates during the breeding season but also serve as a territorial marker to intruding males. Males are known to engage in physical combat with competitors encroaching on their territory, and bouts of grappling and wrestling usually end with one of the males being pushed off a leaf and out of a territory.

Individuals tend to descend closer to the ground when breeding activity commences during the rainy season. Females deposits 50–60 diminutive eggs in a small, single-layered clutch placed on the upper surface of a leaf that overhangs a rapidly flowing stream. The black-and-white eggs, which can be seen clearly through the skin of gravid females, are encased in a layer of gelatinous material once deposited. This gelatinous coating swells tremendously over the course of the first few days by absorbing moisture from the environment, and forms a cone-like mass that droops down from the tip of the leaf. The weight, shape, and location of the egg mass ensure that water running off the

surface of the leaf bathes the eggs, providing adequate moisture for the developing embryos. Upon hatching, the small, black tadpoles wriggle free from the egg mass and drop into the stream below, where they feed on detritus until finishing their larval development. No parental care or guarding of clutches is known for this species.

C. granulosa is quite adaptable; it is still commonly observed in a variety of riparian habitats in Costa Rica, ranging from pristine forests and areas of secondary growth to relatively disturbed habitats with remnant patches of streamside vegetation. The biggest threats to this species likely include excessive loss of riparian habitat and exposure to pesticides and fertilizers, either directly or through pollution of its stream habitat.

C. granulosa has characteristic light blue speckling on its granular skin. Often, individuals also have a few scattered dark blue or black spots on their dorsum.

196

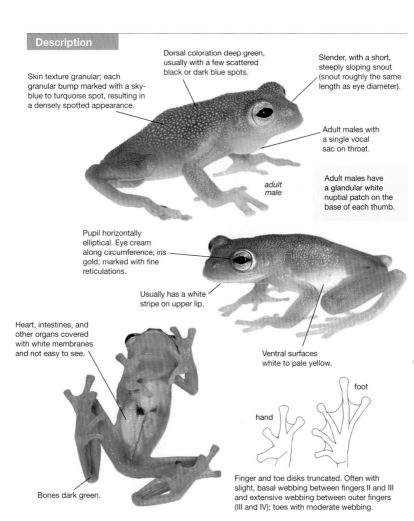

Dorsal coloration deep green, usually with a few scattered black or dark blue spots.

Slender, with a short, steeply sloping snout (snout roughly the same length as eye diameter).

Skin texture granular; each granular bump marked with a sky-blue to turquoise spot, resulting in a densely spotted appearance.

Adult males with a single vocal sac on throat.

Adult males have a glandular white nuptial patch on the base of each thumb.

adult male

Pupil horizontally elliptical. Eye cream along circumference; iris gold; marked with fine reticulations.

Usually has a white stripe on upper lip.

Heart, intestines, and other organs covered with white membranes and not easy to see.

Ventral surfaces white to pale yellow.

foot

hand

Bones dark green.

Finger and toe disks truncated. Often with slight, basal webbing between fingers II and III and extensive webbing between outer fingers (III and IV); toes with moderate webbing.

Cochranella euknemos
(p. 194)

Has a pattern of raised white or yellow spots, a long, sloping snout, and fleshy fringes along the margins of its limbs.

Teratohyla spinosa
(p. 224)

Smaller, has less granular skin, and has a uniform green dorsum without dark spots.

Espadarana prosoblepon
(p. 199)

Has dark spots on its green dorsum but differs in having an intestinal tract that is visible; males have a spur (humeral hook) on each upper arm.

Genus *Espadarana*

This recently created genus includes just three species. *Espadarana pros-oblepon* occurs in Costa Rica, while the other two species are found in South America.

Adult male *Espadarana* are easily distinguished from other glass frogs by the presence of a prominent humeral hook on each upper arm; females lack this feature and can be mistaken for other centrolenids. Additional features that identify *Espadarana* include their green bones and the presence of a ventral parietal peritoneum (a membrane that covers the internal organs) that is mostly white but gradually becomes transparent near the insertion of the hind limbs, leaving sections of the intestinal tract visible.

Male *Espadarana prosoblepon* call from the upper surfaces of leaves. Females deposit their eggs either on the upper or lower surfaces of leaves that are near, or overhanging, a stream.

In older publications, *Espadarana prosoblepon* is referred to as *Centrolenella prosoblepon* or *Centrolene prosoblepon*.

Espadarana prosoblepon
Emerald Glass Frog   **Least concern**

An emerald-green glass frog with small dark dorsal spots, dark green bones, and a white peritoneal membrane that covers much of the ventral surfaces but leaves part of the intestinal tract visible. Adult males have a distinctive, bluntly rounded spur (humeral hook) on each upper arm. Males to 1.06 in (27 mm) in standard length, females to 1.26 in (32 mm).

Ecoregions
1 2 3 4 5

Espadarana prosoblepon occurs from eastern Honduras to northern Colombia and western Ecuador. In Costa Rica it is found in a variety of habitats along the Atlantic and Pacific slopes, but is absent from the dry Pacific northwest; near sea level to 6,225 ft (1,900 m).

Natural history

This common species is regularly encountered in riparian habitats with dense vegetation. Found in a variety of habitats, including streams in pristine cloud forest and remnant vegetation along water courses running through lowland pastures. *E. prosoblepon* usually occupies lower calling sites than other glass frogs and is quite vocal throughout much of the year, making this species relatively easy to detect. Males call from upper surfaces of leaves, often in areas of dense vegetation that reaches close to the water level. The characteristic advertisement call, a sharp series, usually of three notes (*dik-dik-dik*), serves to attract mates and to stake out a male's territory.

Breeding peaks during the rainy season (May–November) but takes place nearly year-round. Amplecting pairs seek out a suitable site for oviposition, usually the upper surface of a leaf, but occasionally a moss-covered rock or branch. The small clutches of about 20 dark brown to black eggs are placed up to 10 ft (3 m) above the water. Parents do not tend to the eggs or guard their clutch; tadpoles

develop approximately 10 days before dropping into the stream below.

Male *E. prosoblepon* are very territorial and use their call to space themselves out along streamside vegetation, thus avoiding territorial disputes; calling males tend to be located about 10 ft (3 m) apart. When an intruder does approach a calling male too closely, physical combat ensues; males will grapple with one another using the humeral hooks on their forearms for leverage and to get a better grip on their opponent. The struggle usually takes place on top of a leaf and ends with one of the males being pushed off; often the two males wrestle with their arms while hanging upside down, dangling by their toes from the underside of a leaf. These territorial scuffles can last upward of 45 minutes.

Populations of *E. prosoblepon* appear stable, although local declines of this species have been reported (e.g., in Monteverde). Habitat alteration is generally the biggest threat to this species, but localized population declines may have been caused by chytrid fungal infections.

199

Eyes large, with silver-gray to golden iris marked with fine bronze specks and reticulations; pupil horizontally elliptical.

Has smooth skin.

Dorsum dark emerald-green, usually with small, dark spots; flanks and venter white to cream; fingers and toes yellow.

A fairly large, slender glass frog, with long limbs and a short, truncated snout.

In adult males, white nuptial patches present on base of inner fingers.

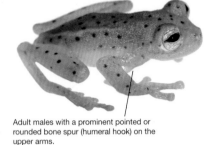

Bones dark green.

Adult males with a prominent pointed or rounded bone spur (humeral hook) on the upper arms.

Internal organs mostly covered by opaque, white membranes, except for an area in the center of the venter, near the insertion of the hind limbs, where part of the intestinal tract is visible.

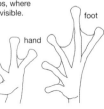

foot

hand

Fingers and toes with truncated disks. A small amount of webbing present between outer fingers (III and IV) only; moderate webbing between all toes.

Females generally lack a humeral hook, but some adult females have a greatly reduced humeral hook.

Sachatamia ilex (p. 219)

Has a lime-green dorsum without dark spots; it also has large silvery-white eyes with a pronounced black reticulum.

Teratohyla spinosa (p. 224)

Considerably smaller. It lacks dark spots on the dorsum and lacks a humeral hook on the upper arm (but note a small prepollical spine at the base of each thumb). A larger area is transparent on its venter.

Genus *Hyalinobatrachium*

The genus *Hyalinobatrachium* includes some of the most commonly encountered glass frogs. Thirty-two species are currently recognized, seven of which occur in Costa Rica.

These small frogs have a flattened look. They are lime-green, with a species-specific pattern of yellow, cream, or white dorsal spots. They also have white bones and lack a humeral spine on the upper arm.

Frogs in this genus count among the most transparent of the glass frogs, with a completely clear peritoneal membrane that leaves the heart and some of the internal organs very visible; usually, however, the digestive tract and liver are covered by a white membrane.

Male *Hyalinobatrachium* call from the underside of leaves. Females deposit a single layer of eggs in a circular clump that is attached to the underside of leaves that overhang a stream. Males care for and guard the eggs.

Hyalinobatrachium chirripoi
Chirripo Bare-hearted Glass Frog **Least concern**

A lime-green glass frog with a faint pattern of minute cream or pale yellow spots. Has white bones. Ventral surfaces transparent, with heart clearly visible (liver and digestive tract enveloped in white membranes). Has significant webbing between fingers II and III, more so than other Costa Rican *Hyalinobatrachium*. Males to 1.02 in (26 mm) in standard length, females to 1.18 in (30 mm).

Ecoregions
1 2 3 4 5

On the Atlanctic slope, *Hyalinobatrachium chirripoi* occurs in southeastern Costa Rica and central Panama; it also is found on the Pacific slope, in eastern Panama; 200–325 ft (60–100 m). There is an isolated population in eastern Honduras. This species also occurs in the Chocó region of western Colombia.

Natural history

Although *H. chirripoi* has been classified as a rare species, it is common at some locations within its limited range. Like other glass frogs, it occurs mainly on streamside vegetation that overhangs the water. It can be found at night, on bushes and trees, either along small streams in forested areas or large rivers passing through pastures and other disturbed areas. This species appears to favor plants with smooth leaves as well as palms.

Adult males' advertisment call is a single, high-pitched, cricketlike buzz, usually made as they perch on the underside of a leaf 3–15 ft (1–5 m) above the ground. During the breeding season, females lay a small round clutch of 65–80 cream to pale green eggs on the underside of a leaf; the male fertilizes the eggs and guards them until they hatch. Males remain near the eggs day and night; they can tend up to three clutches at once. Like other *Hyalinobatrachium*, activity in this species is closely tied to humidity levels. They are most active on warm, humid nights; but both the calls and nocturnal activity cease abruptly in the absence of rainfall.

In 2007, isolated populations of a glass frog in eastern Honduras, formerly described as *H cardiocalyptum*, were classified as *H. chirripoi*. Although no populations of *H. chirripoi* have been identified in the region that lies between Honduras and Costa Rica, further research may reveal that his species is more widespread than is currently understood.

Pair of *H. chirripoi* in amplexus. Note that males are considerably smaller than females.

Adult males with an indistinct glandular, white nuptial patch on the base of each thumb.

Dorsum green with a pattern of minute dark specks, visible upon close examination. The dorsum is also marked with small, faint cream or pale yellow spots.

Each nostril is located on a raised protuberance, giving the snout a notched profile when seen from above.

Iris gold, with fine dark speckling that is most densely concentrated around the horizontally elliptical pupil.

Bones white.

Skin with minute granular bumps.

Ventral surfaces transparent; heart surrounded by pigmentless membrane and clearly visible; liver and intestines covered with opaque, white membrane.

Finger and toe disks truncated; this species has more extensive webbing between the fingers than other *Hyalinobatrachium*, especially between fingers II and III.

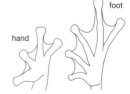

hand   foot

Similar species

H. colymbiphyllum (p. 204)

Very similar in overall appearance but has less webbing between its fingers.

H. dianae (p. 206)

Lacks yellow spots and shows a characteristic small dark spot on its lower back.

H. fleischmanni (p. 208)

The heart, covered by a white membrane, is not visible.

Hyalinobatrachium colymbiphyllum
Bare-hearted Glass Frog   **Least concern**

A lime-green glass frog with small to medium-sized cream or pale yellow spots. Has white bones. Ventral surfaces transparent; heart clearly visible but liver and digestive tract enveloped in white membranes. No webbing between fingers II and III. Males to 1.06 in (27 mm) in standard length, females to 1.22 in (31 mm).

Ecoregions
1 2 3 4 5

On the Atlantic slope, *Hyalinobatrachium colymbiphyllum* occurs in central and southeastern Costa Rica; on the Pacific slope, it ranges from northwestern Costa Rica to northern Colombia; from near sea level to 5,900 ft (1,800 m).

Natural history

This species is common along forested streams. Male *H. colymbiphyllum* are most often located by their call, a sharp, high-pitched *prreeet*. This advertisement call allows males to space their territories and also serves to attract mates. Males generally occupy a preferred calling site, perching upside down on the underside of a leaf located several meters above a fast-flowing stream. They abandon this perch during the daytime but return to the same territory and calling site each night. Resident males will vigorously defend their territory and attack an intruding male by jumping on its back and pressing it against the leaf surface.

Breeding takes place during the rainy season. The female deposits her eggs on the underside of a leaf, generally the same leaf used as a calling site by the male parent. Each clutch is roughly circular in shape and may contain 18–75 pale green eggs; populations at low elevations tend to have a much larger clutch size than populations at higher elevations.

Male *H. colymbiphyllum* tend to their eggs at night, sometimes simultaneously guarding multiple clutches in different stages of development. During the day the male abandons the eggs but he returns again at night. Gastric brooding, in which the male perches on top of the eggs and rehydrates them with moisture from his bladder, has been observed in this species. Predatory wasps have been reported to carry off eggs, one by one, during the day, when the eggs are unattended.

H. colymbiphyllum is most active on humid nights and during, or shortly after, abundant rainfall. On dry nights, males generally call less, or not at all. This species is widespread in a variety of habitats and elevations in Costa Rica. Although it prefers mature undisturbed forest habitat, it appears somewhat tolerant of habitat disturbance and can be found along densely vegetated stream banks bordering pastures and other developed land as long as sizeable tracts of forest are nearby. Populations of this species appear relatively stable throughout the country, and *H. colymbiphyllum* appears to have recovered in areas where it declined significantly in the 1990s.

Egg clutches are circular and attached to the underside of broad-leaved plants overhanging a stream; the tadpoles drop into the water below upon hatching.

Dorsal coloration bright lime-green with a dense pattern of minute dark specks that are visible under magnification or upon very close examination. Dorsum marked with faintly indicated small to moderate-sized cream or yellow spots.

Dorsum minutely granular.

Iris gold, with minute dark brown to black speckling or reticulations, most prominent on either side of the horizontally elliptical pupil.

Adult males with an indistinct white, glandular nuptial patch on the base of each thumb.

Eyes relatively large and protuberant.

Small and relatively stocky with a broad head and a short, bluntly truncated snout. Nostrils located on raised protuberances; resulting indentation between nostrils visible when snout viewed from above.

Bones white.

Underside of limbs white.

Ventral surfaces transparent. Heart surrounded by pigmentless membrane and clearly visible; liver and intestines covered with opaque, white membrane.

foot

hand

In the Talamanca Mountain Range, *H. colymbiphyllum* are large and robust, with a truncated snout, protruding eyes, and strongly raised nostrils.

Pacific lowland populations of *H. colymbiphyllum* typically are smaller, with a more rounded snout and less protruding eyes.

Finger and toe disks truncated. Minimal webbing between outer fingers (III and IV) only; feet with extensive webbing between all toes.

H. chirripoi
(p. 202)

Bears the closest resemblance to *H. colymbiphyllum*; its hands have more extensive webbing and its yellow spots are smaller and more clearly defined.

H. fleischmanni
(p. 208)

Looks similar but its snout is rounded and its heart is not visible.

H. valerioi
(p. 212)

Has larger yellow spots that are surrounded by a dark green reticulum.

H. talamancae
(p. 210)

Differs in having a dark green middorsal stripe.

H. dianae
(p. 206)

Has a uniform green dorsal coloration without yellow spots; it also has silvery-white eyes.

Hyalinobatrachium dianae
Diane's Bare-hearted Glass Frog

Not assessed

Lime-green, without the light spots that characterize some other glass frogs; has a unique oblong black spot on the lower back. Has white bones. Ventral surfaces transparent; heart clearly visible but liver and digestive tract enveloped in white membranes. Males to 1.14 in (29 mm) in standard length, only recorded adult female to 1.10 in (28 mm).

Ecoregions
1 2 3 4 5

This species is endemic to Costa Rica and only known from three sites on the Caribbean slope; these lie between the vicinity of Santa Clara, Heredia Province, and the headwaters of Río Victoria, Limón Province; 1,300 to 2,950 ft (400 to 900 m).

Described in 2015, this is one of the more recent additions to the herpetofauna of Costa Rica. *H. dianae* is most closely related to the other two bare-hearted glass frogs, *H. chirripoi* (p. 202) and *H. colymbiphyllum* (p. 204), and seems to share many traits of their biology. It is nocturnal and lives on streamside vegetation. Adult males usually call while perched on the underside of leaves. Limited observations of this species indicate that males not only call from vegetation that overhangs streams but also from leaves that are located above the forest floor, or above dry gullies. The advertisement call, a single whistle-like note that is long and metallic, differs from that of other glass frogs. Calling activity in this species is apparently quite unpredictable and sporadic. Even when weather conditions seem suitable, males do not always call, which makes detecting this species a challenge.

Like other species of *Hyalinobatrachium*, *H. dianae* lays its eggs on the undersurfaces of leaves. Females produce a roughly circular clutch of 31–68 greenish-yellow eggs, which are fertilized and subsequently tended by a male. Adult males sometimes guard more than one clutch, and may even call to attract mates while guarding developing eggs. The tadpoles of *H. dianae* have thus far not been described.

Because it occurs in remote habitat and because males do not seem to call with any predictable regularity, it is not surprising that this species escaped

detection until very recently. It is likely that *H. dianae* is distributed more widely than is currently known, and possibly also occurs in nearby Panama. Since much of its potential range lies within protected areas, habitat alteration and habitat loss are not likely threats to its survival.

Interestingly, all individuals of *H. dianae* ever observed display a small, dark oblong spot under the skin of their lower back. Although the origin and function of this structure remains uncertain, it is a useful field mark for identifying this species.

Adult male *H. dianae*, calling while guarding a developing egg clutch on the underside of a large leaf.

Iris silvery-white, with minute dark specks and reticulations, especially around the horizontally elliptical pupil.

Dorsum uniform lime-green without spots, although each granular bump on the dorsal skin may be a slightly lighter shade of green.

Dorsal skin coarsely granular.

A unique dark oblong spot is located on the lower back, slightly off-center.

Snout short and truncated; nostrils located on slightly raised protuberances.

adult male

Adult male with a single vocal sac on throat; also note small, glandular nuptial patch on base of inner fingers.

Small, with long slender limbs.

Venter transparent; heart surrounded by pigmentless membrane and clearly visible; liver and intestines covered with opaque, white membrane.

Bones white.

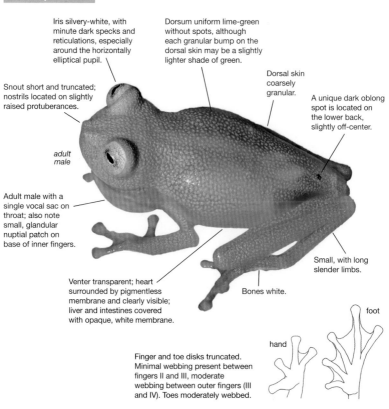

foot

hand

Finger and toe disks truncated. Minimal webbing present between fingers II and III, moderate webbing between outer fingers (III and IV). Toes moderately webbed.

H. chirripoi (p. 202)

Has a pattern of tiny light dots, lacks the black structure on the lower back, and has more extensively webbed hands.

H. colymbiphyllum (p. 204)

Has a pattern of light cream to pale yellow spots on the dorsum and lacks the black structure on the lower back.

H. fleischmanni (p. 208)

Has a pattern of light cream to pale yellow spots on the dorsum, lacks the black structure on the lower back, and its heart is covered with a white membrane.

Hyalinobatrachium fleischmanni
Fleischmann's Glass Frog

Least concern

A small glass frog. Lime-green, with medium-sized yellow spots. Has white bones. Venter transparent, but heart, liver, and digestive tract all enveloped in white membranes. Adult males to 1.02 in (26 mm) in standard length, females to 1.26 in (32 mm).

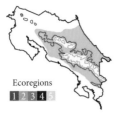

Ecoregions
1 2 3 4 5

Hyalinobatrachium fleischmanni is the most widespread and the northernmost member of the glass frog family. It ranges from Guerrero, Mexico, through Central America, to Ecuador, Colombia, Venezuela, and the Guyanas. In Costa Rica, it occurs in humid habitats throughout much of the country; 325–5,575 ft (100–1,700 m).

Natural history

This frog's call is the easiest means of finding it. On humid nights, it makes a one note, whistle-like *wheeet*, with a rising inflection. If climatic conditions provide sufficient moisture, these frogs call and breed practically year-round. The peak of their reproductive activity, however, coincides with the rainy season. Males call from the underside of leaves, often perched several meters above the water surface. But, in areas with disturbed habitat that lack tall vegetation or trees, calling males may be found close to the ground. Female *H. fleischmanni* actively seek out and choose a potential mate based on his calling behavior. When the female approaches, the male engages in axillary amplexus. She deposits a circular clutch of 18–50 small, cream to pale-green eggs on the underside of a leaf. The oviposition site, generally the same leaf that the male calls from, is located directly above a stream to allow hatching tadpoles to fall into the water below. After oviposition, the pair separates, and the male returns to attend the eggs each night until the larvae hatch. Males will continue to call and attract additional mates while guarding eggs, and it is not uncommon for them to tend to two or three clutches at once. The male performs gastric brooding, in which he perches on top of, or next to, an egg mass and voids moisture from his bladder to keep the embryos from dehydrating. Guarding males also protect developing larvae from parasitic "frog flies," which lay their eggs on the egg clutches and whose predatory larvae attack the developing tadpoles. Male *H. fleischmanni* are highly

territorial and engage in aggressive calling and physical combat to ward off males that encroach on their preferred calling site.

This common and widespread species is likely the most frequently encountered glass frog in Costa Rica. Although it favors humid riverine habitats with dense vegetation, it can tolerate significant habitat disturbance and water pollution. This ability has allowed *H. fleischmanni* to persist in severely altered stream habitats that pass through areas of secondary growth, coffee plantations, pastures, and even human settlements, including the outskirts of the city of San José, as long as the vegetation cover is sufficient to maintain adequate humidity levels. In the 1990s, well-documented populations in Monteverde suffered dramatic declines, possibly as a result of infection by chytrid fungi; these populations are gradually recovering. It has been suggested that the ability to survive in stream corridors contaminated with soaps, detergents, and agricultural runoff might have benefited *H. fleischmanni* because chytrid fungi possibly does not tolerate such conditions. Additional research is needed to corroborate this notion; nevertheless the population status of this species appears to be relatively secure in Costa Rica.

Throughout its extensive distribution range *H. fleischmanni* shows considerable regional variation in color and pattern; this species is possibly composed of a complex of multiple cryptic species, but further study is needed to adequately address this question.

No recorded nuptial patches in adult males of this species.

Ventral surfaces transparent. Heart surrounded by opaque white membrane and not visible; liver and intestines also covered with opaque, white membrane.

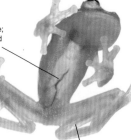

Bones white.

Dorsal coloration quite variable; from yellowish-green to bright lime-green, with varying amounts of faint yellow spots.

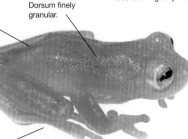

Iris gold, with minute dark brown to black speckling or reticulations, most prominent on either side of the horizontally elliptical pupil.

Toes and fingers yellow.

Dorsum finely granular.

Some individuals can appear almost patternless.

Nostrils barely raised above snout, creating a less truncated snout profile than in similar species (e.g., *H. colymbiphyllum*, p. 204).

foot

hand

Small and stocky, with a broad head and a short, relatively rounded snout.

Finger and toes disks truncated. Extensive webbing between fingers III and IV and between all toes.

Similar species

H. colymbiphyllum (p. 204)

Most similar in appearance but has a truncated snout with prominently raised nostrils and a clearly visible heart.

H. chirripoi (p. 202)

Has more extensive webbing between its fingers and its heart is visible.

H. valerioi (p. 212)

Has a color pattern of large yellow spots surrounded by a dark green reticulum; it also has a visible heart.

Hyalinobatrachium talamancae
Green-striped Glass Frog
Least concern

A small glass frog. Lime-green, with large yellow blotches and a unique, dark green middorsal stripe. Has white bones. Males to 0.94 in (24 mm) in standard length, females to 1.06 (27 mm).

Ecoregions
1 2 3 4 5

Hyalinobatrachium talamancae is known from a few isolated localities on the Atlantic slope of Costa Rica; 1,200–3,650 ft (400–1,116 m). Recently confirmed from Panama, it is likely more widespread than currently thought.

Natural history

H. talamancae can be fairly common in appropriate habitats; it is not easy to find, however, in part because it does not always call. It prefers mature second growth and primary forest habitat, where it inhabits vegetation along rapid brooks and streams. Adult males produce a single, soft whistling call, usually while perched on the underside of a leaf far above the water. These frogs do not call on rainless nights. But even on humid nights, calling is sometimes sporadic, making this species difficult to detect. Calling intensity increases during the wet season, when reproductive activity commences. Females attach small, circular clutches of 20–35 pale green eggs to the underside of a leaf overhanging the water. Male *H. talamancae* guard multiple clutches on the same leaf; they tend to the eggs at night.

For a long time, this species was only known from its type locality near Moravia de Chirripó, but additional field work in recent years has identified it on Tenorio Volcano, Guanacaste Province, in the Tilarán Mountain Range, as well as on slopes of the Central and Talamanca Mountain Ranges. Additional populations are likely awaiting discovery. Because *H. talamancae*

appears to be restricted to relatively mature forest, habitat alteration and loss of forest cover because of logging and agricultural practices are likely its biggest threats.

H. talamancae metamorph. The long, muscular tail will disappear over the course of several days.

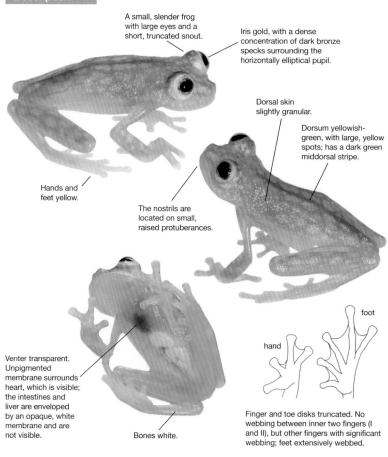

A small, slender frog with large eyes and a short, truncated snout.

Iris gold, with a dense concentration of dark bronze specks surrounding the horizontally elliptical pupil.

Dorsal skin slightly granular.

Dorsum yellowish-green, with large, yellow spots; has a dark green middorsal stripe.

Hands and feet yellow.

The nostrils are located on small, raised protuberances.

Venter transparent. Unpigmented membrane surrounds heart, which is visible; the intestines and liver are enveloped by an opaque, white membrane and are not visible.

Bones white.

foot

hand

Finger and toe disks truncated. No webbing between inner two fingers (I and II), but other fingers with significant webbing; feet extensively webbed.

H. valerioi
(p. 212)

Has a similar color pattern, with large yellow spots on a green background, but lacks the dark green middorsal stripe.

H. vireovittatum
(p. 214)

The dark green middorsal stripe is bordered on either side by a yellow line. Does not co-occur.

H. chirripoi
(p. 202)

Has a pattern of minute cream or yellow spots and lacks a green middorsal stripe.

H. colymbiphyllum
(p. 204)

Lacks the green middorsal stripe and has less webbing between its fingers.

H. fleischmanni
(p. 208)

Lacks the green middorsal stripe; its heart is covered with an opaque, white membrane.

Hyalinobatrachium valerioi
Reticulated Glass Frog

Least concern

A small glass frog with white bones. Has a unique color pattern of large yellow spots surrounded by a green reticulum that is marked with minute dark green to black spots. Heart visible. Males to 0.94 in (24 mm) in standard length, females to 1.02 in (26 mm).

Hyalinobatrachium valerioi occurs in Costa Rica on both the Atlantic and Pacific slopes; it also occurs in Panama, western Colombia, and Ecuador. Although recorded from near sea level to 4,925 ft (1,500 m), this species is usually found below 3,275 ft (1,000 m).

Ecoregions
1 2 3 4 5

Natural history

Like other glass frogs, *H. valerioi* is active on humid nights, in dense streamside vegetation. Since these minute, semi-transparent frogs are difficult to spot, they are most easily detected by their advertisement call. Males produce a relatively soft, single-noted *sheep*, repeated infrequently. Calling activity peaks during the wet season (May–November), when males vocalize from a preferred calling site on the underside of a leaf that overhangs a small stream or seepage area. The call serves to attract mates and to announce territoriality to neighboring males. If a male enters another's territory and is undeterred by increased calling on the part of the resident male, wrestling will ensue, and the resident male will attempt to pin down and subdue the intruder.

Females attracted to a male's call will approach his calling site, where axillary amplexus occurs. The female sticks a circular clutch of 25–35 pale green eggs to the underside of the leaf that the male calls from. Males prefer leaves that are large and smooth, and located above the water. In addition to breeding next to streams, *H. valerioi* also breeds near seepage areas or very small trickles; in such cases, egg masses and calling sites are generally situated on dense bushes and trees close to the ground.

Of all Costa Rican glass frogs, male *H. valerioi* display the most involved parental care and tend to their eggs day and night. As the resident male guards his eggs, he continues to call, and additional females may deposit more egg clutches on the same leaf on consecutive nights. Males have been observed simultaneously guarding as many as seven clutches in different stages of development. At night, the male moves about the leaf, defending the developing eggs against intruding males and engaging in gastric brooding (i.e., releasing moisture from his bladder onto the eggs to keep them from drying out). During the day, he positions himself flat against the leaf surface in a strategic location among the egg masses. Studies have shown that the guarding male's round-the-clock attendance increases larval survivorship. Parasitic drosophilid frog flies lay their eggs in the frog's brood, and their maggots feed on developing eggs and tadpoles. The egglike dorsal pattern of the guarding male may act as a decoy to attract these flies and other egg predators, such as parasitic wasps, affording him the opportunity to eat them or otherwise deter them.

H. valerioi is a relatively uncommon glass frog in Costa Rica, with a discontinuous, patchy distribution. Although it occurs over a wide range of habitats and elevations, it is generally common only in the humid lowlands and foothills of the southern part of the country. This species is somewhat tolerant of habitat alteration and can be found in relatively undisturbed riparian habitats as well as in mature secondary forests, as long as sufficiently dense streamside vegetation persists to maintain adequate moisture levels.

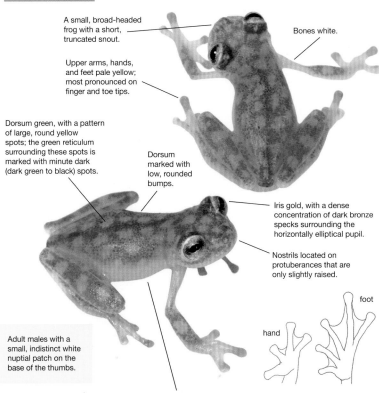

A small, broad-headed frog with a short, truncated snout.

Bones white.

Upper arms, hands, and feet pale yellow; most pronounced on finger and toe tips.

Dorsum green, with a pattern of large, round yellow spots; the green reticulum surrounding these spots is marked with minute dark (dark green to black) spots.

Dorsum marked with low, rounded bumps.

Iris gold, with a dense concentration of dark bronze specks surrounding the horizontally elliptical pupil.

Nostrils located on protuberances that are only slightly raised.

hand   foot

Adult males with a small, indistinct white nuptial patch on the base of the thumbs.

Venter transparent. In most cases, an unpigmented membrane surrounds the heart, which is visible. In some populations, however, white pigment may obscure this view. The intestines and liver are always covered by an opaque, white membrane and are not visible.

Finger and toe disks bluntly rounded to truncated. Minimal webbing between inner three fingers (I-II-III) and moderate webbing between fingers III-IV. Feet with extensive webbing between all toes.

H. talamancae (p. 210)

Has a pattern of large, yellow dorsal spots, but differs in having a distinct dark green middorsal stripe.

H. vireovittatum (p. 214)

Has a green middorsal stripe bordered on each side by a yellow stripe.

H. fleishmanni (p. 208)

Some individuals have enlarged yellow dorsal spots but lack a green reticulum; this species has a white membrane covering the heart.

213

Hyalinobatrachium vireovittatum
Yellow-striped Glass Frog

Data deficient

A small glass frog with white bones. Lime-green, with a pattern of cream to yellow small dots, interrupted by a unique, dark green middorsal stripe that is bordered on both sides by a yellow stripe. Heart visible. Males to 0.91 in (23 mm) in standard length, females to 0.98 in (25 mm).

Ecoregions
1 2 3 4 5

Hyalinobatrachium vireovittatum is endemic to Costa Rica, known from only a few localities along the Pacific slope, primarily between Dominical and San Isidro del General, and the Fila Chonta north of Quepos; 1,225–2,625 ft (375–800 m).

Natural history

H. vireovittatum can be locally common in appropriate habitat. It has been found in relatively cool, humid valleys; it seems to have a preference for high-gradient water courses, where it inhabits the spray zone of rapids and small waterfalls.

Adult male's advertisement call is a single, high-pitched, whistle-like *wheet*. Calling activity reaches its peak on rainy nights during the wet season (July–November); usually no calling activity is detected on dry nights. Reproduction also peaks in the wet season; in one observation, egg masses were found in mid-August on the underside of streamside vegetation. Clutches invariably over-hang a fast-moving rocky stream, 15–25 ft (5–8 m) above the water. The pale yellowish-green eggs are laid in a tight, roughly circular clump, attached to the underside of a leaf.

Male *H. vireovittatum* guard their clutch at night and perform gastric brooding, in which they straddle the eggs and empty their bladder over them to maintain adequate moisture levels. During the day, the male abandons the eggs. Although the tadpole of *H. vireovittatum* remains unknown, it is likely that larvae wriggle free from their protective jelly coat upon hatching and drop into the water below. Presumably, they feed on organic detritus in the streambed substrate until metamorphosis.

Little is known about the exact distribution range of *H. vireovittatum*. Initial records of this species reported from Tenorio Volcano (Guanacaste Province); near Monteverde, in the Tilarán Mountain Range; and from western Panama have recently been shown to represent the similar looking species *H. talamancae*.

Like other species in its genus, male *H. vireovittatum* guard their egg clutches at night and engage in gastric brooding to keep the eggs hydrated.

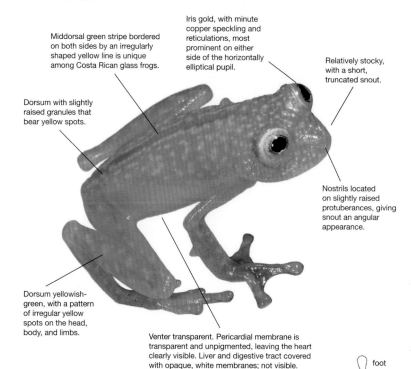

Middorsal green stripe bordered on both sides by an irregularly shaped yellow line is unique among Costa Rican glass frogs.

Iris gold, with minute copper speckling and reticulations, most prominent on either side of the horizontally elliptical pupil.

Relatively stocky, with a short, truncated snout.

Dorsum with slightly raised granules that bear yellow spots.

Nostrils located on slightly raised protuberances, giving snout an angular appearance.

Dorsum yellowish-green, with a pattern of irregular yellow spots on the head, body, and limbs.

Venter transparent. Pericardial membrane is transparent and unpigmented, leaving the heart clearly visible. Liver and digestive tract covered with opaque, white membranes; not visible.

Disks on fingers and toes truncated. No significant webbing between fingers I-II-III; moderate webbing between fingers III and IV. Toes extensively webbed.

hand   foot

H. talamancae (p. 210)

The only other glass frog with a green middorsal stripe, but it lacks the adjacent yellow lines. Does not co-occur.

H. valerioi (p. 212)

Has a pattern of large yellow dorsal spots surrounded by a green reticulum; although this pattern sometimes resembles a green middorsal stripe, it invariably lacks the bordering yellow lines.

Genus *Sachatamia*

A newly created genus that includes three species formerly assigned to other genera (*Centrolene*, *Cochranella*, and *Centrolenella*). Two species of *Sachatamia* occur in Costa Rica, while the third species is restricted to South America.

The two Costa Rican species share these characteristics: they are large and have green bones; all *S. albomaculata* and most *S. ilex* lack a humeral spur on the upper arm (although some adult male *S. ilex* may have a small spine); the ventral parietal peritoneum, a membrane that covers the organs, is white but becomes transparent posteriorly; and the digestive tract is partly visible through the ventral skin, near the insertion of the hind legs.

Male *Sachatamia* call from rocks along streams or from perches on the upper surfaces of leaves above creeks; at these same locations, females deposit their darkly pigmented eggs. There appears to be no parental care in these frogs.

Sachatamia albomaculata
Cascade Glass Frog **Least concern**

Dark green, with numerous white, cream, or yellow spots on the dorsum and limbs. Has a bluntly rounded snout and distinct pale lip stripe. An opaque white membrane covers only the chest region and the anterior part of the venter; posterior part of venter translucent. Bones are dark green. Males to 1.18 in (30 mm) in standard length, females to 1.26 in (32 mm).

Ecoregions
1 2 3 4 5

Sachatamia albomaculata ranges from eastern Honduras to western Colombia, though so far there are no records of this species from Nicaragua. In Costa Rica, it occurs on both the Atlantic and Pacific slopes; from near sea level to 4,925 ft (1,500 m).

Natural history

S. albomaculata is fairly common along forested streams in undisturbed humid forests, in lowlands and foothills. These frogs are generally encountered at night, when they perch on the upper surfaces of leaves in low bushes and trees along stream banks. In rocky, fast-flowing streams this species also occupies perches on spray-soaked boulders and rock walls. Calling males produce a single, short, high-pitched *dik*, which is repeated at intervals of up to several minutes. This call enables males to space their territories and is also an advertisement call to females.

This species seems to be less dependent on humidity and rainfall than other Costa Rican glass frogs, and its breeding activity appears to increase early in the dry season, when water levels are lower. Females deposit a single-layered egg clutch on the upper surface of a large leaf that overhangs the water or on wet rocky surfaces (the walls of small waterfalls, for example). When deposited, the black-and-white eggs are surrounded by a clear jelly; shortly after, the eggs turn dark brown to black, and the jelly becomes clouded. The shape of the clutch is unique among Costa Rican glass frogs. The eggs are typically deposited in a horseshoe pattern. Both parents abandon the eggs once they are in place and there appears to be no parental care or guarding of the developing eggs. Once

they are ready to hatch, tadpoles wriggle free from the jelly coat covering the clutch and drop into the stream below, where they complete their larval development.

When inactive and during the day, individual *S. albomaculata* have been found to retreat into rock crevices located in the spray zone of rapids and waterfalls. Additional observations support the notion that these frogs move away from riparian habitat outside of their breeding season, and individuals have been found considerable distances from the water.

Costa Rican populations of *S. albomaculata*, represent three distinct color morphs. Frogs from the Atlantic region tend to have a more yellowish-green dorsal coloration with moderate-sized, diffuse yellow spots; those from the central and south Pacific region have a dark bluish-green dorsal coloration with small, discrete yellow spots; while individuals from the Tilarán Mountain Range and north Pacific region can have a leaf-green dorsal coloraton and very fine white speckling. Animals from the latter population tend to lack any gold coloration in the eyes and have silver-gray irises.

In older literature, this species is listed as *Cochranella albomaculata* or *Centrolenella albomaculata*.

Males with indistinct white nuptial pads on the base of their thumbs.

Light stripe along upper lip.

Opaque white peritoneal membrane only covers chest and anterior part of venter; posterior half of belly translucent, leaving digestive tract faintly visible.

Iris silver-gray, with dark reticulations; in most populations, area surrounding horizontally elliptical pupil tinged with gold.

Dorsum dark green to bluish-green, with distinct yellow to white spots that are usually located on low tubercles.

Bones dark green.

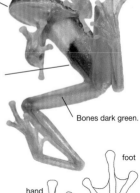

Nostrils, located on low protuberances, slightly raised above snout profile.

foot

hand

Medium-sized, with large, protruding eyes and a short, bluntly rounded snout.

Forearm bears a low, fleshy ridge on trailing edge.

Cream or white stripe present on margins of lower arms and lower legs.

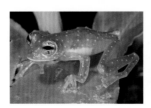

Finger and toe disks truncated. Hands bear webbing between fingers III and IV only; feet with moderate webbing between all toes.

S. albomaculata from Atlantic slope populations tend to have diffuse yellow spots on a yellowish-green to leaf-green dorsum

Individuals from the central and southern Pacific slope are often darker green to bluish-green, with small, discrete bright yellow spots.

Populations from the Tilarán Mountain Range differ in having only a fine dusting of tiny white specks on their leaf-green dorsum.

Similar species

Cochranella euknemos (p. 194)

Has smaller eyes and a distinctively long, sloping snout.

Teratohyla pulverata (p. 222)

Differs in having a clear, transparent peritoneal membrane, more extensively webbed fingers, and a sloping snout profile.

Sachatamia ilex
Ghost Glass Frog

Least concern

A large, uniform leaf-green glass frog. Strikingly large eyes have a silvery-white iris marked with bold black reticulations. Nostrils are prominently raised above the snout profile. Bones are green. The venter is nearly completely covered with a white, opaque membrane. Males to 1.18 in (30 mm) in standard length, females to 1.38 in (35 mm).

Ecoregions
1 2 3 4 5

On the Atlantic slope, *Sachatamia ilex* occurs in scattered localities in eastern Nicaragua, Costa Rica, western Panama, and western Colombia; on the Pacific slope of Costa Rica, a seemingly isolated population reportedly occurs near the Río Grande de Tárcoles; 820–2,950 ft (250–900 m).

Natural history

In suitable habitat, this species can be locally common. Generally prefers vegetation along streams with a significant elevational gradient. Occurs in foothills, in humid, mature secondary forest or pristine primary forest.

Populations of *S. ilex* are most easily detected by the loud, explosive advertisement call, a metallic single-noted *tink* that is repeated at intervals of sometimes several minutes. On humid nights, these calls usually emanate from treetops high overhead; but one most commonly sees non-calling frogs perched on top of a leaf 3–10 ft (1–3 m) above the ground. Interestingly, unlike most glass frogs, *S. ilex* is regularly seen on vegetation that does not directly overhang a stream. Adult males are territorial and engage in combat if an intruding male is not warded off by vocal cues. Wrestling males attempt to dislodge one another from the vegetation and often end up fighting while hanging upside down, suspended from a leaf by their feet.

This species seems more active during the wet season, but calling activity can be observed virtually year-round. Breeding takes place on the upper surface of leaves that overhang moving water. Egg clutches consist of 15–25 dark brown to black eggs that develop into tan or brown larvae. On hatching, the tadpoles drop into the water below, where they complete their development and feed on organic detritus.

During the day, most *S. ilex* retreat into the treetops, although individuals have been found sleeping on exposed leaves 5 ft (1.5 m) above the forest floor—or lower—tucked away in the leaf axils of large elephant ear plants (*Xanthosoma* sp.) or adhering to rock walls in the spray zone of an adjacent waterfall.

In the 1990s, some populations of *S. ilex* experienced dramatic declines, likely in response to the arrival of chytrid fungal pathogens. But this species has recovered in remarkable fashion, and is now common, even in areas where it was reduced to undetectable levels in the late 1990s. Due to its preference for mature forest, *S. ilex* is threatened by lumber extraction and conversion of forest for agriculture or other uses, but the species currently appears stable.

In older publications, this species is referred to as *Centrolenella ilex* or *Centrolene ilex*.

Adult males engage in territorial scuffles, often while hanging upside down, suspended by their feet.

219

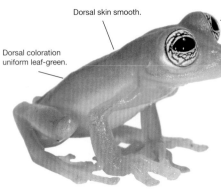

Dorsal skin smooth.

Dorsal coloration uniform leaf-green.

Eyes large and protruding, with characteristic silvery-white iris and bold black reticulum surrounding horizontally elliptical pupils.

Snout short and truncated, with distinctly raised nostrils.

The largest glass frog in Central America.

Adult males with indistinct white nuptial patch on base of thumbs. In some males, a small humeral spine may be visible on each upper arm.

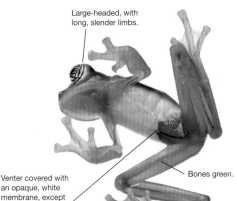

Large-headed, with long, slender limbs.

Venter covered with an opaque, white membrane, except for a small section near the insertion of the hind limbs.

Bones green.

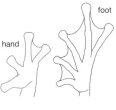

foot

hand

Finger and toe disks truncated. Webbing between fingers minimal or absent except between fingers III and IV. Feet with very extensive webbing between all toes; on some toes, webbing may almost reach disks.

Espadarana prosoblepon (p. 199)

Lacks the striking silver eyes with their black reticulum, and lacks a bluntly truncated snout profile. It generally has a pattern of dark spots on the dorsum; males have prominent humeral hooks.

Teratohyla spinosa (p. 224)

A much smaller species. It usually has golden eyes, with a fine reticulum (sometimes has a bold, dark reticulum). Note its distinct prepollex with a small protruding spine at the base of each thumb.

Genus *Teratohyla*

A recently resurrected genus that currently includes four small glass frog species, two of which occur in Costa Rica. Although quite distinct from one another, both species have pale green to dark green bones and lack a humeral spine on the upper arm.

In *T. spinosa*, the prepollical spine at the base of the thumb is well-developed and protrudes; in *T. pulverata* the spine is less developed and does not protrude. They also differ in the degree to which their internal organs are visible; in *T. spinosa*, the parietal peritoneum that covers the internal organs is white except for a transparent section near the insertion of the hind limbs, while in *T. pulverata* this membrane is completely transparent. The separate membrane that covers the digestive tract is translucent in *T. spinosa* and white in *T. pulverata*.

Male *Teratohyla* call from the upper surfaces of leaves overhanging a stream. Females deposit their eggs along the undersides of leaf margins (in *T. spinosa*) or on the tips of leaves (in *T. pulverata*).

In older literature, *T. spinosa* was referred to as *Cochranella spinosa*; *T. pulverata* was listed as *Hyalinobatrachium pulveratum*.

Teratohyla pulverata
Dusty Glass Frog

Least concern

A medium-sized frog. Leaf-green to dark green, with a dusting of fine white or yellowish spots. Also note sloping, obtuse snout, pale green bones, and extensively webbed hands. Males to 1.14 in (29 mm) in standard length, females to 1.30 in (33 mm).

Ecoregions
1 2 3 4 5

On the Atlantic slope, *Teratohyla pulverata* ranges from east-central Honduras to northern Colombia; on the Pacific slope, it also occurs in southwestern Costa Rica; from near sea level to 3,150 ft (960 m).

Natural history

T. pulverata tends to occupy vegetation along streams that are larger than those preferred by other glass frogs. It also appears better adapted to survive dry climatic conditions than other species and can tolerate moderate habitat alteration. It occurs in relatively undisturbed humid forest habitats, but can persist in seemingly unsuitable open areas with only a strip of trees remaining along a stream bank.

Males call at night from the upper surface of a leafy perch, often high in a tree, making them difficult to spot. Their advertisement call usually consists of a series of three short, high-pitched *tik* notes, not unlike the calls of *Sachatamia albomaculata* or *Espadarana prosoblepon*. Calling males can be heard almost year-round, but activity increases notably during the rainy season (May–November), when breeding takes place. *T. pulverata* eggs are pale green and laid on the upper surface of a leaf overhanging flowing water. Each egg mass contains 30–80 eggs and is generally attached to the drip tip of the leaf. The weight of the clutch tends to make the leaf droop down, which ensures that rainwater and dew on the leaf's surface runs down over the eggs and keeps them hydrated. *T. pulverata* egg masses are unique among Costa Rican glass frogs in that their protective jelly coat absorbs moisture and can swell to almost golf ball size, which may provide additional protection from predation as well as from dehydration in the drier conditions where this frog is found. Unlike the glass frogs in the genus *Hyalinobatrachium*, this species does not appear to guard its eggs or engage in gastric brooding. Its tadpoles drop down from the vegetation after they hatch and complete their development in the sediment at the bottom of the stream.

T. pulverata is relatively common in Costa Rica and its populations seem stable, although it is generally not a frequently observed species due to its habit of calling from the tops of leaves high in trees. Nonetheless, it is regularly seen in the southern Pacific lowlands and on the Osa Peninsula, where it appears to be one of the more common glass frogs.

The gelatinous egg masses of this species are larger than those of other Costa Rican glass frogs and can in extreme cases swell up to almost golf ball size.

Snout elongated, with a sloping, obtuse profile and slightly raised nostrils.

Eyes large and protruding. Iris silvery-gray, with a darker gray fine reticulum; pupils are horizontally elliptical.

Adult males with indistinct, unpigmented nuptial patch on base of each thumb.

A medium-sized glass frog with a flattened body and long, slender limbs.

Dorsal coloration leaf-green to dark green, with a distinctive dusting of white or yellowish specks.

Bones pale green.

Dorsum finely granular.

Lower arm with a white, fleshy ridge along its trailing edge.

Trailing edge of hind limb with a fleshy ridge.

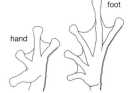

Ventral surfaces transparent; the intestines and heart are covered by white, opaque membranes and are not visible.

Sexual size difference in this species is not as pronounced as it is in other glass frogs. Note that the eggs are visible through the female's skin.

Finger and toe disks truncated. Significant webbing between fingers II and III, as well as between fingers III and IV. Toes with moderate webbing.

Cochranella euknemos
(p. 194)

The only other glass frog with a comparably sloping snout profile; it differs in having less webbing between fingers II and III and a mostly opaque, white ventral surface.

Sachatamia albomaculata
(p. 217)

Some populations may have a similar dorsal pattern but differ in having a short, truncated snout, a white lip stripe, less webbing between the fingers, and part of the venter covered with an opaque, white membrane.

Hyalinobatrachium chirripoi
(p. 202)

May be superficially similar in pattern and amount of finger webbing, but it has a bluntly truncate snout, white bones, and a visible heart.

Teratohyla spinosa
Dwarf Glass Frog

Least concern

A very small glass frog. Uniform green, with no dorsal spots. It can be distinguished by a free prepollex at the base of its thumbs that bears a small spine. Has a rounded to truncate snout and dark green bones. Much of the ventral surfaces is covered by opaque, white membranes, obscuring all organs except the intestinal tract. Males to 0.79 in (20 mm) in standard length, females to 0.95 (24 mm).

Ecoregions
1 2 3 4 5

Teratohyla spinosa occurs in apparently isolated populations along the Atlantic slope of extreme southeastern Honduras and southern Nicaragua; it is widespread on the Atlantic slope of Costa Rica; from near sea level to 2,450 ft (750 m). It also occurs in eastern Panama, Colombia, and Ecuador.

Natural history

Very common locally, in wet forests of the Atlantic lowlands and foothills. Where present, *T. spinosa* tends to occur in high densities. It perches on top of leaves in low vegetation, making this one of the more easily observed glass frogs in the country. Although it can be found along moderate-sized streams, it seems to prefer slow-moving sections of small streams or seeps in relatively undisturbed forest habitats. Males produce a loud, harsh series of usually two or three calls: *creep-creep* or *creep-creep-creep*. They call from the upper surface of a large leaf, often selecting a perch that is covered by a second leaf overhead. Calls can be heard nearly year-round, but intensify after heavy rains and during the wet season (May–November), when their breeding activity increases. *T. spinosa* females lay 14–25 pale green eggs on the underside of vegetation in a single layer of loose jelly. As the larvae develop, the jelly coat starts to droop, facilitating the release of hatching tadpoles. The small larvae drop into the stream below, where they complete their development. Adults of this species do not guard their eggs or keep the clutches hydrated.

T. spinosa is quite common in appropriate habitat but several populations underwent dramatic declines in the early 1990s, presumably in response to the arrival of chytrid fungal pathogens. Even populations that were undetectable throughout much of the 1990s have recovered and are now seemingly thriving. Given its preference for relatively undisturbed habitats, *T. spinosa* is likely most threatened by habitat loss from development and by water pollution as a result of agricultural runoff.

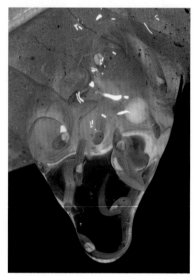

T. spinosa egg mass attached to underside of a leaf, with tadpoles about to drop into a stream below. There is a single, white infertile egg in this clutch.

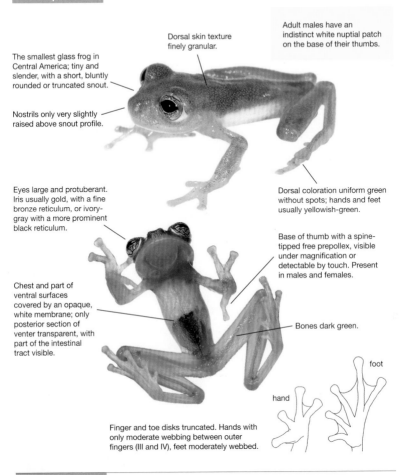

Dorsal skin texture finely granular.

Adult males have an indistinct white nuptial patch on the base of their thumbs.

The smallest glass frog in Central America; tiny and slender, with a short, bluntly rounded or truncated snout.

Nostrils only very slightly raised above snout profile.

Eyes large and protuberant. Iris usually gold, with a fine bronze reticulum, or ivory-gray with a more prominent black reticulum.

Dorsal coloration uniform green without spots; hands and feet usually yellowish-green.

Base of thumb with a spine-tipped free prepollex, visible under magnification or detectable by touch. Present in males and females.

Chest and part of ventral surfaces covered by an opaque, white membrane; only posterior section of venter transparent, with part of the intestinal tract visible.

Bones dark green.

foot

hand

Finger and toe disks truncated. Hands with only moderate webbing between outer fingers (III and IV), feet moderately webbed.

Espadarana prosoblepon (p. 199)

Juveniles can be very similar in appearance but lack the free prepollex and have a white peritoneal sheath that covers all of the internal organs.

Sachatamia ilex (p. 219)

Juveniles can be confused with gray-eyed *T. spinosa* but lack the spine-tipped prepollex and have more prominently raised nostrils.

Family Craugastoridae
(Litter Frogs)

These frogs were previously included in the enormous genus *Eleutherodactylus*, which, until a few years ago, represented the largest genus of any vertebrate, with many hundreds of extremely diverse species. It had been long known that the collection of frogs contained within *Eleutherodactylus* was not reflective of phylogenetic or biogeographic relationships between various species groups, but it was not until about 10 years ago that researchers started the daunting task of trying to make better sense of this assemblage.

Costa Rican species formerly assigned to the genus *Eleutherodactylus* were once all placed in a single family, Leptodactylidae. Today they are divided over three families: Craugastoridae, Eleutherodactylidae, and Strabomantidae. Remaining Costa Rican members of the family Leptodactylidae were considered separate from *Eleutherodactylus*, even prior to the new family arrangement.

Created only in 2008, the exact composition of the family Craugastoridae is still under discussion. It is primarily defined on the basis of molecular traits, and no single, easily recognizable, trait applies to all the species that make up this confusing group. To further complicate the matter, most of the species in this family display tremendous variation in physical appearance.

There are currently 30 recognized species of Craugastoridae in Costa Rica, all placed in the genus *Craugastor*. Most defy easy description.

Genus *Craugastor*

The 30 species of *Craugastor* found in Costa Rica display tremendous variation in body size, shape, coloration, preferred habitat, and other aspects of their biology. This genus includes several commonly observed species and some that are possibly extinct.

In an effort to make sense of the diversity within this genus, similar species have traditionally been grouped into informal species groups whose exact make-up is still the subject of considerable debate. The Costa Rican species of *Craugastor* are assigned to one of 5 groups:

***biporcatus*-species group** (3 species). *C. gulosus*, *C. megacephalus*, and *C. rugosus*

Sometimes colloquially called broad-headed rain frogs because of their strikingly large heads. These are medium-sized to large, toadlike, terrestrial frogs with an exceedingly broad head (about half as wide as their standard length, or wider still). They lack webbing between their fingers and toes and do not have significantly enlarged disks on their digits. Some species have cranial crests and/or distinct dermal ridges on their back, but they never have the parotoid glands seen in true toads (family Bufonidae).

***fitzingeri*-species group** (9 species). *C. andi*, *C. chingopetaca*, *C. crassidigitus*, *C. cuaquero*, *C. fitzingeri*, *C. melanostictus*, *C. phasma*, *C. rayo*, and *C. talamancae*

These are the so-called rain frogs, commonly seen or heard during or immediately after significant rainfall. Characterized by the presence of greatly enlarged truncated disks on the outer two fingers of each hand, while the inner two fingers bear only slightly expanded rounded disks. Their feet have basal to extensive webbing, and the toe tips bear slightly expanded disks that are smaller than the enlarged outer finger disks. Some species are terrestrial, others are arboreal. While some of the species in this group are abundant, others may be extinct.

***gollmeri*-species group** (3 species). *C. gollmeri, C. mimus*, and *C. noblei*

These masked litter frogs are streamlined and have a long snout, long legs, and smooth skin. All have a dark eye mask that covers the sides of the head and extends onto the body. Their coloration is cryptic and resembles the colors of dead leaves; they invariably hide in the leaf litter substrate on the forest floor, relying on their camouflage to avoid detection.

***rhodopis*-species group** (7 species). *C. bransfordii, C. gabbi, C. persimilis, C. podiciferus, C. polyptychus, C. stejnegerianus*, and *C. underwoodi*

A group of tiny to small frogs that inhabit the leaf litter layer of forests throughout Costa Rica. Their small size, stocky body, short legs, and the absence of webbing and disks on the fingers and toes allow one to fairly easily place them in the right species group, but identifying individual species within this group under field conditions is nearly impossible.

***rugulosus*-species group** (8 species). *C. angelicus, C. catalinae, C. escoces, C. fleischmanni, C. obesus, C. ranoides, C. rhyacobatrachus*, and *C. taurus*

These frogs occupy streamside habitats, usually in rocky, cascading streams at middle and high elevations (although *C. ranoides* and *C. taurus* also occupy lowland stream corridors). Once common to abundant in appropriate habitat, all eight species have greatly declined in numbers or disappeared altogether since the late 1980s. The lack of observations and of photographs of living individuals means that their descriptions are often based on minute morphological details visible on preserved museum specimens rather than easily observed field marks, making these frogs very difficult to identify in the field.

Craugastor andi
Yellow-legged Rain Frog **Critically endangered**

A large member of the *fitzingeri*-species group. Characterized by the presence of greatly enlarged truncated disks on the outer two fingers of each hand; the inner two fingers bear only slightly expanded rounded disks. Typically it has bright yellow spots or vertical bars on the dark brown posterior section of the thighs. Yellow ventral surfaces are sometimes suffused with salmon-red. A distinct light stripe runs down the center of its throat. Males to 2.17 in (55 mm) in standard length, females to 3.15 in (80 mm).

Ecoregions
1 2 3 4 5

Craugastor andi is endemic to Costa Rica. It is known from the Atlantic slopes of the Guanacaste, Tilarán, Central, and Talamanca Mountain Ranges in the northern and central parts of the country; 3,275–3,950 ft (1,000–1,200 m).

Natural history

Very little is known about the biology of this rare species. It has been found during the day in arboreal bromeliads several meters above the ground. Adult males reportedly descend from their arboreal retreats during the first heavy rains of the wet season and move towards the banks of nearby streams. Mating and oviposition supposedly takes place in the leaf litter or some other moist microsite in the vicinity of the stream, after which the eggs undergo direct development and hatch as tiny frogs. Adults return to the canopy after the eggs are deposited.

C. andi was once regularly encountered in the Monteverde Cloud Forest Preserve and in other protected areas in the country but has not been reported from anywhere in its range since the early 1990s. In fact, beginning in the late 1980s, many montane amphibians—especially species associated with streams—have disappeared. Given that *C. andi* occurred in protected areas and other places with undisturbed forests, habitat destruction can be ruled out as an explanation for its decline. More likely causes are a chytrid fungal infection and, possibly, climate change. Additional surveys are needed to locate possible surviving poulations and to properly assess this frog's current conservation status, but it is possible that *C. andi* has gone extinct.

Description

Undersurfaces of chest and limbs heavily marked with dark mottling or spots; center line on throat devoid of dark spots, appearing as a light midgular stripe.

Distinct sexual size dimorphism; females significantly larger than males.

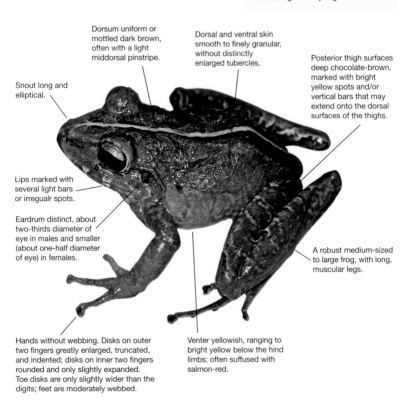

Dorsum uniform or mottled dark brown, often with a light middorsal pinstripe.

Dorsal and ventral skin smooth to finely granular, without distinctly enlarged tubercles.

Posterior thigh surfaces deep chocolate-brown, marked with bright yellow spots and/or vertical bars that may extend onto the dorsal surfaces of the thighs.

Snout long and elliptical.

Lips marked with several light bars or irregualr spots.

Eardrum distinct, about two-thirds diameter of eye in males and smaller (about one-half diameter of eye) in females.

A robust medium-sized to large frog, with long, muscular legs.

Hands without webbing. Disks on outer two fingers greatly enlarged, truncated, and indented; disks on inner two fingers rounded and only slightly expanded. Toe disks are only slightly wider than the digits; feet are moderately webbed.

Venter yellowish, ranging to bright yellow below the hind limbs; often suffused with salmon-red.

C. fitzingeri (p. 246)

Shares many diagnostic features, but has rugose dorsal skin, lacks indented finger disks, is smaller, and sometimes has a broad middorsal stripe (but never a pinstripe).

C. cuaquero (p. 243)

NO PHOTO

Has a mottled groin pattern, is smaller, and has less webbing between its toes.

C. crassidigitus (p. 240)

Is smaller; the posterior section of its thighs is uniform brown (without yellow spots or vertical bars).

Craugastor angelicus
Yellow-bellied Streamside Frog **Critically endangered**

A member of the *rugulosus*-species group. The frogs in this group are extremely difficult to distinguish under field conditions as they often differ only in very subtle ways. Geographic provenance, the presence of only slightly expanded finger and toe disks, and its colorful ventral surfaces may help in identifying this species. Males to 1.81 in (46 mm) in standard length, females to 2.95 in (75 mm).

Ecoregions
1 2 3 4 5

Craugastor angelicus is endemic to Costa Rica. It has been reported from isolated locations on Cerro Cacao in the Guanacaste Mountain Range, from the Tilarán Mountain Range (3,950–5,250 ft [1,200–1,600 m]), and on the Atlantic slope of Poás Volcano, in the Central Mountain Range (1,975–5,575 ft [600–1,700 m]).

Natural history

C. angelicus is nocturnal and terrestrial. It occurs in forests at middle and high elevations, where it was most frequently seen on the steep banks of clear, cascading streams. A female was reported laying a clutch of 77 eggs in a hole, covering the eggs with sand, and then guarding the clutch for several days. The larvae of *Craugastor* species skip a free-swimming tadpole stage and complete their development inside the egg.

This species was formerly abundant throughout its range but has declined dramatically since the 1980s. The last known sighting of *C. angelicus* was in 1994, in the Tilarán Mountains, but in spite of repeated searches in several localities where it was known to occur, there are no more recent observations, and this species may be extinct. *C. angelicus* vanished from several well-protected, undisturbed sites, and thus habitat alteration or degradation is clearly not the main reason for its decline. In Mesoamerica, global warming, airborne pollution, and chytridiomycosis are all possible factors in the demise of this and several other montane amphibians associated with streams. Additional surveys are urgently needed to determine if this species still persists in parts of its historic range and to assess its current conservation status.

Dorsum dark brown, olive, or gray, often with obscure, irregular dark markings. On large adults, dorsum and top of head sometimes show red blotches.

Relatively smooth-skinned, with only minimally enlarged tubercles or dermal ridges on the dorsum.

The snout is moderately long; it appears ovoid when viewed from above and rounded in profile. The canthus rostralis is rounded.

Legs long and slender.

Posterior thigh surfaces dark, usually marked with yellow spots.

A distinct supratympanic fold follows the upper curve of the eardrum, which is clearly visible.

Ventral surfaces bright yellow. In some populations and in large, old individuals the limbs, throat, and edges of the belly may be bright orange-red to red.

Medium-sized to large; females dramatically larger than males. Adult males have nuptial pads. They lack vocal slits.

Finger I is equal in length or only slightly longer than finger II. Fingers and toes bear slightly expanded (approximately 1.5x the width of the digit), rounded disks. Feet are only basally webbed, and the unwebbed sections of the toes, as well as all fingers, bear weak lateral keels.

C. escoces (p. 244)

Has larger finger disks and has vocal slits.

C. fleischmanni (p. 249)

Has barely expanded finger and toe disks and usually a pale yellow or cream venter.

C. ranoides (p. 274)

Lacks expanded finger and toe disks and does not have a brightly colored venter.

Craugastor bransfordii
Bransford's Litter Frog

Least concern

A member of the *rhodopis*-species group. These tiny, stocky leaf litter frogs are characterized by their short limbs and the absence of both interdigital webbing and enlarged finger and toe disks. All species in this group are extremely variable in coloration, pattern, and skin texture and can be almost impossible to identify in the field. *Craugastor bransfordii* typically is small, has projecting, pointed tubercles on the soles of its hands and feet, and often has reddish coloration in the groin area. Adult males have a white nuptial patch at the base of each thumb. Males to 0.91 in (23 mm) in standard length, females to 1.02 in (26 mm).

Ecoregions
1 2 3 4 5

Craugastor bransfordii occurs on the Atlantic slope, from eastern Nicaragua to central Costa Rica. Although known to occur at higher elevations in Nicaragua, all Costa Rican populations occur between 200–2,885 ft (60–880 m). An isolated population in central and eastern Panama currently assigned to *C. bransfordii* represents a distinct but undescribed species.

Natural history

In many lowland areas, *C. bransfordii* is the most abundant leaf litter frog. It is a very successful colonizer of disturbed areas; even in pristine forests, it prefers edge situations, trails, and open areas over the forest interior. The density of this species is normally highest at the end of the dry season, when the ground is covered with a thick layer of dead leaves that provides many hiding places and a plethora of small invertebrate prey.

During the day, *C. bransfordii* hops across the layer of fallen leaves on the forest floor. Because of its small size and cryptic coloration, it is easily overlooked. If startled, these frogs will remain motionless and rely on their camouflage for protection. When approached too closely, in rapid succession they make two or three short hops and dive under the leaf litter. This species is very active, spending most of the day foraging for insect prey. A food generalist, *C. bransfordii* eats a wide variety of invertebrates, including ants, spiders, mites, beetles, and insect larvae.

C. bransfordii reproduces at night, although its mating call, a soft *pew*, can also be heard on rainy, overcast days. Males call from under leaf litter, from grassy roadsides, or other areas with dense vegetation. The eggs are laid on the forest floor, in a moist, shaded spot near the calling site. The larvae undergo direct development; they complete their metamorphosis inside the egg and hatch as miniature versions of the adults.

C. bransfordii and other members of the *rhodopis*-species group are notorious for their extreme variation in skin texture and color pattern. This group includes many cryptic species that can only be reliably distinguished based on biochemical characteristics, since their morphological features—such as size, color, pattern, and skin texture—overlap between species. Secondary sex characteristics in males (such as presence or absence of nuptial patches) may be helpful for identification, but females and juveniles are impossible to positively identify that way under field conditions. In some cases, geographic provenance may help in reaching a tentative identification, but especially within the range of *C. bransfordii* even that becomes challenging.

The confusing taxonomic situation of *C. bransfordii* makes it difficult to assess the numbers and conservation status of this species. The fact that these frogs undergo dramatic fluctuations in population size in response to seasonal changes in depth of the leaf litter further complicates the matter. However, *C. bransfordii* appears common throughout its range and is probably not at risk.

Upper lip marked with alternating dark and light bars or spots.

A short, diagonal row of pink or salmon tubercles generally present between the bottom of the eardrum and the insertion of the front limbs.

female

Skin texture extremely variable. Can be smooth, irregularly granular, or warty; sometimes, also has dorsal ridges (in a variety of patterns) superimposed on the skin.

Soles of hands and feet with distinctly projecting, pointed tubercles.

Snout short and bluntly rounded.

In males, tympanum (eardrum) roughly equal in size to the eye; distinctly smaller in females.

male

Adult males have a white nuptial patch at the base of each thumb.

Arms and legs relatively short and stocky.

Venter and undersides of limbs dull yellow; throat usually dirty white, with minute dark speckling.

foot

hand

In many individuals, the thighs and flanks are suffused with orange or red.

Inner fingers (I and II) of equal length. Lacks webbing between the digits; tips of fingers and toes are not expanded into disks.

234

As these seven images show, coloration is extremely variable in *C. bransfordii*; usually shows a cryptic pattern of tan, gray, cream, or brown, with a pattern of spots, chevrons, or stripes. Sometimes has a pair of cream, dorsolateral stripes or, less frequently, a narrow cream middorsal stripe.

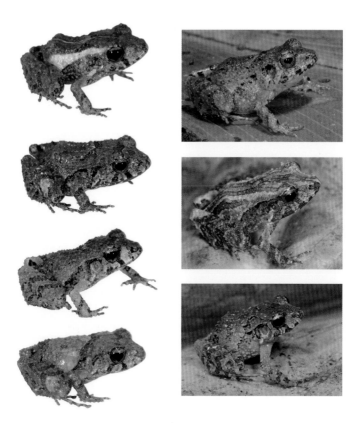

Similar species

C. persimilis (p. 265)

Smaller and has rounded tubercles on the soles of its hands and feet. Finger II is longer than finger I. Adult males lack nuptial patches.

C. polyptychus (p. 272)

Is larger, and tends to occur at lower elevations. Adult males lack nuptial patches.

C. underwoodi (p. 289)

Is larger, has rounded, less projecting, tubercles on the soles of its hands and feet, and tends to occur at higher elevations.

C. podiciferus (p. 269)

Has several enlarged heel tubercles and often has a dark eye mask. It occurs at higher elevations.

Craugastor catalinae
Montane Streamside Frog   **Critically endangered**

A member of the *rugulosus*-species group. These frogs are extremely difficult to identify under field conditions, as the eight species differ only in very subtle ways. Nevertheless, geographic provenance, the presence of nuptial patches in adult males, and lack of colorful ventral surfaces may help in identifying this species. Males to 1.77 in (45 mm) in standard length, females to 2.95 (75 mm).

Ecoregions
1 2 3 4 5

Craugastor catalinae is known from just a small area on the Pacific slope, in extreme southwestern Costa Rica and adjacent western Panama; 4,000–5,900 ft (1,219–1,800 m).

Natural history

This is a relatively new addition to the *rugulosus*-species group. It was previously included in *Craugastor punctariolus* (as currently understood, *C. punctariolus* only occurs in the Panamanian highlands).

C. catalinae is a terrestrial, riparian species. It inhabits stream banks and streambeds, where it lives under and between moss-covered rocks.

It was formerly common, but known populations have declined dramatically in the last decades. The areas from which it disappeared include several pristine areas where habitat loss and alteration could not have been a factor in its decline; possibly chytridiomycosis or other environmental factors have played a role. Repeated searches for this species in Panama resulted in no recent sightings, and additional surveys are urgently needed to ascertain whether this species still persists anywhere in its known range. Although a few other species of the *rugulosus*-species group have been rediscovered in recent years, the majority are likely extinct.

Supratympanic fold is fairly prominent and passes over top of the distinctly visible eardrum.

An interorbital dark bar, bordered anteriorly by a light bar, is usually present.

Several low tubercles present on upper eyelids.

Dorsal skin smooth to granulate, with scattered large tubercles. Glandular ridges on the dorsum are weakly developed or absent.

Posterior thigh surfaces dark brown, with yellow mottling.

Snout short and elliptical when viewed from above. Has a rounded canthus rostralis.

Dark bands on the hind limbs and dark markings on the sides of the head are often difficul to see due to the overall dark dorsal coloration.

Ventral surfaces white to pale cream, with brown mottling on the throat and chest.

Dorsal surfaces dark brown, usually bordered on the flanks by a paler hue (gray, cream, tan). Occasionally with a light middorsal pinstripe.

Adult males have nuptial pads on the base of their thumbs; they also have paired vocal slits.

Finger I longer than finger II. All fingers bear weak lateral keels and only slightly enlarged rounded disks (about 1.5x the width of the digit). Toes with small round disks that are similar in size to those on the fingers. Feet are moderately webbed; sections of toes that are not webbed bear lateral fringes.

C. ranoides (p. 274)

This is the only similar species that may co-occur. Males lack nuptial pads and vocal slits. Tends to have yellow or orange on the posterior ventral surfaces and the undersides of its hind limbs.

Craugastor chingopetaca
Río San Juan Rain Frog

Not assessed

An enigmatic species in the *fitzingeri*-species group. Characterized by the presence of greatly enlarged truncated disks on the two outer fingers of each hand, while the two inner fingers bear only slightly expanded rounded disks. It has uniform brown posterior thigh surfaces, indented finger and toe disks, and moderate webbing between its toes; it lacks a light stripe on the throat. Standard length of the only known individual, an adult female, is 1.50 in (38 mm).

Ecoregions
1 2 3 4 5

Craugastor chingopetaca is known only from the type locality of Chingo Petaca, on the Nicaraguan side of the Río San Juan, which forms the border with Costa Rica; found at 130 ft (40 m). This species is currently not known from any Costa Rican localities, but given the proximity of the type locality it should be expected in Barro del Colorado and Tortuguero.

Natural history

C. chingopetaca is known from just a single individual, discovered in Nicaragua (near the Río San Juan); it was found during the day, in the leaf litter of a lowland rainforest. In spite of repeated searches in the area, no other representatives of this species have been found, although two other species in the *fitzingeri*-species group (*C. fitzingeri* and *C. talamancae*) were present in the area. *C. chingopetaca* is very similar in appearance to *C. crassidigitus*, a species not currently known from the area near the Río San Juan, but which might conceivably occur there, in low numbers. There is no doubt that cryptic species are contained within the current definition of *C. crassidigitus*, *C. fitzingeri*, and *C. talamancae*, but it is also true that frogs in the genus *Craugastor* are notoriously variable in their coloration and morphology, and they often show considerable overlap in their diagnostic characteristics. Until additional individuals of *C. chingopetaca* are identified, its status will remain uncertain.

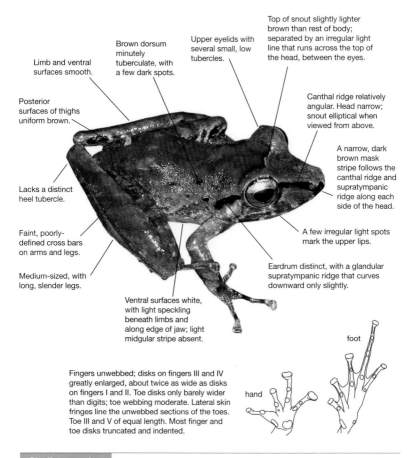

Limb and ventral surfaces smooth.

Brown dorsum minutely tuberculate, with a few dark spots.

Upper eyelids with several small, low tubercles.

Top of snout slightly lighter brown than rest of body; separated by an irregular light line that runs across the top of the head, between the eyes.

Posterior surfaces of thighs uniform brown.

Canthal ridge relatively angular. Head narrow; snout elliptical when viewed from above.

A narrow, dark brown mask stripe follows the canthal ridge and supratympanic ridge along each side of the head.

Lacks a distinct heel tubercle.

A few irregular light spots mark the upper lips.

Faint, poorly-defined cross bars on arms and legs.

Eardrum distinct, with a glandular supratympanic ridge that curves downward only slightly.

Medium-sized, with long, slender legs.

Ventral surfaces white, with light speckling beneath limbs and along edge of jaw; light midgular stripe absent.

foot

Fingers unwebbed; disks on fingers III and IV greatly enlarged, about twice as wide as disks on fingers I and II. Toe disks only barely wider than digits; toe webbing moderate. Lateral skin fringes line the unwebbed sections of the toes. Toe III and V of equal length. Most finger and toe disks truncated and indented.

hand

C. fitzingeri (p. 246)

Has a light midgular stripe and cream spots on the otherwise dark posterior thigh surfaces.

C. talamancae (p. 284)

Has uniform brown posterior thigh surfaces, but differs in having only minimal toe webbing; also note that toe III is longer than toe V.

C. crassidigitus (p. 240)

Has uniform brown posterior thigh surfaces and often a light midgular stripe, but its supratympanic ridge curves downward noticeably (only slightly in *C. chingopetaca*) over the eardrum.

Craugastor crassidigitus
Slim-fingered Rain Frog

Least concern

A member of the *fitzingeri*-species group. Characterized by greatly enlarged truncated disks on the two outer fingers of each hand, while the two inner fingers bear only slightly expanded rounded disks. The combination of uniform reddish-brown to orange posterior thigh surfaces, substantial webbing between the toes, and absence of a white lip stripe is unique. Males to 1.18 in (30 mm) in standard length, females to 1.89 in (48 mm).

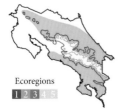

Ecoregions
1 2 3 4 5

Craugastor crassidigitus is widespread along both coasts of Costa Rica, throughout Panama, and in extreme northwestern Colombia; near sea level to 6,550 ft (2,000 m).

Natural history

C. crassidigitus is a very common species in many parts of Costa Rica. It is mostly seen at night, when it perches on low vegetation. Individuals are also often seen on the ground, or on rocks bordering small, rocky waterways. During the day, these frogs retreat into the leaf litter, where they are often flushed out by hikers. The closely related species *C. fitzingeri* occupies a very similar ecological niche, and the two species sometimes occur together.

Male *C. crassidigitus* produce a two-toned *chuckle*, which serves both to advertise their presence and mark territory. They call with the greatest intensity at dusk or early evening, especially after heavy afternoon rains; males also occasionally call on overcast, drizzly days. Call activity is most pronounced during the rainy season, which is presumably the peak breeding season. Females deposit their eggs in moist microsites in the leaf litter or in shallow depressions in the soil, often under a log or rock. The larvae complete their entire development and metamorphosis inside the eggs; they hatch as small froglets.

C. crassidigitus occurs in pristine forests but is also common in secondary vegetation, plantations, and other disturbed habitats. It is an adaptable and abundant species that seems able to survive—and thrive—in areas with tree cover and associated leaf litter substrate remain relatively intact.

In recent studies, the author has encountered *C. crassidigitus* that were infected with the chytrid fungal pathogen *Bd*. In some areas those frogs appeared not to be affected by the pathogen, whereas in other areas dead or dying individuals were found that showed signs of chytridiomycosis, the disease caused by *Bd*. Even in areas where individual *C. crassidigitus* succumb to *Bd*, their overall population status appears to be relatively stable. This species is still commonly observed throughout its range, but additional research is needed to better understand the transmission dynamics of *Bd* in amphibian communities, as well as the factors that keep potentially devastating disease outbreaks and amphibian population declines in check in areas where *Bd* is present.

This is a common species that people are likely to encounter but will have a difficult time identifying, as it is highly variable in appearance.

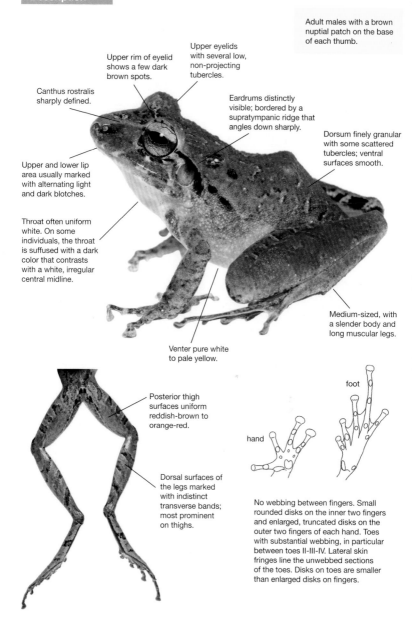

Adult males with a brown nuptial patch on the base of each thumb.

Upper eyelids with several low, non-projecting tubercles.

Upper rim of eyelid shows a few dark brown spots.

Canthus rostralis sharply defined.

Eardrums distinctly visible; bordered by a supratympanic ridge that angles down sharply.

Dorsum finely granular with some scattered tubercles; ventral surfaces smooth.

Upper and lower lip area usually marked with alternating light and dark blotches.

Throat often uniform white. On some individuals, the throat is suffused with a dark color that contrasts with a white, irregular central midline.

Medium-sized, with a slender body and long muscular legs.

Venter pure white to pale yellow.

Posterior thigh surfaces uniform reddish-brown to orange-red.

foot

hand

Dorsal surfaces of the legs marked with indistinct transverse bands; most prominent on thighs.

No webbing between fingers. Small rounded disks on the inner two fingers and enlarged, truncated disks on the outer two fingers of each hand. Toes with substantial webbing, in particular between toes II-III-IV. Lateral skin fringes line the unwebbed sections of the toes. Disks on toes are smaller than enlarged disks on fingers.

Color pattern of *C. crassidigitus* is extremely varied; generally consists of shades of gray or brown with dark spots demarcating the dorsal and dorsolateral tubercles. A broad middorsal light stripe may be present. Many individuals have a pink or reddish cast on their dorsal coloration, which is particularly visible at night. Occassionally with irregular green mottling on the dorsum.

Females generally more robust than males.

Populations from high elevations in the Talamanca Mountain Range differ from lowland populations in size, coloration, and head shape. *C. crassidigitus* likely consists of a complex of several species.

Juveniles from several highland populations have a significant amount of green in their coloration.

Adults occasionally show irregular metallic green mottling on head and dorsum.

Similar species

C. fitzingeri (p. 246)

Has dark brown or black posterior thigh surfaces, marked with distinct, pale yellow spots. It also has more rugose dorsal skin, marked with scattered tubercles and ridges, and moderate webbing between its toes.

C. talamancae (p. 284)

Has uniform reddish-brown posterior thigh surfaces, but only has basal webbing between its toes. It is usually easily distinguished by the presence of a white lip stripe that extends onto the body.

C. chingopetaca (p. 238)

Has indented disks on its outer fingers and its toes and lacks a white stripe on the throat.

C. rayo (p. 276)

Has uniform dark purplish-brown posterior thigh surfaces and a distinct tubercle on each heel.

242

Craugastor cuaquero
Tilarán Rain Frog

Data deficient

A member of the *fitzingeri*-species group. Characterized by the presence of greatly enlarged truncated disks on the two outer fingers of each hand; the two inner fingers bear only slightly expanded rounded disks. It has small cream to yellow spots on the dark brown surfaces of the posterior section of the thighs; these spots are sometimes fused to form irregular vertical lines. The enlarged outer finger disks are indented. *Craugastor cuaquero* typically has only basal webbing between its toes. The only known male measures 1.53 in (39 mm) in standard length, females reach at least 1.89 in (48 mm).

Ecoregions
1 2 3 4 5

Craugastor cuaquero is endemic to Costa Rica and is known only from the type locality near Monteverde, in the Tilarán Mountain Range; 4,985–5,250 ft (1,520–1,600 m).

Natural history

This rare species is only known from two females and one male collected in the early 1960s. Nothing is known about the biology of *C. cuaquero*. It inhabits lower montane rainforest and has only been found within the Monteverde Cloud Forest Preserve. There are no recent sightings of this species, in spite of repeated searches. It is easily mistaken for other, much more common forms like *C. fitzingeri*. It is possible that *C. cuaquero* is a canopy species, which would only add to the difficulty of observing it. Nevertheless, the three known individuals were collected on low herbaceous vegetation. Apart from a single report from the early 1980s, this species has not been observed since its original discovery in 1964. Additional research is needed to accurately assess the validity of this species, as well as its population and conservation status. *C. cuaquero* had not been observed for quite some time before chytrid fungal infections caused dramatic declines in the amphibian populations of the Monteverde region in the late 1980s, and it is unclear whether chytridiomycosis played a role in its perceived absence.

Description

- Medium-sized, slender, and long-limbed. Disks on inner two fingers only slightly expanded and rounded; disks on outer two fingers distinctly enlarged, indented, and truncated. Toe disks smaller than largest finger disks. Hands unwebbed; feet only basally webbed.
- Dorsum mottled brown; limbs with transverse bands.
- Posterior thigh surfaces dark brown with distinct cream to yellow spots that occasionally fuse to form vertical stripes.
- Ventral surfaces and undersides of limbs yellow; throat heavily mottled with brown pigment, except for a midgular light stripe.

Similar species

C. fitzingeri (p. 246)

Very similar in overall appearance but has more toe webbing and slightly differently shaped disks on its fingers.

C. andi (p. 229)

Is a much larger species with more extensive toe webbing.

Craugastor escoces
Red-bellied Streamside Frog

Extinct

A member of the *rugulosus*-species group. These frogs are extremely difficult to identify under field conditions because they often differ only in very subtle ways. Geographic provenance, the presence of nuptial patches in adult males, the relatively uniformly colored smooth dorsum, and its red ventral surfaces may help in identifying this species. Males to 1.81 in (46 mm) in standard length, females to 2.83 in (72 mm).

Ecoregions
1 2 3 4 5

Craugastor escoces is endemic to Costa Rica. It is known from the slopes of Barva, Irazú, and Turrialba Volcanoes, in the Central Mountain Range; 3,600–6,890 ft (1,100–2,100 m).

Natural history

C. escoces is nocturnal and terrestrial. It occurs in pristine forests, where it frequents the banks of cascading mountain streams. This species breeds through direct development; the larvae lack a free-swimming tadpole stage and complete their development inside the egg. The eggs are laid in a moist site on the ground.

Like other members of the *rugulosus*-species group that occur at middle elevations, this species was once common throughout its range; however no live *C. escoces* have been seen since the mid-1980s, in spite of repeated searches. It has disappeared from several undisturbed sites within Braulio Carillo National Park and other well-protected areas, thus habitat alteration or degradation is clearly not the main reason for its decline. Global warming, airborne pollution, and chytridiomycosis are all possible factors in the demise of this species, which may now be extinct. Additional surveys of known sites are urgently needed to assess whether remnant populations of *C. escoces* still survive.

A distinct supratympanic fold follows the upper curve of the distinctly noticeable eardrum.

Dorsum uniform brown to dark olive, sometimes with a few diffuse dark spots, but generally patternless.

Posterior thigh surfaces dark brown; if present, light spots are densely suffused with dark pigment and appear indistinct.

Snout moderately long. Oval in dorsal profile and rounded in lateral profile; the canthus rostralis is rounded.

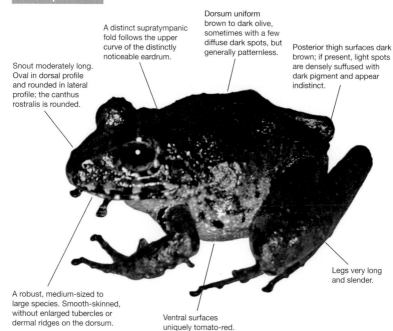

Legs very long and slender.

A robust, medium-sized to large species. Smooth-skinned, without enlarged tubercles or dermal ridges on the dorsum.

Ventral surfaces uniquely tomato-red.

Females substantially larger than males. Adult males have nuptial pads and paired vocal slits.

Juveniles tend to have a uniform pale yellow venter, which changes to bright red by the time sexual maturity is reached.

Finger I is longer than finger II. All fingers bear well-developed lateral keels and moderate-sized rounded disks. Toe tips slightly or moderately expanded. Feet only basally webbed; the unwebbed sections of the toes bear weak lateral keels.

C. angelicus (p. 231)

Is a closely related, similar species, but it normally lacks a red venter. It also has smaller finger disks that are only slightly expanded.

C. fleischmanni (p. 249)

Lacks a red venter and its finger and toe disks are barely expanded.

C. ranoides (p. 274)

Has pale yellow undersurfaces and a pattern of light spots on its dark posterior thigh surfaces.

Craugastor fitzingeri
Common Rain Frog

Least concern

A member of the *fitzingeri*-species group, characterized by the presence of greatly enlarged truncated disks on the two outer fingers of each hand; the two inner fingers bear only slightly expanded rounded disks. This is one of the more commonly observed frogs in the country, but one that often puzzles observers due to its extreme variability. It has rugose, warty skin, dark brown posterior thigh surfaces marked with cream or yellow spots, and a light stripe on the center of its otherwise mottled throat. Males to 1.38 in (35 mm) in standard length, females to 2.32 in (59 mm).

Ecoregions
1 2 3 4 5

Craugastor fitzingeri is abundant and widespread. It occurs from northeastern Honduras on the Atlantic slope, and northwestern Costa Rica on the Pacific slope, to northwestern Colombia; from near sea level to 4,985 ft (1,520 m).

Common throughout its range—and abundant near human settlements—this species is among the frogs in Costa Rica that are most commonly seen by the casual observer. It prefers edges of clearings and other open areas such as the banks of roads and rivers, ideally in areas surrounded by humid forest. During the day, it retreats into the leaf litter in shaded parts of the forest floor; at night, these frogs perch on low vegetation, or sit on exposed logs or rocks in or near streams, rivers, or road cuts.

When approached, *C. fitzingeri* remains still until the last second, then makes a rapid series of long jumps, escaping into the safety of a brushy tangle or pile of leaf litter. Along water courses, it may dive from its perch into the water and hide on the bottom. These frogs often fall prey to common frog-eating snakes, including young fer-de-lance (*Bothrops asper*), cat-eyed snakes (genus *Leptodeira*), and blunt-headed vine snakes (e.g., *Imantodes inornatus*).

Male *C. fitzingeri* make a clucking call that resembles the sound of two pebbles rapidly being tapped together; you can produce a fair imitation of the call by using your tongue to make a clucking call (which might elicit a response). The call is most frequently heard at dusk, the early evening, and also during—and right after—strong afternoon rains. The call serves to assert a male's territorial claims and to attract females. Even when calling, the males are exceedingly difficult to locate since they are usually concealed in leaf litter or in dense clumps of vegetation. They call sporadically, perhaps to avoid giving predators a means of tracking them. At least one predator, the fringe-lipped bat (*Trachops cirrhosus*), is known to locate these frogs based on their calls. And indeed when they are perched at night on an exposed spot, they rarely call.

C. fitzingeri employs direct development: the larvae skip a free-swimming tadpole stage and complete their metamorphosis inside the egg; they hatch as miniature froglets. Clutch sizes of direct developing frogs tend to be smaller than those depositing their eggs in the water, and *C. fitzingeri* broods have been found to range from 39 to at least 85 eggs (usually 65–70). It places its relatively large eggs in a moist terrestrial site away from standing water; generally, the eggs are deposited in a shallow burrow underneath a rock or log, in holes in tree buttresses, or in a nest depression below the leaf litter. Some species of *Craugastor* exhibit parental care and tend to their eggs, and adult female *C. fitzingeri* have been observed sitting on egg clutches until they hatch. It is not known whether they hydrate the eggs (gastric brooding) or exhibit other forms of parental care other than guarding their offspring from predators and parasites. In one study, attending females were observed with egg clutches throughout the year, both during dry and wet periods. This suggests that *C. fitzingeri* has a prolonged breeding season. Hatchlings of this species measure on average about 0.28 in (7 mm) and tend to stay near their nest for the first day of their life, presumably to resorb the remaining yolk sac before dispersing into the habitat.

C. fitzingeri is ubiquitous in many lowland settings. It is more common in disturbed areas than it is in primary forests and is often heard or seen in gardens, plantations, and farms. Although it does well in disturbed habitat, this does not mean that it is impervious to wholesale loss of habitat or pollution caused by agrochemicals. Individual *C. fitzingeri*

have tested positive for the chytrid fungal pathogen *Bd.* Sometimes these animals do not show any obvious detrimental effects of this infection, but occasional dead or sick frogs that display signs of chytridiomycosis are encountered. More research is needed to better understand the dynamics of disease transmission in tropical amphibian communities and the environmental factors that may suppress potentially devastating outbreaks of these dangerous pathogens and the diseases they cause.

Description

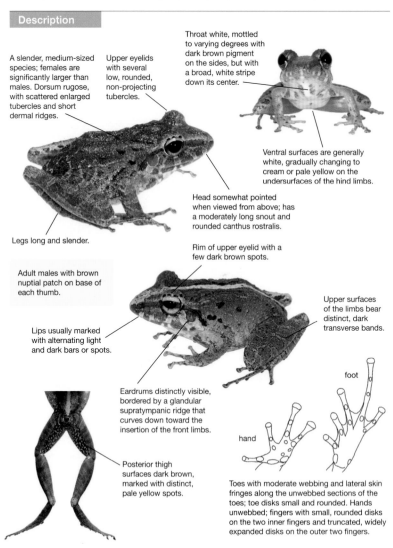

Throat white, mottled to varying degrees with dark brown pigment on the sides, but with a broad, white stripe down its center.

A slender, medium-sized species; females are significantly larger than males. Dorsum rugose, with scattered enlarged tubercles and short dermal ridges.

Upper eyelids with several low, rounded, non-projecting tubercles.

Ventral surfaces are generally white, gradually changing to cream or pale yellow on the undersurfaces of the hind limbs.

Head somewhat pointed when viewed from above; has a moderately long snout and rounded canthus rostralis.

Legs long and slender.

Rim of upper eyelid with a few dark brown spots.

Adult males with brown nuptial patch on base of each thumb.

Upper surfaces of the limbs bear distinct, dark transverse bands.

Lips usually marked with alternating light and dark bars or spots.

Eardrums distinctly visible, bordered by a glandular supratympanic ridge that curves down toward the insertion of the front limbs.

foot

hand

Posterior thigh surfaces dark brown, marked with distinct, pale yellow spots.

Toes with moderate webbing and lateral skin fringes along the unwebbed sections of the toes; toe disks small and rounded. Hands unwebbed; fingers with small, rounded disks on the two inner fingers and truncated, widely expanded disks on the outer two fingers.

male

female

The size of the tympanum is sexually dimorphic: in males, approaches diameter of eye; in females, about two-thirds the diameter of the eye.

As these four photographs show, the dorsum is extremely variable in coloration. Usually consists of shades of gray, beige, or brown, often marked with a broad yellow or orange middorsal stripe, or with a dark spot either in the shape of an hourglass or a W.

Similar species

C. crassidigitus
(p. 240)

C. talamancae
(p. 284)

C. cuaquero
(p. 243)

C. andi
(p. 229)

C. ranoides
(p. 274)

NO PHOTO

Differs most notably in having uniform orange to brown posterior thigh surfaces, smoother dorsal skin, and more extensive webbing between its toes.

Has uniform brown or red posterior thigh surfaces, and usually a white lip stripe that extends onto the body.

An exceedingly rare species with a limited distribution range. It has more yellow underneath its hind limbs and the light spots on its posterior thigh surfaces are often fused to form irregular vertical bars.

A very rare species that is larger and has relatively smooth skin. Its finger disks are indented and the light spots on its posterior thigh surfaces are often fused to form irregular vertical bars.

Lacks greatly expanded, truncated disks on the two outer fingers.

Craugastor fleischmanni
Fleischmann's Streamside Frog
Critically endangered

A member of the *rugulosus*-species group. These frogs are extremely difficult to identify under field conditions as they often differ only in very subtle ways. Geographic provenance, the presence of nuptial patches in adult males, the barely expanded finger and toe disks, and presence of yellow ventral surfaces may help in identifying this species. Males to 1.77 in (45 mm) in standard length, females to 2.95 in (75 mm).

Ecoregions
1 2 3 4 5

Craugastor fleischmanni is a Costa Rican endemic. Historically, it was widespread in the Central Mountain Range, throughout the Central Valley, on the Pacific slopes of Poás and Barva Volcanoes, and on the Atlantic slopes of Irazú and Turrialba Volcanoes, as well as the northern slopes of the Talamanca Mountain Range; 3,445–7,500 ft (1,050–2,286 m). The only recent (2010) record for this species is from the headwaters of the Río Ciruelas, on Poás Volcano.

Natural history

C. fleischmanni is a nocturnal species. It occurs in middle and high elevation rainforests, where it inhabits stream corridors, although it has been found considerable distances from water. It forages for invertebrate prey in or near streambeds. Like other species in its genus it reproduces through direct development; eggs are deposited in a moist terrestrial site, where their larvae develop entirely inside the egg. The larvae hatch as small froglets, skipping a free-swimming tadpole stage.

Somewhat tolerant of habitat alteration, *C. fleischmanni* has suffered a decline in the Central Valley, both from significant loss of habitat as a result of urbanization and from the effects of pollution. However, until a dramatic population decline occurred in the mid- to late 1980s, it was still commonly observed in moderately disturbed habitats, as well as in several pristine, well-protected areas. In spite of repeated searches in known sites, *C. fleischmanni* was not found from 1987 to 2010, when a single adult individual was sighted on Poás Volcano. No new observations have been made since then, but biologists hold out hope that *C. fleischmanni* is not extinct.

Additional surveys are urgently needed to assess the status of the population on Poás Volcano, to search for survivors in other areas, and to better understand the forces behind the collapse of this species in the 1980s and 1990s. Chytrid fungal infections have been confirmed within the range of this species; other likely causes are climatic or environmental changes.

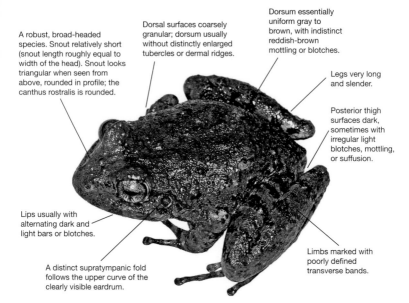

A robust, broad-headed species. Snout relatively short (snout length roughly equal to width of the head). Snout looks triangular when seen from above, rounded in profile; the canthus rostralis is rounded.

Dorsal surfaces coarsely granular; dorsum usually without distinctly enlarged tubercles or dermal ridges.

Dorsum essentially uniform gray to brown, with indistinct reddish-brown mottling or blotches.

Legs very long and slender.

Posterior thigh surfaces dark, sometimes with irregular light blotches, mottling, or suffusion.

Lips usually with alternating dark and light bars or blotches.

A distinct supratympanic fold follows the upper curve of the clearly visible eardrum.

Limbs marked with poorly defined transverse bands.

Venter bright yellow; may become suffused with red in large adults.

Finger I is longer than finger II. Fingers with only barely expanded, rounded disks and weakly indicated lateral keels. Toe tips barely expanded. Feet only basally webbed, while the unwebbed sections of the toes bear weak lateral ridges.

Adult males have white nuptial patches on the base of their thumbs, but lack vocal slits.

C. angelicus (p. 231)

Has slightly expanded finger and toe disks (about 1.5x the width of the digits) and a yellow-orange to orange venter.

C. escoces (p. 244)

Has distinctly expanded finger and toe disks and a red venter (pale yellow in juveniles).

C. ranoides (p. 274)

Has only slightly expanded finger and toe disks and pale yellow to yellow ventral surfaces; it generally occurs at lower elevations.

Craugastor gollmeri
Red-eyed Masked Litter Frog   **Least concern**

A member of the *gollmeri*-species group. Characterized by its smooth skin, long snout and legs, a distinct dark eye mask that extends onto the flanks, and coloration that mimics that of dead leaves. *Craugastor gollmeri* is distinguished from the two other Costa Rican species in this group by: disks on the fingers and toes that are only slightly enlarged, basal webbing between the toes, enlarged heel tubercles, and a dark red iris. Males to 1.46 in (37 mm) in standard length, females to 2.13 in (54 mm).

Ecoregions
1 2 3 4 5

On the Atlantic slope, *Craugastor gollmeri* occurs from northern Costa Rica to eastern Panama; from near sea level to 2,790 ft (850 m). It has also been found on the Pacific slopes of several volcanoes in Guanacaste Province; up to 4,985 ft (1,520 m).

Natural history

This diurnal frog is most active at dawn and dusk. Though relatively common in appropriate habitat, *C. gollmeri* is easily overlooked because of its cryptic coloration and habits. It occurs in the leaf litter of lowland and foothill forests. Individuals are usually seen when they are inadvertently spooked by hikers; they tend to jump just as a foot is moving down on them, making long leaps—in a zigzag pattern—to escape, then suddenly freezing, becoming nearly invisible amid the dead leaves.

Not much is known about the biology of these beautiful frogs. Because they do not have a vocal sac, they were long thought to be mute, but recent observations indicate that they are capable of producing a short, croaking call. Significant sexual size difference suggests that this species engages in amplexus prior to breeding. Their eggs are likely deposited in the leaf litter, where they undergo direct development and develop into tiny froglets.

The population density of *C. gollmeri* tends to fluctuate with the depth of the leaf litter layer. Numbers often increase as the leaf litter layer becomes thicker and provides more hiding places, more suitable microclimates for reproduction, and likely a higher population density of invertebrate prey. Therefore, these frogs are most often seen during drier periods, when trees drop their leaves with greater frequency.

C. gollmeri is fairly tolerant of habitat alteration and is common in regenerating secondary forests and shade-grown plantations that often have a thick leaf litter layer. Apart from the outright removal of tree cover from its habitat, no other specific threats to this species are known.

In areas where *C. gollmeri* co-occurs with the related species *C. mimus*, it tends to occupy higher elevations ranges, while the latter species favors lower elevations.

When sitting motionless in leaf litter, *C. gollmeri* is easily overlooked.

251

Upper half of the iris deep red.

Eardrums distinctly visible. On males, eardrums approach the size of the eye; on females, eardrums are about half the size of those of males.

Has a long, pointed snout.

Dorsal surfaces mostly smooth, but scattered low tubercles are present on the head, body, and limbs.

Dark eye mask extends onto the body.

male

Ventral surfaces uniform white.

Anterior thigh surfaces marked with a dark brown stripe that extends along the margin of the lower limb.

Often a pair of circular dorsal blotches is present on the middle of dorsum.

A few enlarged tubercles usually located above the shoulder blades and near the tympanum.

female

Generally tan, gray, burnt-orange, or brown, with a cream or pale yellow middorsal pinstripe.

Heel marked with one or two enlarged tubercles.

A short, inconspicuous glandular ridge is present behind each eye, usually curving downward over top of eardrums.

Posterior thigh surfaces uniform rust-brown.

foot

Upper surfaces of the limbs marked with distinct transverse bars.

hand

Medium-sized and slender, with long, slender legs.

juvenile

Very small juveniles lack the red eyes and have a gold to bronze iris instead.

Disks on the digits only marginally expanded. Hands unwebbed; feet only have basal webbing.

C. mimus (p. 259)

Similar in general appearance but has a bronze or copper iris, has more toe webbing, and lacks enlarged heel tubercles.

C. noblei (p. 261)

Has distinctly enlarged disks on the two outer fingers, a pair of glandular dorsolateral skin folds, and lacks enlarged heel tubercles.

C. talamancae (p. 284)

Its dark eye mask is generally bordered below by a white line that extends along the upper lip and onto the body.

252

Craugastor gulosus
Montane Broad-headed Litter Frog

Endangered

A member of the *biporcatus*-species group. Characterized by the very broad head and unwebbed hands and feet. Though large and toadlike, this species has smooth skin and cranial crests; it lacks distinct ridges on its dorsum. It never has parotoid glands in the neck region (a trait of true toads). Males reach a standard length of 2.36 in (60 mm), females are much larger and can grow to 4.06 in (103 mm).

Ecoregions
1 2 3 4 5

Craugastor gulosus occurs in the Talamanca Mountain Range, in extreme southeastern Costa Rica and adjacent northwestern Panama; 2,590–6,150 ft (790–1,873 m).

Natural history

This rare species has been observed just a handful of times in Costa Rica; it is currently only known from a small mountainous region near the Panamanian border. It occurs at middle and high elevations in humid pristine forests. Like other broad-headed litter frogs in its genus, this species is a nocturnal sit-and-wait predator that spends most of its time sitting in or near the entrance of its burrow, waiting for suitable prey to pass within striking distance; it likely feeds on relatively large prey. Although no observational data exists, this species is presumed to reproduce by direct development.

C. gulosus has not been observed for many years in Costa Rica, but relatively recent sightings of this species in neighboring Panama (from La Amistad International Park) suggest that it likely still survives in Costa Rica as well. The secretive habits of this cryptic leaf litter frog, along with the inaccessibility of its preferred habitat, perhaps explain its perceived rarity.

Well-developed cranial crests present on top of head, but it lacks distinct ridges on the dorsum (typically seen in related species *C. megacephalus* [p. 255] and *C. rugosus* [p. 280]).

Has a pronounced supratympanic skin fold behind each eye that curves over the top of eardrum toward the corner of the mouth. Usually bordered below by dark markings that cover the small tympanum.

Iris dark brown to nearly black; pupil horizontally elliptical.

Dorsum generally smooth, but on adults it sometimes bears scattered low warts.

Generally a dark irregular spot present below each eye.

In adults, dorsum reddish-brown, grayish-brown, or nearly black. Males usually have discrete dark spots on the dorsum; females tend to be uniformly colored.

Upper lip shows a dirty white to cream lip stripe, usually marked with dark blotches.

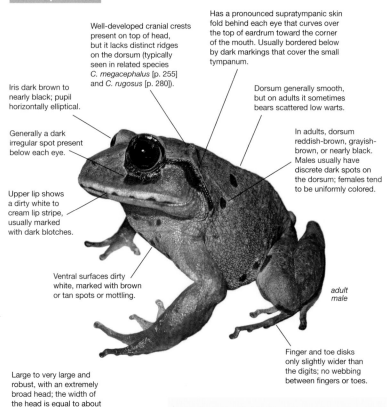

Ventral surfaces dirty white, marked with brown or tan spots or mottling.

adult male

Large to very large and robust, with an extremely broad head; the width of the head is equal to about half of the standard length!

Finger and toe disks only slightly wider than the digits; no webbing between fingers or toes.

Juveniles reportedly differ from adults in coloration and skin texture; descriptions of their appearance mention that the dorsal skin is covered with many small, white-tipped pustules, and that it is mottled light brown in coloration.

Similar species

C. gulosus is toadlike in overall appearance but it lacks the paired parotoid glands seen in true toads of the genera *Incilius*, *Rhaebo*, and *Rhinella*.

C. megacephalus (p. 255)

Similar in morphology but has distinct ridges on its dorsum.

Strabomantis bufoniformis (p. 511)

Has rugose skin, distinct tubercles on the upper eyelids, and webbing between its toes.

Craugastor megacephalus
Atlantic Broad-headed Litter Frog

Least concern

A member of the *biporcatus*-species group. Characterized by the very broad head and unwebbed hands and feet. Medium-sized to large. Toadlike, with a pair of distinct cranial crests, a series of hourglass-shaped ridges on the shoulder region, and a V-shaped ridge on the lower back. It lacks parotoid glands. Males 1.69 in (43 mm) in standard length, females to 2.91 in (74 mm).

Ecoregions

Craugastor megacephalus is widespread in the Atlantic lowlands and foothills from extreme southeastern Honduras to western Panama; from near sea level to 3,940 ft (1,200 m).

Natural history

C. megacephalus is a relatively common species. Adults are most often seen at night, as they perch in or near the entrance of their burrow or in the leaf litter. When discovered, they either freeze—relying on their superb camouflage—or flee rapidly with a series of powerful short hops and retreat into their burrow. Juveniles tend to be active during the day, when they can be spotted in deep leaf litter. Adults are only seen at night and presumably retreat into their burrow during daylight hours.

These frogs are sit-and-wait predators and hunt by ambushing passing prey. Large spiders seem to be a favorite prey, although millipedes, beetles, and other relatively large prey items have also been found in their stomach. The broad head and big mouth allow *C. megacephalus* to eat prey (including small frogs and lizards) that is not accessible to many frogs that are smaller and have narrower heads.

C. megacephalus remains common throughout its Costa Rican range. Although it prefers relatively undisturbed forests, it persists in areas with minor habitat alteration as long as sufficient canopy cover and a thick leaf litter layer persist. It is not dependent on bodies of water for its reproduction and breeds through direct development. Presumably egg clutches are deposited in an adult's burrow or in a moist microclimate within the leaf litter.

C. megacephalus was previously thought to be part of the wider ranging species *Eleutherodactylus biporcatus* (from which *C. rugosus* and several other forms were also split off). Given the morphological variation that exists in *C. megacephalus*, it is possible that this species still includes additional cryptic forms; more research is needed to adequately assess its taxonomy. As a result of these taxonomic changes, this species is referred to as *Eleutherodactylus megacephalus* or *Eleutherodactylus biporcatus* in some older literature.

The broad head and robust body of *C. megacephalus* give it a toadlike appearance, but it lacks parotoid glands and warty skin.

Dorsum with a series of hourglass-shaped ridges on shoulder region and a V-shaped ridge on the lower back.

Dorsal coloration generally reddish, gray, tan, or dark brown; the dorsal ridges and tubercles are often outlined with dark shading.

Head with paired cranial crests.

Posterior thigh surfaces dark brown to black, with tiny cream to yellow spots or mottling.

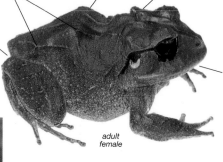

Robust, with a broad head (the width of the head is equal to about half of the standard length!). Females significantly larger than males. Males, however, have a much larger tympanum than females; it approaches the diameter of the eye in males and is about half that size in females.

adult female

In large adults, dorsum becomes relatively smooth; on juveniles, dorsum shows numerous scattered tubercles. Several pointed tubercles adorn top of eyelids; these are most pronounced in young individuals.

No vocal slits or vocal sac; presumably does not call.

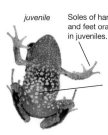

Iris gold to dark bronze, with fine black speckling and reticulations; pupil horizontally elliptical.

juvenile Soles of hands and feet orange in juveniles.

juvenile

Ventral color pattern distinct: venter is dark brown, with white polka dots or mottling; these white marks fade with age. Young individuals often also have yellow, orange, or red on the posterior sections of their venter and on the undersides of the hind limbs.

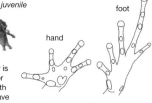

foot

hand

Digits slender with only slightly expanded tips; no webbing between fingers or toes.

Although somewhat toadlike in general appearance, *C. megacephalus* lacks the paired parotoid glands seen in true toads of the genera *Incilius*, *Rhaebo*, and *Rhinella*.

C. gulosus (p. 253)

Similar in morphology but is larger and lacks the dorsal ridge pattern.

Strabomantis bufoniformis (p. 511)

Lacks a dorsal ridge pattern and has very tuberculate dorsal skin. It also has some webbing between the toes.

Craugastor melanostictus
Banded-thighed Rain Frog

Least concern

A member of the *fitzingeri*-species group, characterized by the presence of greatly enlarged truncated disks on the two outer fingers of each hand; the two inner fingers bear less expanded, rounded disks. *Craugastor melanostictus* has a distinctly enlarged, projecting tubercle on each upper eyelid and lacks toe webbing. It also has a characteristic color pattern of alternating dark vertical bars and contrastingly colored (yellow, orange, red, pink, or green) interspaces on its posterior thigh surfaces. Males to 1.69 in (43 mm) in standard length, females to 2.20 in (56 mm).

Ecoregions
1 2 3 4 5

Craugastor melanostictus occurs in isolated populations in the foothills and mountains of Costa Rica and adjacent western Panama; 3,775–8,850 ft (1,150–2,700 m).

Natural history

C. melanostictus is uncommon, rarely seen, and poorly understood. Striking morphological variation exists between populations found at different elevations in the same mountain range, while large differences exist between isolated populations found at high elevations on different mountain ranges. Thus, it is very likely that several undescribed, cryptic species are included in the definition of *C. melanostictus*. Not enough information is currently available, however, to split this enigmatic species in any meaningful way.

This nocturnal, arboreal species occurs in humid forest habitats at middle and high elevations. At night, it has been observed perched in low vegetation, 3–13 ft (1–4 m) above the ground. During the day, it retreats into arboreal bromeliads and underneath the bark of standing trees, although individuals have also been found on the forest floor, under rocks and logs.

C. melanostictus primarily inhabits pristine cloud forests; it does not appear to tolerate changes to its habitat. And, because it occurs at middle and high elevations, it is likely susceptible to infection by the chytrid fungal pathogen *Bd*. Although population declines have been noted in a few areas, it is not frequently found even in areas with seemingly robust populations, and it is therefore difficult to evaluate how chytridiomycosis affects this species. Consequently, additional research is needed to better understand the geographic distribution, taxonomy, biology, and conservation status of *C. melanostictus*.

The rugose skin texture in some populations of *C. melanostictus*, combined with the presence of an enlarged tubercle on the upper eyelid, presence of enlarged heel tubercles, and absence of webbing between the toes can lead to possible confusion with highland populations of a distantly related species, *Pristimantis cruentus* (family Strabomantidae), p. 497. The latter species lacks the alternating dark and brightly colored bars on the posterior thigh surface and its outer toe (toe V) is much longer than toe III. Proper identification and reporting of sightings of these rarely seen montane species will ultimately lead to a better understanding of their distribution patterns and their biology.

Small juveniles tend to be green; they generally lack the greatly enlarged finger and toe disks of adults.

A medium-sized, fairly robust frog. Females larger than males. Dorsal skin smooth to finely rugose; in some populations, individuals have scattered tubercles and a few indistinct, low dermal ridges.

A pronounced, glandular supratympanic fold follows the outline of the eardrum and angles downward toward the insertion of the front limbs.

Upper eyelids with a prominently enlarged, protruding tubercle, surrounded by a series of small, low tubercles.

Snout fairly short; ovoid when seen from above. Canthus rostralis usually sharply angled.

Has a distinctly enlarged projecting tubercle on each heel; sometimes additional enlarged tubercles are present along trailing edge of tarsus.

Usually a pattern of dark and light bars present on each side of the head, as though dark bands were radiating out from the eye. Lips marked with light blotches between dark bars.

Dorsal coloration variable; generally shows shades of tan or brown, marked with reddish, orange, yellow, or green marbling and spots. A paler, broad dorsal field or thin light middorsal pinstripe may be present.

female

Tympanum large and round in males and small and oval in females.

hand

foot

Adult males with white nuptial patch on base of thumb.

Limbs distinctly banded with dark bars. Dark bands on thighs extend onto posterior thigh surfaces and form a distinctive pattern of dark brown to black vertical bars, alternating with brightly colored interspaces that can be yellow, orange, pink, purple, red, or green.

Finger I shorter than finger II. Fingers with very large disks, most prominently enlarged on fingers III and IV, and to a far lesser degree on finger II. Finger disks on outer two fingers truncated and indented. Toe disks enlarged but smaller than finger disks; some toe disks indented. Hands and feet unwebbed.

C. rayo (p. 276)

Has moderate webbing between its toes, lacks the unique color pattern on the posterior thigh surfaces, and does not have enlarged tubercles on the upper eyelid.

C. crassidigitus (p. 240)

Has uniform orange to brown posterior thigh surfaces and extensively webbed feet.

Pristimantis cruentus (p. 497)

Usually has mottled brown posterior thigh surfaces and a pattern of bright yellow to orange spots in the groin. Its toe III is much shorter than toe V.

Craugastor mimus
Masked Litter Frog

Least concern

A member of the *gollmeri*-species group. Characterized by its smooth skin, long snout and long legs, a distinct dark eye mask that extends onto the flanks, and coloration that mimics that of dead leaves. *Craugastor mimus* is distinguished from the two other Costa Rican species in this group by a bronze or copper iris, substantial webbing between the toes, and the absence of expanded toe and finger disks. Males to 1.46 in (37 mm) in standard length, females to 2.28 in (58 mm).

Ecoregions
1 2 3 4 5

Craugastor mimus occurs in the lowlands and foothills of the Atlantic slope, from eastern Honduras to southeastern Costa Rica; near sea level to 3,075 ft (940 m).

Natural history

C. mimus is a diurnal, leaf litter inhabitant. It often passes unnoticed due to its cryptic coloration and because it remains motionless when observed. If approached too closely, it escapes in long bounds into the leaf litter or dense vegetation. These frogs appear to be most active at dawn and dusk.

Because they are hard to find, little is known about their biology. Like other species in the *gollmeri*-species group, these frogs were long thought to be mute. In recent years, however, calls have been confirmed in *C. gollmeri* and *C. noblei*, thus it is likely that *C. mimus* will eventually be found to produce an advertisement call as well. Even though an extendable vocal sac is absent in this species, the presence of very large eardrums in males may indicate that vocalizations are an important form of infraspecific communication. Like other species in the genus, *C. mimus* likely reproduces through direct development, with its eggs being deposited in a moist microclimate in the leaf litter.

In terms of their size and habits, *C. mimus* and *C. gollmeri* are apparently very similar species whose distribution ranges overlap in some areas. Wherever this happens, *C. mimus* tends to inhabit lower elevation sites than does *C. gollmeri*. Nevertheless, it appears that the two species have very similar ecological requirements, which raises the question of how they avoid competition over scarce resources in the areas where they co-occur. In some places, *C. mimus* appears to favor wet,

undisturbed forests, while *C. gollmeri* tends to occur in drier habitats and in areas of regenerating secondary growth. Nevertheless, *C. mimus* has also been found in disturbed edge habits and may be tolerant of minor habitat alteration.

As a leaf litter inhabitant, this species probably shows population increases and declines in response to changes in the depth of the leaf litter. This phenomenon is generally driven by climatic changes and seasonality; during dry periods, more trees will drop their leaves and add to the forest floor's substrate, thus creating more room in which frogs and their prey can hide, breed, and feed.

Like all frogs in the *gollmeri*-species group, *C. mimus* has smooth skin, a long snout, long legs, and a distinct, dark eye mask that extends onto the flanks.

259

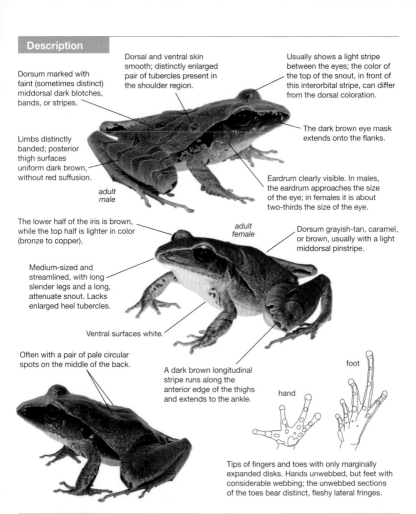

Dorsum marked with faint (sometimes distinct) middorsal dark blotches, bands, or stripes.

Dorsal and ventral skin smooth; distinctly enlarged pair of tubercles present in the shoulder region.

Usually shows a light stripe between the eyes; the color of the top of the snout, in front of this interorbital stripe, can differ from the dorsal coloration.

Limbs distinctly banded; posterior thigh surfaces uniform dark brown, without red suffusion.

adult male

The dark brown eye mask extends onto the flanks.

Eardrum clearly visible. In males, the eardrum approaches the size of the eye; in females it is about two-thirds the size of the eye.

The lower half of the iris is brown, while the top half is lighter in color (bronze to copper).

adult female

Dorsum grayish-tan, caramel, or brown, usually with a light middorsal pinstripe.

Medium-sized and streamlined, with long slender legs and a long, attenuate snout. Lacks enlarged heel tubercles.

Ventral surfaces white.

Often with a pair of pale circular spots on the middle of the back.

A dark brown longitudinal stripe runs along the anterior edge of the thighs and extends to the ankle.

hand

foot

Tips of fingers and toes with only marginally expanded disks. Hands unwebbed, but feet with considerable webbing; the unwebbed sections of the toes bear distinct, fleshy lateral fringes.

C. gollmeri (p. 251)

Has a deep red iris, basal toe webbing only, and enlarged heel tubercles.

C. noblei (p. 261)

Has distinctly enlarged disks on the two outer fingers and a pair of glandular dorsolateral skin folds.

C. talamancae (p. 284)

Resembles members of the *gollmeri*-species group and also has a dark eye mask, but mask is bordered below by a white line that extends along the upper lip and onto the body.

Craugastor noblei
Noble's Masked Litter Frog

Least concern

A member of the *gollmeri*-species group. Characterized by its smooth skin, long snout and legs, a distinct dark eye mask that extends onto the flanks, and coloration that mimics that of dead leaves. *Craugastor noblei* is distinguished from the two other Costa Rican species in this group by a yellowish-gold iris, greatly enlarged disks on its two outer fingers, a short dorsolateral skin fold behind each eye, and basal webbing between its toes. Males to 1.89 in (48 mm) in standard length, females to 2.60 in (66 mm).

Ecoregions
1 2 3 4 5

On the Atlantic slope, *Craugastor noblei* occurs from northeastern Honduras to western Panama, and on the Pacific slope from central Costa Rica to west-central Panama; near sea level to 3,935 ft (1,200 m).

Natural history

This species inhabits the floor of humid forests, in lowlands and foothills. Although it is diurnal, individuals have been encountered at night sitting on thin twigs in low bushes, up to 2 m above the ground. Their cryptic coloration enables them to blend in perfectly with the dead leaves on which they sit. When threatened, they rely on their camouflage and remain motionless, pressing their body close against the ground. However, when approached too closely, they launch themselves at the last second and, with their powerful legs, bound to safety in the underbrush or dive into the leaf litter layer in a zigzag pattern of long leaps.

Like other species in its genus, *C. noblei* presumably deposits its eggs in moist microsites within the leaf litter, and its larvae undergo direct development inside the eggs, hatching as tiny froglets. Lacking a vocal sac and vocal slits, this species was long thought to be mute. However, recent observations indicate that it produces a high-pitched, two note call, primarily in late afternoon and only at the height of the rainy season (November–December). Males call individually or form choruses of up to five individuals; they usually call from hidden locations in the leaf litter, inside burrows, or beneath fallen logs. The confined spaces from which they call might function as a resonating chamber, serving

to amplify their song in the absence of a vocal sac. It is not known whether the vocalizations of *C. noblei* serve to defend a territory, attract a mate, or both, but the presence of very large eardrums in the males of this species hints at the importance of sounds as a means of communicating.

Though relatively uncommon, *C. noblei* is widely distributed throughout Costa Rica. It tends to occur in mostly undisturbed lowland and foothill forests, but also tolerates edge situations and minor habitat alteration.

The characteristic eye mask, seen in all species in the *gollmeri*-species group, is not always obvious in some color forms of *C. noblei*.

Snout long, triangular, and pointed when seen from above.

Dorsum often with a dark brown hourglass-shaped spot (sometimes outlined by a light border).

Cryptically colored; generally a shade of tan, bronze, or brown, with a thin light middorsal line.

Backs of the thighs characteristically uniform brown, suffused with red to scarlet.

Lacks enlarged heel tubercles.

Dorsum shows weak granulation and a pair of glandular dorsolateral ridges that extend from behind the eye towards the groin.

Side of head marked with a dark brown masklike eye stripe that extends well onto the body.

male

Eardrum distinctly visible. In males, the eardrum approaches the size of the eye; in females, it is about half that size.

Eyes large, with a golden iris and a horizontally elliptical pupil.

juvenile

Ventral surfaces smooth, uniform white or cream, but may be marked with dark pigment on the underside of the limbs and throat.

foot

hand

C. noblei is quite variable in its coloration. Some color forms have a boldly lined pattern, with a series of dark stripes on the limbs, dorsum, and flanks.

Hands unwebbed; feet with basal webbing. Disks on all toes. Some fingers only slightly expanded; only those on the two outer fingers are distinctly enlarged.

C. gollmeri (p. 251)

Has a deep red iris, basal toe webbing only, and lacks distinctly enlarged disks on the two outer fingers.

C. mimus (p. 259)

Similar in appearance but has a bronze or copper iris, has more toe webbing, and lacks the enlarged disks on the two outer fingers.

C. talamancae (p. 284)

Has a dark eye mask that is bordered below by a white line that extends along the upper lip and onto the body.

Craugastor obesus
Atlantic Torrent Frog **Endangered**

A member of the *rugulosus*-species group. These frogs are extremely difficult to identify under field conditions as they often differ only in very subtle ways. *Craugastor obesus* has rugose dorsal skin covered with tubercles and glandular ridges; moderately enlarged, rounded finger and toe disks; extensively webbed feet; and skin fringes along the unwebbed portions of its toes. Males to 1.93 in (49 mm) in standard length, females 3.39 in (86 mm).

Ecoregions
1 2 3 4 5

Craugastor obesus occurs in southeastern Costa Rica and northwestern Panama, on the Atlantic slope of the Talamanca Mountain Range; 1,310–4,750 ft (400–1,450 m).

Natural history

In spite of the tree-frog-like disks on its hands and feet, this is a terrestrial frog found in or near steeply cascading mountain streams. *C. obesus* inhabits stream banks and streambeds, where it lives under and between moss-covered rocks. It is nocturnal and has been observed at night as it perches on rocks in the middle of moderate-sized streams or in the spray zone of small waterfalls. Although its distribution is still poorly understood, to date this species has only been reported from undisturbed medium-sized streams in foothill forests.

C. obesus is a rare species. In Costa Rica, it has been found in just two areas in the Río Lari watershed (Limón Province), where it was last seen in 1984. Recent searches have uncovered no new individuals. It also occurs in Panama, including in several protected areas, but populations have also disappeared there. Loss of suitable habitat due to development and agriculture may be a factor in unprotected areas, but likely chytrid fungal infection played a role in the disappearance of this species. Most riparian *Craugastor* in the *rugulosus*-species group have declined dramatically or have disappeared entirely, and the population status of *C. obesus* is uncertain. Additional efforts are needed to locate surviving individuals of this species.

C. obesus was previously included in what was once considered a wide-ranging form, *Eleutherodactylus punctariolus*, and until recently it was listed under that name. Due to the very subtle morphological differences between the species in this complex, direct comparison is needed to idenitify individual animals. Because this is not feasible in the field, geographic provenance can be a useful aid in establishing this species' identity.

Adult males have a nuptial patch on the base of each thumb; they also have paired vocal slits.

Robust, medium-sized to large. Females significantly larger than males.

Dorsum rugose, with scattered tubercles and short glandular ridges.

Snout short and rounded in profile; has steeply sloped sides and a sharply edged canthus rostralis.

Dorsum generally orange-brown to dark olive-brown, mottled with even darker blotches.

Hidden, posterior surfaces of thighs black with many yellow spots.

Iris gold, with delicate dark reticulations; pupil horizontally elliptical.

Tympanum small but clearly visible; bordered above by a distinct supratympanic ridge that looks swollen and tuberculate.

Dark bands on the hind limbs and dark markings on the sides of the head are not always easy to detect.

Ventral surfaces usually immaculate pale yellow, although throat and chest sometimes moderately to heavily suffused with dark pigment.

Finger I longer than finger II. All fingers bear enlarged, rounded to bluntly truncated disks and distinct lateral skin fringes. The toes are extensively webbed and bear medium-sized round disks. Distinct lateral skin fringes adorn the unwebbed portions of the toes.

C. ranoides (p. 274)

Similar in overall morphology and biology but has only minimal webbing between its toes.

Craugastor persimilis
Similar Litter Frog

Vulnerable

A member of the *rhodopis*-species group. These tiny, stocky leaf litter frogs are characterized by their short limbs and the absence of both interdigital webbing and enlarged finger and toe disks. All species in this group are extremely variable in coloration, pattern, and skin texture and can be almost impossible to identify in the field. *Craugastor persimilis* differs from related species in that finger I is shorter than finger II. It often has a dark, downward curving mark on the eardrum. Adult males to 0.71 in (18 mm) in standard length, females to 0.91 in (23 mm).

Ecoregions
1 2 3 4 5

Craugastor persimilis is endemic to Costa Rica. It occurs locally in central and southeastern Costa Rica, on the Atlantic slope; 125–3,935 ft (39–1,200 m).

Natural history

This is a small, diurnal leaf litter frog. It was previously included in *C. bransfordi*; confusion over taxonomy, combined with the difficulty of identifying *C. persimilis* under field conditions, makes it difficult to obtain biological information about this species. However, it likely has a very similar biology to other species in the *C. rhodopis* group and very likely breeds through direct development.

As with other leaf litter species, the population levels fluctuate with the depth of the forest floor substrate. During dry spells, when trees tend to drop their leaves, the leaf litter layer increases in depth and provides additional hiding spaces and food resources for these frogs, causing their populations to increase. Conversely, during prolonged wet periods, their numbers may go down dramatically. This species appears stable and is still commonly encountered in low and middle elevation rainforests; it also occurs in plantations and other altered habitats with sufficient tree cover and leaf litter.

C. persimilis is frustratingly variable in its coloration and is easily confused with other members of the *rhodopis*-species group.

Skin texture variable but generally either relatively smooth or irregularly rugose; dorsum usually lacks distinctly enlarged tubercles, although a pattern of low, finely textured dorsal ridges may be present.

Coloration varies but is always cryptic; often relatively uniform tan to mottled brown or gray. If patterned, usually marked with irregular blotches or a light middorsal pinstripe.

adult male

Upper lip marked with alternating dark and light bars or spots.

Arms and legs relatively short and stocky.

A short, diagonal row of white, cream, or pinkish tubercles is generally present between the bottom of the eardrum and the insertion of the front limbs.

Soles of hands and feet with distinctly projecting, but relatively rounded, tubercles.

Often has a downward-curving, dark mark on the upper half of the eardrum. A similarly dark spot in front of each eye may create the appearance of a dark eye mask.

adult female

Snout longer than wide; elliptical when seen from above.

In males, eardrum roughly equal in size to the eye; in females, the eardrum is distinctly smaller. Adult males lack a nuptial patch on the base of each thumb.

Venter and undersides of limbs cream or dirty white; chest and throat marked with dark pigment.

foot

hand

Innermost finger (finger I) shorter than finger II. Fingers and toes without webbing. Tips of fingers and toes only marginally expanded; disks on finger and toe tips sometimes pointed.

C. bransfordii (p. 233)

Fingers I and II are about the same length. Has projecting, pointed tubercles on the soles of the hands and feet, and adult males have a white nuptial patch at the base of the thumb.

C. underwoodii (p. 289)

Fingers I and II are about the same length. Has low, rounded tubercles on the soles of the hands and feet, and adult males have a white nuptial patch at the base of the thumb.

C. polyptychus (p. 272)

A slightly larger species. Fingers I and II are of roughly equal length.

Craugastor phasma
Ghost-like Rain Frog

Data deficient

A member of the *fitzingeri*-species group. Characterized by the presence of greatly enlarged truncated disks on the two outer fingers of each hand, while the two inner fingers bear only slightly expanded, rounded disks. This is the only *Craugastor* that is grayish-white. It is long-legged, has only basal webbing between its toes, and has indented disks on some of its fingers and toes. Only known from a single female, which measured 1.89 in (48 mm) in standard length; males unknown, but presumably smaller.

Ecoregions
1 2 3 4 5

Craugastor phasma is a Costa Rican endemic. It is known only from its type locality on the Río Cotón, in the Las Tablas Protected Zone, located in the southern Talamanca Mountain Range, Puntarenas Province; 6,075 ft (1,850 m).

Natural history

A single individual of *C. phasma* was found on the rocky banks of the Río Cotón, in a middle elevation rainforest, near the Panama border. It was discovered during a rainy afternoon in an area that had been regularly surveyed over a two-year period. In the 1990s, that area was severely affected by the fungal disease chytridiomycosis; the frog was found in association with dead and dying frogs and was thought to be sick. No additional specimens of *C. phasma* have ever been found. Although this enigmatic species occurs within the protected La Amistad Biosphere Reserve, disease may have taken a toll.

Due to the lack of additional comparative material, it is difficult to adequately assess the taxonomic placement of this species. Based on its morphology it is tentatively assigned to the *Craugastor fitzingeri*-species group, but more study is needed to evaluate its phylogeny and population status.

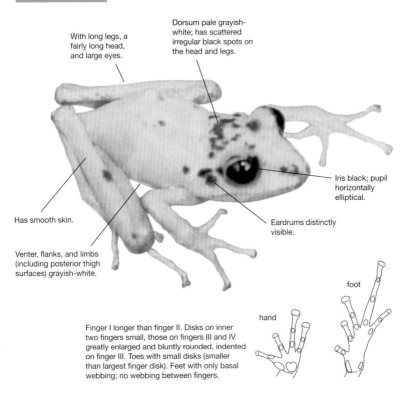

With long legs, a fairly long head, and large eyes.

Dorsum pale grayish-white; has scattered irregular black spots on the head and legs.

Iris black; pupil horizontally elliptical.

Has smooth skin.

Eardrums distinctly visible.

Venter, flanks, and limbs (including posterior thigh surfaces) grayish-white.

foot

hand

Finger I longer than finger II. Disks on inner two fingers small, those on fingers III and IV greatly enlarged and bluntly rounded, indented on finger III. Toes with small disks (smaller than largest finger disk). Feet with only basal webbing; no webbing between fingers.

Similar species

No other *Craugastor* has a similar color pattern. In addition to its coloration, it differs from *C. crassidigitus* (p. 240) and *C. fitzingeri* (p. 246) in having longer legs and less toe webbing.

Craugastor podiciferus
Highland Litter Frog   **Near threatened**

A member of the *rhodopis*-species group. These tiny, stocky leaf litter frogs are characterized by their short limbs and the absence of both interdigital webbing and enlarged finger and toe disks. All species in this group are extremely variable in coloration, pattern, and skin texture and can be almost impossible to identify in the field. *Craugastor podiciferus* differs from other members in the group in having relatively smooth skin, often a series of dorsal and dorsolateral ridges, a dark eye mask, and a series of one to three tubercles on each heel. Males to 1.10 in (28 mm) in standard length, females to 1.58 in (40 mm).

Ecoregions
1 2 **3** 4 5

In Costa Rica, *Craugastor podiciferus* occurs in isolated populations along both slopes of the Tilarán, Central, and Talamanca Mountain Ranges; it also occurs in adjacent western Panama; 3,575–8,695 ft (1,090–2,650 m).

Natural history

C. podiciferus is a leaf litter inhabitant of middle and high elevation rainforests. It generally occurs at elevations that far exceed the range of other Costa Rican members of the *rhodopis*-species group, which can facilitate identification in the field. However, note that it may co-occur with *C. underwoodi* at lower elevations.

Like related species, *C. podiciferus* likely breeds in moist terrestrial microsites; its eggs undergo direct development, meaning that the larvae complete their development entirely inside the egg and hatch as small froglets. Both males and females produce a soft, squeaky *pweet* advertisement call.

The conservation status of this species is not very well known. Local declines have been reported from several areas and chytridiomycosis has been implicated as a possible culprit. Nevertheless, *C. podiciferus* now persists in seemingly healthy numbers in several areas where chytrid fungal pathogens are still present and where many other amphibian species have disappeared.

C. podiciferus shows significant differences in size and coloration from population to population; it likely represents a complex of multiple cryptic forms that are distributed over the isolated mountain ranges it inhabits.

The contrasting brown and tan markings in this *C. podiciferus* are a form of disruptive camouflage and effectively break up the outline of the frog's body.

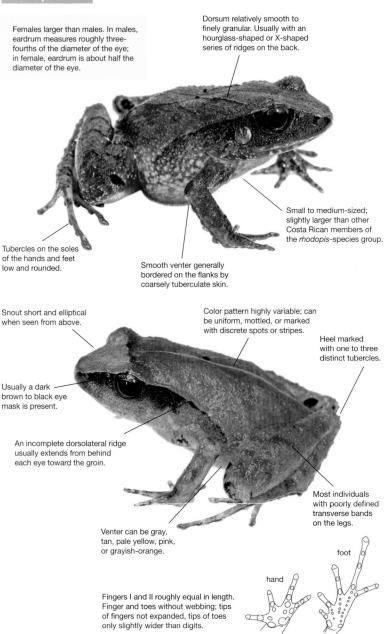

Females larger than males. In males, eardrum measures roughly three-fourths of the diameter of the eye; in female, eardrum is about half the diameter of the eye.

Dorsum relatively smooth to finely granular. Usually with an hourglass-shaped or X-shaped series of ridges on the back.

Small to medium-sized; slightly larger than other Costa Rican members of the *rhodopis*-species group.

Tubercles on the soles of the hands and feet low and rounded.

Smooth venter generally bordered on the flanks by coarsely tuberculate skin.

Snout short and elliptical when seen from above.

Color pattern highly variable; can be uniform, mottled, or marked with discrete spots or stripes.

Heel marked with one to three distinct tubercles.

Usually a dark brown to black eye mask is present.

An incomplete dorsolateral ridge usually extends from behind each eye toward the groin.

Most individuals with poorly defined transverse bands on the legs.

Venter can be gray, tan, pale yellow, pink, or grayish-orange.

foot

hand

Fingers I and II roughly equal in length. Finger and toes without webbing; tips of fingers not expanded, tips of toes only slightly wider than digits.

Like other members of the *rhodopis*-species group, *C. podiciferus* shows a dazzling variety of color patterns and skin textures. Whatever its pattern, it is always well-camouflaged. The following seven images are all of *C. podiciferus*.

Similar species

C. underwoodi (p. 289)

Lacks heel tubercles; also does not have a dark eye mask.

C. gollmeri (p. 251)

Has a longer snout, basal webbing between its toes, and deep red eyes.

Craugastor polyptychus
Coastal Plain Litter Frog **Least concern**

A member of the *rhodopis*-species group. These tiny, stocky leaf litter frogs are characterized by their short limbs and the absence of both interdigital webbing and enlarged finger and toe disks. All species in this group are extremely variable in coloration, pattern, and skin texture and can be almost impossible to identify in the field. *Craugastor polyptychus* has prominently projecting, pointed tubercles on the soles of its hands and feet. It often has red suffusion on the hidden surfaces of the hind limbs and the groin. Males to 1.06 in (27 mm) in standard length, females to 1.14 in (29 mm).

Ecoregions
1 2 3 4 5

Craugastor polyptychus occurs in the lowlands of the Atlantic slope, from southeastern Nicaragua to northwestern Panama; near sea level to 195 ft (60 m).

Natural history

C. polyptychus is a small, diurnal species found in leaf litter. It was formerly included in *C. bransfordii* but recently split off based on biochemical characteristics. Like *C. bransfordii*, this frog shows extreme variation in color, pattern, and skin texture, and these two species are difficult to distinguish under field conditions.

The life history of this species has not been studied, but it is assumed to be similar to that of the closely related *C. bransfordii*. Males produce a single, soft squeaky *pew* call, mostly at dusk and in the early evening. Presumably this call serves to attract a mate, but it might also have a territorial function. Eggs are deposited in a moist terrestrial site and larvae develop into small froglets inside the egg, as is the case in other, related species.

C. polyptychus occurs at very low elevations in moist and wet forests; it also persists in secondary vegetation, plantations, and other open areas with sufficient shade and leaf litter. Its populations seem stable. The biggest threats to this species are habitat loss and pollution; in the lowland habitat of *C. polyptychus*, much of the original forest has been replaced with agricultural land and plantations.

Frogs in the genus *Craugastor* reproduce through direct development. Their larvae skip a free-swimming tadpole stage and leave the eggs as tiny versions of the adults.

One of the largest Costa Rican members of the *rhodopis*-species group. Robust, with stocky limbs.

Coloration variable; cryptic with shades of brown and tan. Usually with a pattern of scattered blotches, spots, or stripes.

Dorsal skin sometimes marked with a series of longitudinal dorsal ridges.

adult female

Lips with several dark brown to black spots or bars, alternating with cream or dirty-white blotches.

Posterior surfaces of thighs, hidden surfaces of hind limbs, and groin suffused with red.

Head, body, and limbs rugose.

Soles of the hands and feet have distinctly projecting, pointed tubercles.

Head longer than broad; snout elliptical when seen from above.

adult male

hand   foot

Limbs marked with poorly defined alternating dark and light bands.

Adult males lack nuptial patches on their thumbs.

There is a distinct difference in the size of the tympanum; the eardrum is approximately two-thirds the size of the eye in males and about half the size of the eye in females.

Two inner fingers (fingers I and II) roughly equal in size. Hands and feet lack webbing. Finger tips not expanded; toes bear only slightly enlarged, pointed disks.

C. bransfordii (p. 233)

A smaller species that generally occurs at slightly higher elevations. Adult male has a white nuptial patch on the base of each thumb.

C. persimilis (p. 265)

Smaller. Finger I is shorter than finger II (both fingers are roughly equal in size in *C. polyptychus*).

C. underwoodii (p. 289)

Generally occurs at considerably higher elevations. Has low, rounded tubercles on the soles of its hands and feet (projecting and pointed in *C. polyptychus*).

273

Craugastor ranoides
Common Streamside Frog

Critically endangered

A member of the *rugulosus*-species group. These frogs are extremely difficult to identify under field conditions as they often differ only in very subtle ways. Lacks distinctly enlarged finger and toe disks; dorsum with fairly uniform coloration; posterior thigh surfaces dark, marked with small, discrete light spots (which may be obscured by dark pigment). Males to 1.77 in (45 mm) in standard length, females to 2.91 in (74 mm).

Ecoregions
1 2 3 4 5

Historically, *Craugastor ranoides* occurred on the Atlantic slope from southeastern Nicaragua, and on the Pacific slope from northwestern Costa Rica, southward to extreme western Panama; from near sea level to 4,000 ft (1,220 m). Currently, in Costa Rica this species is only known to survive in a small watershed within Guanacaste National Park. It has also been reported in recent years from Escudo de Veraguas Island, Panama.

Natural history

C. ranoides was once common in Costa Rica, along rocky streams in low and middle elevation rainforests. These frogs were most often observed at night; they would perch and hunt on rocks in streams. During the day, they retreat under boulders or hide in leaves that accumulate between rocks on the wrack line. Like other species in its genus, *C. ranoides* reproduces through direct development. It deposits small clutches of eggs in moist sites in the stream bank and its larvae complete their metamorphosis inside the egg. The surviving population in Guanacaste National Park is carefully monitored. Preliminary data suggests that it is holding steady and producing offspring; young juveniles and recent metamorphs have been found as recently as 2015.

Like most amphibians, *C. ranoides* is affected by habitat loss and alteration, though this does not explain its disappearance from pristine habitats. Chytridiomycosis has affected many highland and middle elevation populations of related forms, but in general lowland forms are much less severely impacted and the factors behind the disappearance of *C. ranoides* are somewhat of a mystery still. Interestingly, the surviving population inhabits a region characterized by dry forests and high temperatures; conditions that are not favored by the chytrid fungal pathogen *Bd*. In comparison, recent surveys in the cooler streams of the nearby Guanacaste Mountain Range, located at a slightly higher elevation and once home to *C. ranoides*, have not resulted in any sightings. The only other known surviving population of this species (on the opposite end of its historic range, on Escudo de Veraguas Island, in Panama) is also found in a habitat with high temperatures.

A surviving population of the only other lowland form in the *rugulosus*-species group in Costa Rica, *C. taurus*, was recently re-discovered in a hot, dry microclimate on the edge of its historic range. Further research is urgently needed to examine the relationship between habitat and climate and amphibian diseases and declines.

The recent discovery of very young juvenile *C. ranoides* in the single surviving population in Costa Rica indicates that individuals are reproducing, offering a glimmer of hope for the future of this species.

Snout fairly short and rounded when seen from above, with a rounded canthus rostralis.

Weak supratympanic fold follows curve of the relatively small tympanum.

Dorsum either uniform or mottled dark olive-green to olive-brown, sometimes with a few dark spots or a cream, yellow, or orange thin middorsal stripe. It is relatively smooth or finely granular, with a few scattered tubercles, and low short ridges.

adult

Usually shows a faint pattern of alternating dark and light blotches on the lips.

Limbs with well-developed transverse dark bands, which may be difficult to see on dark individuals.

Throat and chest often immaculate white or cream, but may be heavily mottled with dark pigment in some individuals.

An indistinct thin interorbital dark stripe is occasionally present on top of the head, between the eyes. It often separates a lighter, brown to grayish brown snout coloration from the darker dorsal color pattern.

Ventral surfaces usually immaculate pale yellow to bright golden-yellow (brightest yellow observed in southernmost populations).

Small juveniles have burnt-orange or reddish on venter.

Adult males lack nuptial thumb pads and vocal slits.

foot

hand

Posterior thigh surfaces very dark brown, marked with small light spots or swirls that may be obscured by dark pigment and indistinct in older individuals.

Finger I longer than finger II. All fingers bear weak lateral keels. Finger tips only slightly expanded into rounded disks (disks measure about 1.5x the diameter of the digit). Toe tips slightly expanded, and bear rounded disks. Feet with basal to moderate webbing; unwebbed sections of toes bear weak lateral keels.

Similar species

C. fitzingeri (p. 246)

Superficially similar in appearance but differs in having greatly enlarged, truncated disks on its outer fingers (fingers III and IV).

C. taurus (p. 287)

Has more widely expanded disks on its fingers and only barely expanded toe disks.

C. angelicus (p. 231)

Usually has a bright yellow, orange, or reddish venter and tends to occur at higher elevations.

C. fleischmanni (p. 249)

Has only barely expanded finger and toe disks, lacks the light markings on the posterior thigh surfaces, and generally occurs at higher elevations.

Craugastor rayo
Talamanca Rain Frog

Data deficient

A member of the *fitzingeri*-species group. Characterized by the presence of greatly enlarged truncated disks on the two outer fingers of each hand; the two inner fingers bear only slightly expanded, rounded disks. This poorly known species differs from other members of its species group in having uniform purplish-brown posterior thigh surfaces, an enlarged heel tubercle, and only moderate webbing between its toes. Males to 1.77 in (45 mm) in standard length, females to 2.79 in (71 mm).

Ecoregions

Craugastor rayo is endemic to Costa Rica. It is known from just a few isolated localities in the Talamanca Mountain Range; 4,855–5,975 ft (1,480–1,820 m).

Natural history

The biology of this rare species is poorly understood. It appears to be an arboreal species that descends to the forest floor to breed in the vicinity of forest streams. Like other species of *Craugastor*, it displays direct development and does not lay its eggs in water. However it does seem to have a loose association with riparian habitats during the breeding season. Outside the reproductive season, *C. rayo* may be a canopy species, and therefore rarely seen.

The holotype of *C. rayo* was collected in 1964; it was observed in daylight, calling from an arboreal bromeliad during a period of heavy rainfall. Only a few other individuals of this species were seen from 1972 to 1976, mostly as they hopped through leaf litter near streams.

There have been no confirmed recent sightings. Due to limited knowledge of the morphology, phylogeny, and biology of *C. rayo*, its taxonomic status remains unclear. Increased survey efforts in the few known sites are needed to ascertain whether this species still persists; other middle elevation species of *Craugastor* that are associated with streams have been severely affected by chytrid fungal pathogens.

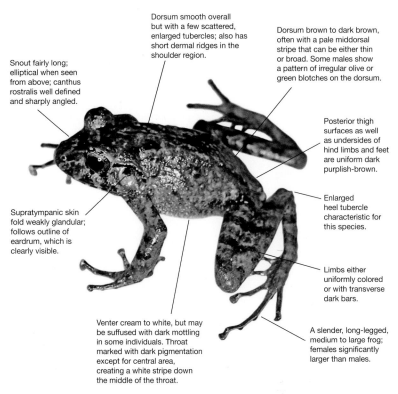

Dorsum smooth overall but with a few scattered, enlarged tubercles; also has short dermal ridges in the shoulder region.

Dorsum brown to dark brown, often with a pale middorsal stripe that can be either thin or broad. Some males show a pattern of irregular olive or green blotches on the dorsum.

Snout fairly long; elliptical when seen from above; canthus rostralis well defined and sharply angled.

Posterior thigh surfaces as well as undersides of hind limbs and feet are uniform dark purplish-brown.

Supratympanic skin fold weakly glandular; follows outline of eardrum, which is clearly visible.

Enlarged heel tubercle characteristic for this species.

Limbs either uniformly colored or with transverse dark bars.

Venter cream to white, but may be suffused with dark mottling in some individuals. Throat marked with dark pigmentation except for central area, creating a white stripe down the middle of the throat.

A slender, long-legged, medium to large frog; females significantly larger than males.

Outer two fingers (III and IV) with expanded, truncated disks; the inner two fingers bear rounded narrowly expanded disks. Hands lack webbing; feet with moderate webbing.

C. crassidigitus (p. 240)

Lacks a heel tubercle, has substantial webbing between its toes, and the posterior surfaces of its thighs are orange or red.

C. melanostictus (p. 257)

Lacks a heel tubercle and webbing between its toes. The posterior surfaces of its thighs, are marked with black bars that alternate with brightly colored (e.g. red, yellow, orange) bars.

Craugastor rhyacobatrachus
Talamanca Torrent Frog

Endangered

A member of the *rugulosus*-species group. These frogs are extremely difficult to identify under field conditions as they often differ only in very subtle ways. *Craugastor rhyacobatrachus* has very rugose dorsal skin covered with scattered tubercles and glandular ridges. Also note enlarged, rounded finger and toe disks, extensively webbed feet, and skin fringes along the unwebbed portions of the toes. In spite of its tree-frog-like hands and feet, this species is a terrestrial frog found only in (or near) steeply cascading mountain streams. Males to 1.97 in (50 mm) in standard length, females to 3.19 in (81 mm).

Ecoregions
1 2 3 4 5

Craugastor rhyacobatrachus was last reported in 1964. Historically it occurred in streams on the southern slopes of the Talamanca Mountain Range, in both Costa Rica and adjacent Panama; 3,150–5,900 ft (950–1,800 m).

Natural history

Craugastor rhyacobatrachus inhabits clear cascading mountain streams in humid middle elevation forests. It is generally associated with rocky sections of torrential streams and the splash zones of waterfalls. This nocturnal species has mostly been observed at night, when it perches on rocks in small to large mountain streams (other streamside frogs in the genus *Craugastor* tend to perch on the banks of streams). The characteristically enlarged finger and toe disks may help these frogs cling to the slippery wet rock surfaces. Based on historic descriptions, *C. rhyacobatrachus* was once relatively common in appropriate habitats, but in spite of recent searches it has not been seen in Costa Rica in more than 50 years.

C. rhyacobatrachus is a cryptic frog that was recently split from a species that was formerly called *Eleutherodactylus punctariolus* (now *Craugastor punctariolus*), which is restricted to western Panama, but was once thought to be more wide-ranging. A second, closely related form, *C. obesus* (p. 263), was split off at the same time as *C. rhyacobatrachus*. Both species are very similar in morphology and likely share a similar biology. Due to taxonomic confusion and the difficulty of identifying these cryptic species in the field, it is currently not known what the status of *C. rhyacobatrachus* is in the Panamanian section of its presumed range; although it resides in La Amistad International Park and other protected areas with pristine habitat, those very same areas were severely affected in the 1990s by the fungal disease chytridiomycosis. Most other species in the *rugulosus*-species group that inhabit streams have declined or disappeared altogether as a result of this disease, and *C. rhyacobatrachus* may be extinct in Costa Rica and possibly in Panama too.

- A robust frog. Medium-sized to large; females are significantly larger than males.
- Dorsum rugose, with scattered large tubercles and dermal ridges.
- Upper eyelids have many medium-sized tubercles.
- A well-developed supratympanic fold passes over the top of the eardrum, which is distinctly visible.
- Snout is moderately short, with a sharply angled canthus rostralis.
- Finger I longer than finger II. All fingers with low lateral skin fringes and large, rounded disks. Feet extensively webbed. All toes have lateral skin fringes on the unwebbed sections and bear large rounded disks, similar in size to those on the fingers.
- Adult males have paired vocal slits and bear nuptial pads on the base of each thumb.
- Dorsum tan to olive-brown, with dark mottling.
- Posterior thigh surfaces mottled pale yellow and brown.
- Ventral surfaces pale yellow, heavily marked with brown. Dark pigmentation is most pronounced in females and on the throat of large adults.

C. fleischmanni (p. 249)

Similar in overall appearance but lacks distinctly enlarged, rounded finger and toe disks and has only basal webbing between its toes.

C. catalinae (p. 236)

Has small, rounded finger and toes disks and slightly enlarged disks on toes III and IV only; it has a moderate amount of toe webbing.

Craugastor rugosus
Pacific Broad-headed Litter Frog

Least concern

A member of the *biporcatus*-species group. Characterized by a very broad head and unwebbed hands and feet. Medium-sized to large. Toadlike, with paired cranial crests and an extremely rugose dorsum that bears a series of hourglass-shaped ridges on the shoulder region. It lacks parotoid glands. Adult males to 1.73 in (44 mm) in standard length, females to 2.72 in (69 mm).

Ecoregions
1 2 **3** 4 5

Craugastor rugosus occurs on the Pacific slope, in central and southwestern Costa Rica and in adjacent Panama; from near sea level to 4,000 ft (1,220 m).

Natural history

C. rugosus is a relatively common, but little studied, species. Like its closely related Atlantic slope counterpart, *C. megacephalus* (p. 255), this species inhabits the leaf litter of lowland and foothill moist forests. Adults are nocturnal and can be seen in or near the entrance of their burrow as they wait for prey to pass within reach. Occasionally adults stray from the burrow entrance and can be found in deep leaf litter. When discovered, they usually freeze, relying on their cryptic coloration to avoid detection; but when further pressed, they escape with a series of powerful short hops, return to their burrow and hide underground.

Because of their large head and powerful physique, these frogs are able to feed on relatively large prey. Juveniles can be found active during the day as well as at night and inhabit the moist microclimates that exist within the leaf litter that covers the forest floor. Like other *Craugastor*, this species breeds by direct development; their larvae develop into small froglets before leaving the egg. Clutches are likely deposited in the leaf litter or inside an adult's burrow.

C. rugosus is somewhat tolerant of habitat alteration and can persist in disturbed situations, secondary forests, and plantations—provided that a suitable thick leaf litter layer is present. The camouflaged color pattern and sedentary nocturnal lifestyle make this a difficult species to detect, which may contribute to its perceived rarity in some areas.

This species was previously included in what was once considered a more wide-ranging species, *Eleutherodactylus biporcatus*, and is reported as such in older literature.

Description

Head with paired cranial crests; crests not always evident in small juveniles.

Dorsum generally olive-gray, tan, or dark brown, often with some dark shading bordering the ridges and tubercles.

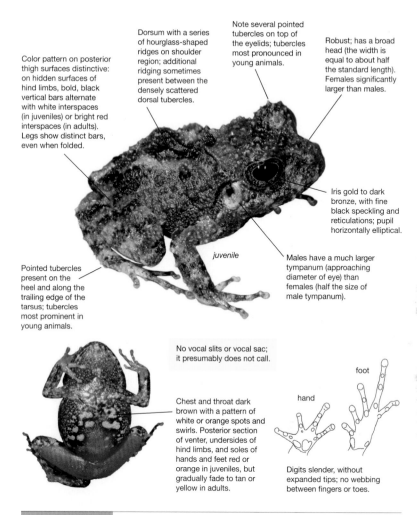

Color pattern on posterior thigh surfaces distinctive: on hidden surfaces of hind limbs, bold, black vertical bars alternate with white interspaces (in juveniles) or bright red interspaces (in adults). Legs show distinct bars, even when folded.

Dorsum with a series of hourglass-shaped ridges on shoulder region; additional ridging sometimes present between the densely scattered dorsal tubercles.

Note several pointed tubercles on top of the eyelids; tubercles most pronounced in young animals.

Robust; has a broad head (the width is equal to about half the standard length). Females significantly larger than males.

Iris gold to dark bronze, with fine black speckling and reticulations; pupil horizontally elliptical.

juvenile

Pointed tubercles present on the heel and along the trailing edge of the tarsus; tubercles most prominent in young animals.

Males have a much larger tympanum (approaching diameter of eye) than females (half the size of male tympanum).

No vocal slits or vocal sac; it presumably does not call.

Chest and throat dark brown with a pattern of white or orange spots and swirls. Posterior section of venter, undersides of hind limbs, and soles of hands and feet red or orange in juveniles, but gradually fade to tan or yellow in adults.

hand
foot

Digits slender, without expanded tips; no webbing between fingers or toes.

Similar species

Engystomops pustulosus (p. 451)

Although somewhat toadlike in appearance, *C. rugosus* lacks the paired parotoid glands seen in true toads of the genus *Incilius* (family Bufonidae), p. 159.

Small juveniles may be confused with the rugose-skinned *Engystomops pustulosus* (family Leptodactylidae), but that species lacks dorsal and cranial ridges and has a narrow head.

Craugastor stejnegerianus
Pacific Litter Frog

Least concern

A member of the *rhodopis*-species group. These tiny, stocky leaf litter frogs are characterized by their short limbs and the absence of both interdigital webbing and enlarged finger and toe disks. All species in this group are extremely variable in coloration, pattern, and skin texture and can be almost impossible to identify in the field. *Craugastor stejnegerianus* is the only species in this confusing group that occurs widely on the Pacific slope, and thus location is perhaps the best key to identification. Adult males to 0.71 in (18 mm) in standard length, females to 0.87 in (22 mm).

Ecoregions
1 2 3 **4** 5

Craugastor stejnegerianus occurs in the Pacific lowlands and foothills, from northwestern Costa Rica to western Panama; it also ranges locally into the western Central Valley; from near sea level to 4,365 ft (1,330 m). The area in red indicates populations assigned to the newly described species *Craugastor gabbi*; 3,600–4,200 ft (1,100–1,280 m).

Natural history

This very common leaf litter species was previously included in *C. bransfordii*, but was relatively recently separated based on biochemical characteristics. The extreme variation in color, pattern, and skin texture in the Costa Rican members of the *rhodopis*-species group generally make identification in the field a challenge. Fortunately, *C. stejnegerianus* is the only species in this group that ranges widely throughout the Pacific slope and it can usually be identified based on its geographic occurrence. Only in the Atlantic foothills of the Tilarán Mountain Range, in the vicinity of Lake Arenal, does *C. stejnegerianus* co-occur with *C. bransfordii*, and possibly *C. underwoodi*. The latter species definitely co-occurs with *C. stejnegerianus* in the Las Cruces area, in the foothills of the southern Talamanca Mountain Range.

C. stejnegerianus has long been included in treatises of *C. bransfordii*, and the taxonomic confusion that surrounds this cryptic species complex makes it difficult to extract biological information pertinent to *C. stejnegerianus*. However, it presumably has a very similar biology to the other species. It reproduces through direct development, and its eggs are deposited in moist microhabitats in the leaf litter, or in small burrows below logs or rocks. Males produce a soft, squeaky, single-noted *pew* call at irregular intervals, usually at dusk or in the early evening.

C. stejnegerianus is widespread, adaptable, and sometimes extremely common locally. As with other leaf litter frogs, its population density fluctuates with the depth of the leaf litter layer, which often varies with seasonal changes in rainfall and temperature. It occurs in a variety of habitats, from dry or humid lowland forest to streamside gallery forests and foothill forests. It also persists in disturbed habitat types, including areas with secondary vegetation, plantations, and other open areas with sufficient shade tree cover and leaf litter. Its populations are stable; the biggest threats to this species include clear cutting and other forms of habitat loss, as well as pollution.

Middle elevation populations of *C. stejnegerianus*, located in the southern Talamanca Mountain Range and nearby Fila Costeña, were described as a new species, Gabb's Litter Frog (*Craugastor gabbi*), as this book was going to press. It differs from related species primarily in molecular characteristics, but may be identified by the presence of dark pigment on its venter (*C. stejnegerianus* has a white venter) and the absence of heel tubercles (a characteristic of *C. podiciferus*). *C. gabbi* likely co-occurs with *C. underwoodi* in the Las Cruces area, but these two species are difficult to distinguish based on morphological traits. *C. gabbi* is apparently common in the leaf litter of undisturbed forests, plantations, and a variety of other habitats.

Skin on head, body, and limbs rugose, often with scattered enlarged tubercles or ridges.

Head roughly as broad as it is long; snout elliptical when seen from above.

female

In males, the eardrum measures approximately three-fourths the diameter of the eye; in females it is about half the size of the eye.

Tubercles on soles of hands and feet prominent and rounded.

Often with a dark supratympanic mark that covers the top of the eardrum.

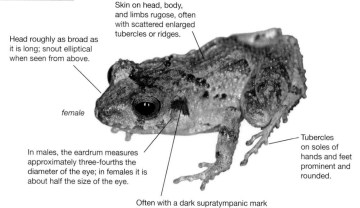

Dorsal coloration extremely variable. Generally cryptically colored in shades of gray or brown.

male

Lips usually marked with alternating dark and light bars or blotches.

A short, diagonal row of white, cream, or orange tubercles located between the bottom of the eardrum and the insertion of the front limbs.

Venter uniform white.

Tiny and stocky, with short limbs.

foot

hand

The two inner fingers (fingers I and II) are of roughly equal size. The hands and feet lack webbing; finger tips are not expanded and the toes bear only slightly expanded disks.

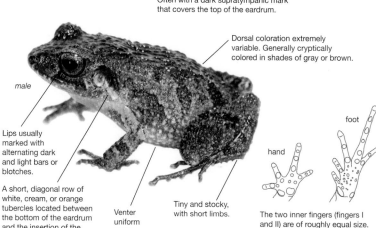

C. bransfordii (p. 233)

Smaller. The tubercles on the soles of its hands and feet are even more prominent (but pointed). Adult males have a white nuptial patch on the base of each thumb.

C. underwoodi (p. 289)

Larger. Has low, rounded tubercles on the soles of its hands and feet.

283

Craugastor talamancae
White-lipped Rain Frog

Least concern

A member of the *fitzingeri*-species group. Characterized by the presence of greatly enlarged truncated disks on the two outer fingers of each hand; the two inner fingers bear only slightly expanded, rounded disks. It is distinguished from other species in this group by having only minimal webbing between its toes and uniform reddish-brown to red posterior thigh surfaces. A defining trait, found in most individuals, is a white stripe or series of spots on the lip that extends onto the flanks. Males to 1.18 in (30 mm) in standard length, females to 1.97 in (50 mm).

Ecoregions
1 2 3 4 5

Craugastor talamancae occurs in lowland forests on the Atlantic slope, from central Nicaragua to eastern Panama; near sea level to 2,330 ft (710 m).

Natural history

C. talamancae is a nocturnal species of humid lowland forests. It is most frequently encountered on humid nights, on low vegetation; during the day, it retreats into the leaf litter, where it is sometimes flushed out by passing hikers.

This species reproduces through direct development; it probably deposits its eggs under leaves on the ground, or in shallow depressions in the soil. At night, males call from an elevated perch, producing a loud, catlike *mew* at irregular intervals. Where populations are dense, calls are often answered by neigboring males, indicating that this call may have a territorial spacing function, as well as serve to attract mates. *C. talamancae* feeds on a variety of insects and other arthropods, but large adults are known to occasionally consume hatchling lizards, small frogs, and other small vertebrates.

This species appears to be restricted mostly to undisturbed habitats and old-growth secondary forest. It probably does not tolerate habitat alteration as

successfully as its close relatives *C. crassidigitus* and *C. fitzingeri*. The presence of a thick layer of leaf litter on the forest floor appears to be a critical factor in determining the population density of *C. talamancae*. Like other litter frogs, its numbers tend to fluctuate with seasonal changes in the depth of the leaf litter, and numbers go up during dry spells, when more trees are apt to drop their leaves.

C. talamancae is relatively common throughout its range, and its populations appear to be stable. Some lowland populations of *C. talamancae* have persisted in spite of infection by the chytrid fungal pathogen *Bd* and do not show any adverse effects from this infection. This may be related to the prevailing high temperatures, which are not optimal for the fungus. Apart from outright loss of its forest habitat due to agriculture, development, or logging, as well as the possible detrimental effects from agrochemicals, *C. talamancae* does not appear to face any major threats.

Dorsum finely tuberculate without distinctly enlarged tubercles or dermal ridges.

A distinct white labial stripe is clearly visible in juveniles and subadults, but may be suffused with dark pigment in large adults to form an interrupted series of light labial spots.

subadult male

Dorsum uniform grayish-brown to olive-brown, occasionally with a pattern of irregular blotches or dark markings that are hourglass-shaped or arranged in a linear pattern. Most individuals show a dark eye mask that extends posteriorly to incorporate the tympanum.

Eardrums distinctly visible, borderd above by a short, downward-curving supratympanic fold.

Head is somewhat pointed when viewed from above, showing a long snout and a sharply angular canthus rostralis.

Adult males with a brown nuptial patch on the base of each thumb.

Throat, chest, belly, and undersurfaces of the hind limbs are uniform white to cream; in some adults, venter may be suffused with red. Throat never shows a white midgular stripe (as is seen in *C. fitzingeri* [p. 246]).

adult female

Females are significantly larger than males.

adult female

Dorsal surfaces of the limbs are marked with dark transverse bands.

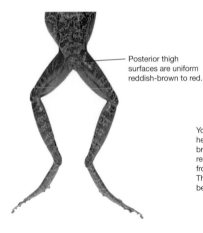

Posterior thigh surfaces are uniform reddish-brown to red.

young juvenile

Young juveniles have a distinctive color pattern: head and dorsum chocolate-brown; limbs reddish-brown; one to three oblique white stripes on flanks, reminiscent of the color pattern seen in rocket frogs of the genus *Silverstoneia* (Dendrobatidae). The light stripes often darken gradually, and become less prominent as these frogs age.

Toes with basal webbing; lateral skin fringes line the unwebbed sections of the toes. Toe disks small, barely expanded. Hands lack webbing; fingers bear small rounded disks on the two inner fingers and truncated, widely expanded disks on the outer two fingers. Finger I is longer than finger II.

hand

foot

Similar species

C. crassidigitus
(p. 240)

Has extensive webbing between its toes, uniform orange to brown posterior thigh surfaces, lacks a white lip stripe.

C. fitzingeri
(p. 246)

Has dark brown to black posterior thigh surfaces marked with yellow spots, lacks a white lip stripe, and has a rugose dorsum marked with tubercles and ridges.

C. chingopetaca
(p. 238)

Has indented finger and toe disks, slightly more webbing between its toes, and lacks a white lip stripe.

C. gollmeri
(p. 251)

This species, *C. mimus* (p. 259), and *C. noblei* (p. 261) have smooth skin and a dark eye mask, but all lack greatly enlarged, truncated disks on their outer fingers (fingers III and IV).

Silverstoneia flotator
(p. 310)

Resembles some juvenile *C. talamancae* in coloration and pattern, but lacks enlarged, truncated disks on the outer fingers.

Craugastor taurus
South Pacific Streamside Frog

Critically endangered

A member of the *rugulosus*-species group. These frogs are extremely difficult to identify under field conditions as they often differ only in very subtle ways. This species is the only member of the group with slightly expanded finger disks that lacks expanded toe disks. Males to 1.73 in (44 mm) in standard length, females to 3.15 in (80 mm).

Ecoregions
1 2 **3** 4 5

Craugastor taurus historically occurred in the lowlands and foothills of southwestern Costa Rica and adjacent Panama; from near sea level to 1,725 ft (525 m). It is currently only known to survive in the Punta Banco area in extreme southwestern Costa Rica, and in neighboring Punta Burica in extreme western Panama.

Natural history

C. taurus was once widespread throughout southwestern Costa Rica, where it inhabited rocky stream corridors. This is a nocturnal species; individuals were formerly seen at night on rocks and in streambeds. Like other species in the genus, it breeds through direct development, and its eggs are deposited in depressions under or between rocks and logs along the stream bank. The eggs develop directly into small froglets, and skip a free-swimming tadpole stage.

Though somewhat tolerant of habitat alteration, this species is predominantly found in undisturbed humid lowland forests. Even in pristine habitats, chytridiomycosis has been implicated in the demise of many populations of Costa Rican amphibians during the second half of the 1980s. Middle elevation and highland populations of stream-associated amphibians (including all species of the *rugulosus*-species group) were severely impacted by this disease, while lowland populations appear to

be less vulnerable. In late 2011, an extensive search for this species revealed only two surviving populations, in the Río Banco area. More recently, *C. taurus* was also confirmed to persist in small populations in nearby Punta Burica, across the border with Panama.

The last surviving populations of *C. taurus* persist on the fringe of the species' original distribution range, in an area that is relatively warmer and drier than the core of its historic range. Recent tests have shown that a significant percentage of the surviving *C. taurus* are infected with the chytrid fungal pathogen *Bd*, but it appears that the prevailing climatic conditions, which are suboptimal for *Bd*, may prevent an outbreak of chytridiomycosis. Additional research is needed to better understand the dynamics of disease transmission and pathology and to explore how these frogs survived the dramatic population crashes that decimated other populations of *C. taurus*.

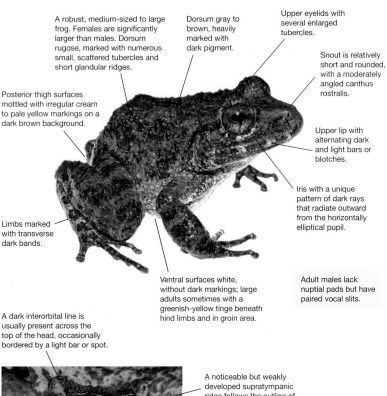

A robust, medium-sized to large frog. Females are significantly larger than males. Dorsum rugose, marked with numerous small, scattered tubercles and short glandular ridges.

Dorsum gray to brown, heavily marked with dark pigment.

Upper eyelids with several enlarged tubercles.

Snout is relatively short and rounded, with a moderately angled canthus rostralis.

Posterior thigh surfaces mottled with irregular cream to pale yellow markings on a dark brown background.

Upper lip with alternating dark and light bars or blotches.

Iris with a unique pattern of dark rays that radiate outward from the horizontally elliptical pupil.

Limbs marked with transverse dark bands.

Ventral surfaces white, without dark markings; large adults sometimes with a greenish-yellow tinge beneath hind limbs and in groin area.

Adult males lack nuptial pads but have paired vocal slits.

A dark interorbital line is usually present across the top of the head, occasionally bordered by a light bar or spot.

A noticeable but weakly developed supratympanic ridge follows the outline of the smallish eardrum.

Finger I longer than finger II. All fingers bear distinctly enlarged, rounded disks; those on fingers III and IV are largest, measuring about twice the diameter of the digit. Hands lack webbing, but fingers are lined with fleshy fringes. Toe disks only barely expanded. Feet moderately webbed; unwebbed sections of toes bear distinct fleshy fringes.

C. ranoides (p. 274)

Has only slightly expanded finger and toe disks, a distinct pattern of light spots on its dark brown posterior thigh surfaces, and is generally pale yellow underneath the hind limbs and on the venter.

Craugastor underwoodi
Underwood's Litter Frog

Least concern

A member of the *rhodopis*-species group. These tiny, stocky leaf litter frogs are characterized by their short limbs and the absence of both interdigital webbing and enlarged finger and toe disks. All species in this group are extremely variable in coloration, pattern, and skin texture and can be almost impossible to identify in the field. *Craugastor underwoodi* differs from other species in this group in having only a few low, rounded tubercles on the soles of its hands and feet. Adult males to 1.02 in (26 mm) in standard length, females to 1.18 in (30 mm).

Ecoregions
1 2 3 **4** 5

Craugastor underwoodi occurs at middle elevations on both the Atlantic and Pacific slopes of the Tilarán, Central, and Talamanca Mountain Ranges, as well as in mountains along the Pacific coast; 3,000–5,225 ft (920–1,600 m).

Natural history

This is a small, diurnal leaf litter species previously included in *C. bransfordii*; it was fairly recently separated based on biochemical characteristics. *C. underwoodi* is difficult to distinguish from other Costa Rican members of the *rhodopis*-species group due to its extreme variation in color, pattern, and skin texture. Not much pertinent information is available on the biology of *C. underwoodi*, in part due to the difficulty of positively distinguishing it from related species.

C. underwoodi is a common species of middle elevation humid forests, and it likely also occurs in shaded coffee plantations and other moderately altered habitats, as long as adequate canopy cover and sufficient leaf litter are present. Like related species, it breeds through direct development, and its eggs are deposited in a humid terrestrial site. The larvae complete their development entirely inside the egg and hatch as miniature versions of the adults. *C. underwoodi* produces a soft, two-toned squeaky advertisement call that differs from the single *pew* emitted by most other species in the *rhodopis*-species group.

This species can show marked changes in population density in response to seasonal variation in leaf litter depth. Such dynamic population levels, combined with the difficulty of identifying this species, make it hard to adequately assess the conservation status of *C. underwoodi*. Nevertheless, its continued presence in a variety of protected and managed areas suggest that it is currently not under any severe threat.

C. underwoodi can be difficult to see when it sits still in leaf litter.

Adult males with a small, white nuptial patch at the base of each thumb.

Skin texture of dorsum varies; it can be relatively smooth, with scattered tubercles or ridges, or entirely rugose.

Head broad, with a relatively short, rounded snout.

Lips with a pattern of alternating cream and dark brown blotches or bars.

Tympanum clearly visible. In males, it is roughly equal in size to the diameter of the eye; in females, it is about three-fourths the diameter of the eye.

Posterior thigh surfaces uniform reddish-brown.

Yellowish ventral surfaces sometimes suffused with red in the groin area.

Coloration cryptic and extremely variable. Usually brown or tan, with a variety of blotches, spots, and mottling; occasionally with a light middorsal pinstripe.

Hands and feet lack webbing. Finger tips not expanded; toes bear only slightly expanded disks. Two inner fingers (fingers I and II) are approximate equal in length. Tubercles on the soles of the hands and feet are relatively low and smooth.

Limbs marked with alternating dark and light bands.

C. bransfordii
(p. 233)

Smaller and has projecting, pointed tubercles on the soles of its hands and feet.

C. persimilis
(p. 265)

Smaller; finger I is shorter than finger II.

C. podiciferus
(p. 269)

Has a series of one to three tubercles on the heel and usually a dark eye mask.

C. stejnegerianus
(p. 282)

Smaller and has projecting, pointed tubercles on the soles of its hands and feet.

Family Dendrobatidae
(Poison Dart Frogs)

Poison dart frogs are small to very small frogs that inhabit the rainforests of southern Central America and South America. More than 300 species are assigned to this family, but its taxonomy is under discussion and it may eventually be split into two separate families: Dendrobatidae and Aromobatidae. Costa Rica is home to eight species of poison dart frog, placed in five genera.

The family's common name is derived from the skin toxins found in frogs in several dendrobatid genera. Frogs in the genus *Phyllobates*, for example, have skin toxins that can be especially strong; laboratory tests have shown that the skin of a South American species, *Phyllobates terribilis*, can contain toxins sufficient to kill 20,000 mice, which amounts to the hypothetical equivalent of 6–7 adult humans. This species' skin secretions are so potent that it should be only handled while wearing gloves. Most poison dart frogs, however, are far less toxic, and many species even lack specialized alkaloids and other toxins altogether.

Although the term poison dart frog refers to all species in the family, only three species—*Phyllobates aurotaenia*, *Phyllobates bicolor*, and *Phyllobates terribilis*, all highly toxic—were regularly used to poison the tips of blow darts. Indigenous people from the Chocó region in Colombia capture the frogs, carefully holding them inside a folded leaf to avoid getting poison on their skin. The tribal hunters simply rub the tips of their blowgun darts and arrows against a frog's back to transfer its skin secretions. If they use less toxic species, the frog is sometimes held over a fire or pierced with a sharp stick to cause it to secrete more of its defensive toxins. The skin secretions carry a cocktail of chemicals that includes strong neurotoxins that may cause paralysis, or even cardiac arrest, within minutes. The darts are used to bring down monkeys, sloths, or other animals that usually live high in the forest canopy; if a dart merely grazes an animal and some of the frog's toxins enters its bloodstream, it will succumb quickly and fall paralyzed to the forest floor.

Interestingly, the alkaloids and other chemical compounds responsible for the toxicity of the skin secretions appear to be derived from the food that these frogs eat. Although other frogs feed on the same kinds of small ants, mites, beetles, and other invertebrates that make up the diet of poison dart frogs, only species in the family Dendrobatidae are able to extract toxic compounds from their food and subsequently sequester them in their skin. Poison dart frogs in zoos and other captive settings are maintained on a diet of fruit flies and other prey items that generally do not occur in their natural habitat and that lack the necessary chemical precursors to produce the toxic skin secretions; therefore captive poison dart frogs lose their toxicity.

Frogs in this family differ morphologically from all other frogs by having two shieldlike scutes on the top of the tips of the fingers and toes; it is sometimes hard to see these scutes without a magnifying glass. Some Costa Rican poison dart frogs (genera *Dendrobates*, *Oophaga*, and *Phyllobates*) are easily recognized by their bright coloration, which is used to advertise their toxicity. However, species in the genera *Allobates* and *Silverstoneia* are cryptically colored and not poisonous.

All Costa Rican poison dart frogs display elaborate parental care behavior. Breeding dendrobatids lay their eggs in leaf litter on the forest floor. One of the parents guards the clutch until the eggs hatch, and, at least in *Dendrobates*, *Oophaga* and *Phyllobates*, the male parent regularly moistens the eggs by emptying his bladder on them. Upon hatching, the tadpoles slither onto the back of one of the parents, who subsequently carries them to a small pool, a water-filled bromeliad, or a small stream, where they complete their development. The tadpoles of *Allobates, Dendrobates, Oophaga* and *Phyllobates* are generally placed in a small body of water—a small puddle on the forest floor, the water held between the leaf axils of arboreal bromeliads or other plants, or some other water-filled object. *Allobates* tadpoles, however, are sometimes found in quiet pools in streambeds. *Silverstoneia* tadpoles are usually deposited in small streams.

Since these beautiful frogs are diurnal, they are among the more easily observed Costa Rican amphibians. During the day, males call from fixed, elevated sites (e.g., trunks, logs, or rocks) to attract females and to keep other males at a distance. They are extremely territorial and literally wrestle intruders out of their territory.

Although skin toxins are present in Costa Rican species of *Dendrobates*, *Oophaga*, and *Phyllobates*, these frogs are not dangerous to humans. It is always advisable to wash your hands after handling any amphibian, however, as the effects of some of their skin secretions are largely unknown and some people can react more strongly to these chemicals than others.

Genus *Allobates*

A diverse group of 50 cryptically colored, nontoxic species of poison dart frogs that were formerly included in the genus *Colostethus*. All inhabit South America, except *Allobates talamancae*, which ranges into Central America, as far north as southern Nicaragua. The widely separated populations of this species likely represent a complex of several cryptic species.

Some biologists assign the genus *Allobates* and a few other genera of nontoxic frogs to the family Aromobatidae, and consider it a sister group to Dendrobatidae. Other biologists, including the author, prefer to view this group as a sub-family within the family Dendrobatidae.

Frogs in the genus *Allobates* are well-camouflaged inhabitants of leaf litter that are brown and marked with light stripes. Members of the genus *Silverstoneia* have similar coloration; in fact, members of both genera were formerly placed in the genus *Colostethus*.

Allobates talamancae
Striped Rocket Frog   **Least concern**

A small terrestrial frog. On each side of the body, a bronze, tan, or copper dorsolateral stripe runs from the tip of the snout to the vent; this line separates the brown dorsum from chocolate-brown flanks. Also has a pattern of white lines (none of which run diagonally) on the sides. Males to 0.87 in (22 mm) in standard length, females to 0.94 (24 mm).

Ecoregions
1 2 3 4 5

On the Atlantic slope, *Allobates talamancae* occurs in extreme southeastern Nicaragua, Costa Rica, and eastern Panama; on the Pacific slope, it occurs in southwestern Costa Rica, eastern Panama, Colombia, and northern Ecuador; from near sea level to 2,300 ft (700 m).

Natural history

A. talamancae lives in the thick leaf litter of undisturbed forests with a closed canopy; it usually occurs close to streams, though individuals sometimes stray some distance from water. When startled, these frogs launch headfirst into the water.

Although this frog is strictly diurnal, it prefers periods of low light. Males call in early morning, late afternoon, or during cloudy spells. The advertisement call is a rapid, high-pitched trill, with a pause before the fourth beat: *peet-peet-peet---peet* (repeated steadily in bursts of 10 to 20 seconds). It serves to attract mates but also as a territorial marker against neighboring males.

These frogs reproduce throughout the year, but with greatest intensity in wetter months. There is no amplexus but rather a mated pair will sit vent to vent on the forest floor. The female deposits her eggs in moist leaf litter, where they are fertilized by the male. Like all other Costa Rican dendrobatids, *A. talamancae* displays elaborate parental care. Upon hatching, the male carries the 8–29 newly emerged tadpoles on his back and deposits them in small bodies of standing water such as water-filled depressions in logs and rocks, shallow puddles in stream banks, or occasionally in cans or bottles that have filled with water. The larvae of *A. talamancae* lack the umbrella-shaped mouth that characterizes *Silverstoneia* tadpoles (p. 312), and do not feed at the water surface.

This is a common species in lowland and foothill forests with relatively high levels of rainfall. Seemingly healthy populations of this species are known from secondary forest types and shaded plantations, indicating that *A. talamancae* is somewhat tolerant of habitat alteration in forests with canopy cover and high humidity.

It appears that *A. talamancae* represents a complex of cryptic forms, and it may eventually be split into multiple, isolated species.

Male *Allobates talamancae* carrying a large number of tadpoles to a suitable body of water, where the larvae will complete their development.

In females, the throat and venter are white, cream, or yellow; males have a black throat.

Body, head, and limbs finely granular.

A broad bronze, tan, or copper dorsolateral stripe extends along each side of the body, from the tip of the snout to the vent, separating brown dorsum from chocolate-brown flanks.

Snout moderately long, bluntly rounded.

Eardrums indistinct but visible.

Small and stocky, with relatively short limbs.

Hind limbs with faint dark transverse bars.

On the lower flank, a white, irregularly shaped line runs from the groin to the upper lip; it is bordered above and below by chocolate brown coloration.

Venter smooth.

Adult males with a single, external vocal sac on throat; when not in use, recognizable as a dark throat patch.

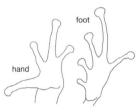

foot

hand

adult male

No webbing or significantly expanded disks on hands or feet; a pair of shieldlike flaps on top of each finger and toe. Adult males do not have a distinctly swollen third finger (unlike male *Silverstoneia*).

Silverstoneia flotator (p. 310)

Co-occurs and is similar in coloration and behavior. It differs in having a diagonal white stripe on its sides, between the groin and the upper eyelid. This diagonal stripe may be incomplete in some populations on the Atlantic slope of Costa Rica.

Silverstoneia nubicola (p. 312)

Likely does not co-occur. It also has a diagonal pale stripe on each side of its body.

295

Genus *Dendrobates*

After the taxonomy of this genus was recently revised, it now contains just five species. One, *Dendrobates auratus*, occurs in Central America; it is widespread in Costa Rica.

Frogs in the genus *Dendrobates* are diurnal and display aposematic (warning) coloration. The expanded disks on each finger and toe bear a pair of dorsal scutes.

These frogs have the ability to sequester lipophilic alkaloids from the food they eat; these toxins are stored in their brightly colored skin. Although the skin secretions of *Dendrobates auratus* are not dangerous to humans, it is always best to avoid touching eyes, mouth or other mucous membranes after handling any frogs; and you should not handle them at all if you have cuts or scrapes on your fingers. Washing with soap and water should remove the toxins from your hands.

Dendrobates auratus
Green-and-Black Poison Dart Frog

Least concern

In Costa Rica, this is the only lime-green or bluish-green frog with a pattern of bold, irregular black swirls and spots. Males to 1.57 in (40 mm) in standard length, females to 1.65 in (42 mm).

Ecoregions
1 2 3 4 5

On the Atlantic slope, *Dendrobates auratus* ranges from southeastern Nicaragua to Colombia, and on the Pacific slope from southwestern Costa Rica to Colombia; from near sea level to 3,275 ft (1,000 m). It has also been introduced to Hawaii.

Natural history

Throughout its range, *D. auratus* shows extensive geographic variation in coloration and pattern. Colombian populations can be uniform golden-yellow, while some Panamanian populations are brown with a pattern of cream, bronze, or copper markings. In comparison, Costa Rican populations are relatively less showy and display some variety of a green and black color pattern. Nevertheless, there certainly are noticeable regional variants; for example, southern Caribbean populations tend to be bluish-green with relatively limited black markings, usually reduced to small spots, while Pacific slope populations usually have a more yellowish-green color with extensive black markings.

This is generally a common to abundant frog. It prefers leaf litter in humid and wet forest, in lowlands and foothills. Although it primarily occurs in forests, *D. auratus* is fairly adaptable and survives in dense secondary growth, gardens, and plantations. Throughout its range, it is especially common around trees with buttressed roots and in areas with leaf or stick piles; population densities can also be very high in areas with dense concentrations of potential breeding sites.

Although abundant, *D. auratus* is quite shy and not always amenable to sustained observation. It is strictly diurnal, and most active early in the morning, especially after a nighttime rain. Males are in constant motion, moving with a characteristic series of short hops, pausing briefly at regular intervals, and occasionally calling between hops. Its call, a fairly high-pitched, insectlike buzz (*cheez-cheez-cheez*) serves to attract females and establish territoriality. Both males and females sometimes act aggressively towards members of the same sex,

which can lead to physical altercation. Agitated *D. auratus* rapidly twitch their toes and will pounce on another individual's head, trying to pin it down and subdue it.

Courtship in *D. auratus* is relatively brief, after which a female selects a calling male as a potential mate. No amplexus takes place in dendrobatid frogs. Instead, both partners sit vent to vent, facing opposite directions. Female *D. auratus* lay 3–13 eggs in a moist location in the leaf litter; the male fertilizes the eggs as they emerge from the female's cloaca. A male attempts to mate with multiple females within his territory; he tends and guards all egg clutches until the eggs hatch. Male parental care includes repeated visits to each clutch to hydrate and rotate the eggs, and remove any moldy or otherwise non-viable eggs. On hatching, the male transports each tadpole on his back to a suitable body of water, usually one tadpole at a time. Larvae are deposited in small water-filled tree cavities, bromeliads, empty coconut shells or husks, or other water-filled basins. Preference is given to terrestrial breeding sites, but *D. auratus* is a skillful climber, and individuals will deposit their tadpoles in arboreal bromeliads located at a significant distance above the ground, if other, more easily accessed, basins are not available.

Larval *D. auratus* are omnivorous and eat algae, detritus, microscopic aquatic organisms, and insect larvae. When eggs do not hatch at the same time, considerable size differences can develop between tadpoles of the same clutch, and cannibalism within a brood is not uncommon. Seven to 15 weeks after being deposited in their basin, the tadpoles complete metamorphosis and emerge as

297

dull-colored replicas of their parents. Captive individuals of this species have lived to the impressive age of eight years, but life expectancy in the wild is likely considerably lower.

Loss of habitat, habitat alteration, and pollution are the principal threats to this species. Though these frogs are popular in the pet trade, it is illegal to collect them in the wild, and international trade in this species is regulated through CITES (Convention on International Trade in Endangered Species). Costa Rican *D. auratus* populations appear to be stable.

A population of this species persists on the island of Oahu, Hawaii, where it was introduced in the early 1930s in an (unsuccessful) attempt to control an introduced insect pest.

Description

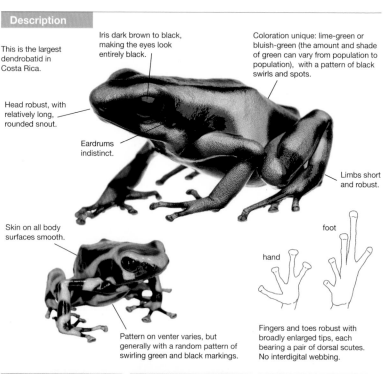

This is the largest dendrobatid in Costa Rica.

Iris dark brown to black, making the eyes look entirely black.

Coloration unique: lime-green or bluish-green (the amount and shade of green can vary from population to population), with a pattern of black swirls and spots.

Head robust, with relatively long, rounded snout.

Eardrums indistinct.

Limbs short and robust.

Skin on all body surfaces smooth.

foot

hand

Pattern on venter varies, but generally with a random pattern of swirling green and black markings.

Fingers and toes robust with broadly enlarged tips, each bearing a pair of dorsal scutes. No interdigital webbing.

Pacific populations often display more black than those on the Atlantic coast.

This individual from the Quepos area on the central Pacific coast is yellow-green, with extensive black swirls.

Individuals from the southern Atlantic coast (Cahuita) tend to be bluish-green, with relatively small black spots.

Similar species

No frog in Costa Rica resembles *Dendrobates auratus*.

Genus *Oophaga*

This group of brightly colored poison dart frogs was formerly included in the genus *Dendrobates*. The nine species of *Oophaga* are grouped in the same genus primarily on the basis of the similar morphology of their tadpoles and on certain behavioral characteristics that are related to their parental care.

Members of the genus *Oophaga* deposit their eggs in leaf litter on the forest floor. On hatching, the female carries the tadpoles to an arboreal, water-filled leaf axil, where the larvae complete their development. The tadpoles are obligatory egg feeders (*oophaga* means egg-eating), and solely eat nutritive infertile eggs that the female parent provides at regular intervals when she repeatedly visits the leaf axils where she deposited her tadpoles.

Two species of *Oophaga* occur in Costa Rica. *Oophaga pumilio* is an iconic frog that is commonly found throughout much of the Atlantic lowlands, while *O. granulifera* is restricted to the Pacific southwest of the country. Both species are capable of sequestering lipophilic alkaloids in their skin and are toxic, but not dangerous to humans. All amphibians secrete a variety of chemical substances from their skin, and it is always advisable to avoid touching your eyes or mouth and to cover cuts and scrapes when handling frogs. A quick wash with water and soap will easily remove any residual toxins from your hands.

Oophaga granulifera
Granular Poison Dart Frog   **Vulnerable**

A small poison dart frog with very granular skin. Head and dorsum red, with light blue to turquoise hind legs; in some populations, the red dorsal coloration is replaced with olive-green, yellow, or orange-tan. Both sexes reach 0.91 in (23 mm) in standard length.

Ecoregions
1 2 **3** 4 5

Oophaga granulifera is known from southwestern Costa Rica and adjacent southwestern Panama; sea level to 1,300 ft (400 m). There may be a disjunct population in extreme southeastern Costa Rica (in the Río Sandbox area in Limón Province), likely introduced by humans.

Natural history

This frog is strictly diurnal. It inhabits moist leaf litter, preferring the steeply sloping banks of fast flowing streams that run through humid, undisturbed forests. Males call from fixed calling sites, which are invariably elevated, usually about 8–60 in (20–150 cm) above the forest floor. Each male uses one to three fixed calling sites during the rainy months (May–December). The advertisement call is an insectlike *buzz-buzz-buzz*, lasting several seconds. Calling activity is typically bimodal, with two distinct activity periods. The first period starts at sunrise and lasts until mid-morning; a second bout of activity takes place in late afternoon and ends at dusk. Calling during the afternoon session is generally less intense than in the morning.

The reproductive biology of this species is similar to that of *O. pumilio* (p. 302), including the elaborate parental care. These frogs do not engage in amplexus; instead both sexes sit with their vents touching and the male fertilizes the eggs as the female deposits them on the forest floor. The male tends to the developing eggs and repeatedly hydrates the clutch with water from his bladder in order to prevent desiccation of the eggs. On hatching, the female transports her tadpoles to water-filled leaf axils of arboreal bromeliads and repeatedly revisits her offspring during the larval development period to feed them unfertilized nutritive eggs.

With the onset of the dry season, in January, the male abandons his home range, returning to his calling sites when the next rainy season commences. The calls of male *O. granulifera* may

serve both to attract females and to ward off the males that invade their territories. When an intruding male ignores the acoustic signals and trespasses into another male's territory, the resident male invariably tries to remove the intruder from his domain.

Populations vary significantly in color pattern. In the south, in the Osa Peninsula, individuals have a red head and dorsum; in the north, the head and dorsum are green; moving north to south, populations change from yellow-green to tan, orange, or (approaching the southern end of its range) orange-red. In a few localities, populations can be polymorphic, with different color morphs present. A recent study has shown that the red 'southern' *O. granulifera* are genetically isolated from the green and polymorphic 'northern' populations, with noticeable differences in both calls and morphology. Individuals from the Osa Peninsula, for example, tend to be smaller and produce longer calls than those from the northern part of the range. Many populations of *O. granulifera* are isolated, a situation that is exacerbated by the destruction of suitable habitat between populations, and this will likely lead to further genetic divergence between morphs. The evolutionary advantage of losing the bright red warning coloration in some populations is still under investigation, but the results from one study suggest that green, cryptic *O. granulifera* may be more toxic than red individuals.

O. granulifera is still regularly encountered in appropriate habitat throughout its range but, as deforestation proceeds, this species is increasingly

300

forced into disturbed or cultivated areas. Some populations are known to persist in secondary forest and even plantations, but their population density is generally lower than in undisturbed forest habitat. This species occurs in a few protected areas, including Corcovado National Park.

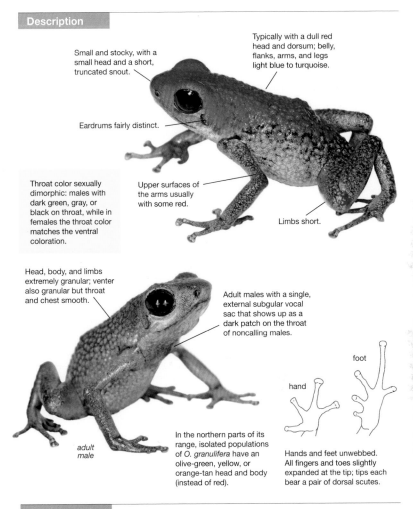

Small and stocky, with a small head and a short, truncated snout.

Typically with a dull red head and dorsum; belly, flanks, arms, and legs light blue to turquoise.

Eardrums fairly distinct.

Throat color sexually dimorphic: males with dark green, gray, or black on throat, while in females the throat color matches the ventral coloration.

Upper surfaces of the arms usually with some red.

Limbs short.

Head, body, and limbs extremely granular; venter also granular but throat and chest smooth.

Adult males with a single, external subgular vocal sac that shows up as a dark patch on the throat of noncalling males.

foot

hand

adult male

In the northern parts of its range, isolated populations of *O. granulifera* have an olive-green, yellow, or orange-tan head and body (instead of red).

Hands and feet unwebbed. All fingers and toes slightly expanded at the tip; tips each bear a pair of dorsal scutes.

O. pumilio (p. 302)

O. granulifera closely resembles *O. pumilio* but differs in having very granular skin on its upper surfaces. The two species do not co-occur, except perhaps in the Río Sandbox area, Limón Province, where an introduced population of *O. granulifera* may still persist.

301

Oophaga pumilio
Strawberry Poison Dart Frog

Least concern

Small, with smooth skin. A gemlike red frog; hind limbs and forearms usually blue, purplish-blue, or black. Males and females reach a standard length of 0.94 in (24 mm).

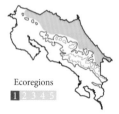

Oophaga pumilio occurs in eastern central Nicaragua, Costa Rica, and northwestern Panama, including many islands in Bocas del Toro; from sea level to 3,225 ft (980 m).

Ecoregions
1 2 3 4 5

Natural history

The blue legs of this species give rise to its nickname, blue jeans frog. Although some color and pattern variation exists in Costa Rican populations, most are red or orange, sometimes with blue forearms and legs. In the Bocas del Toro region of Panama, just south of the Costa Rican border, *O. pumilio* occurs in an extremely diverse combination of colors and patterns, including blue, green-yellow, white, and brown, often with bold round spots. Many islands in the Bocas del Toro Archipelago are each home to a unique color morph of this species.

This is a common to abundant species throughout its range. Because it is diurnal—and often occurs in altered habitats close to human settlements—it is fairly easy to see. Males can be heard calling incessantly throughout the day, and sometimes even on bright moonlit nights. They produce a harsh, insectlike *buzz-buzz-buzz* that serves to advertise their presence and mark out their territory; after sunrise and as the day heats up, this call can be heard at locations throughout its range.

There is no amplexus in Costa Rican dendrobatids; male and female mate vent to vent. Oviposition takes place in moist leaf litter; females generally lay 3–5 eggs on the forest floor. The eggs are immediately fertilized by the male, who then guards the clutch, occasionally emptying his bladder onto the eggs to prevent them from dehydrating. Individual males can tend to as many as three clutches at the same time, usually located on the same leaf. After about one week, the eggs hatch, and the female returns to the clutch, waiting for one of the tadpoles to wriggle onto her back and attach itself using its oral disk. The female carries the tadpoles, 1–4 at a time, to the water-filled leaf axil of a bromeliad or other suitable plants such as *Dieffenbachia* or a large aroid.

The offspring of a single clutch are not always placed in the same plant, but can be distributed among different plants. *O. pumilio* tadpoles feed exclusively on unfertilized eggs provided by their mother. On her feeding trips, the female climbs a plant containing one of her tadpoles and starts by searching for the appropriate crevice. The resident tadpole indicates its presence by vibrating its tail, and the female backs into the leaf axil and deposits 1–5 nutritional eggs. Females only feed their own tadpoles and can recognize their offspring by the manner in which they vibrate their tail. Smaller tadpoles suck out the egg yolk through a little hole they bite in the egg's jellylike coating, whereas larger tadpoles often eat the entire egg. Until their metamorphosis into miniature frogs, some 43–52 days after hatching, the female parent feeds the tadpoles

In *Oophaga pumilio*, females usually transport only one to two tadpoles at a time.

every 4 days. On average a tadpole eats 20–50 nutritive eggs before reaching metamorphosis.

Male *O. pumilio* are very territorial and can remain in the same area for several weeks at a time; the territory typically measures 54–320 ft² (5–30 m²). An ideal territory includes several logs or rocks or other elevated calling sites, potential nest sites, plants with water-filled leaf axils for depositing tadpoles, and usually a buffer zone, which the male vigorously defends against intruding males. In experiments, two-thirds of males that were taken from their territory and moved several meters away returned to their territory.

O. pumilio eats large numbers of small invertebrates, particularly tiny ants, which comprise up to 86% of recorded prey. Ants contain high concentrations of alkaloids, which are important components of the skin toxin in dendrobatids—the diet of *O. pumilio*
is at least partly responsible for its toxicity. The bright red aposematic coloration in this species undoubtedly serves to scare off potential attackers, and *O. pumilio* appears to have few predators.

O. pumilio is an abundant, adaptable species that is more common in disturbed open habitats with a thick leaf litter layer than in undisturbed forest. In densely forested locations, it is usually found in tree fall areas and edge situations. Local populations, especially those on islands in the Bocas del Toro Archipelago, are susceptible to habitat loss due to development for tourism. These populations also fall prey to commercial collectors who collect for the international pet trade. Trade in *O. pumilio* is regulated through the Convention on International Trade in Endangered Species (CITES), but more research is needed to adequately assess the impact of commercial collecting on populations.

Description

Skin on head, body, limbs, and venter smooth, sometimes with fine granulations.

Orange-red to deep scarlet, often marked with diminutive dark streaks or spots.

Eardrums indistinct.

In some populations, the hind limbs and the forearms are red, but in most cases they are patterned with bright blue or purple spots on a black background.

Limbs short.

adult male

hand

foot

Adult males with a single, external, subgular vocal sac, recognizable as a brown patch on the throat when not calling.

Hands and feet unwebbed. All fingers and toes slightly expanded; tips of digits bear a pair of dorsal scutes.

Similar species

O. granulifera (p. 300)

No similar-looking species occur within its range, with the possible exception of a population of *O. granulifera* that was possibly introduced to extreme southeastern Costa Rica (Río Sandbox, Limón Province). *O. granulifera* invariably has coarse, granular skin.

Genus *Phyllobates*

The five species in the genus *Phyllobates* are among the most toxic amphibians in the world. Three South American species have been famously used by the Emberá in the Chocó region of Colombia to poison the tips of their blow darts. The skin toxins of the two Costa Rican *Phyllobates* are not quite as potent as those found in South American species, but care should be taken when handling these animals.

Frogs in the genus *Phyllobates* have bright, aposematic coloring; small, narrow disks on the fingers and toes; and lack webbing between their fingers and toes. The Costa Rican *Phyllobates* are both jet black, with a bold pattern of yellow, orange, or red longitudinal stripes (a pattern shared with a non-toxic mimic, *Pristimantis gaigei*, p. 500).

Phyllobates lugubris
Striped Poison Dart Frog

Least concern

A small frog. Glossy-black, with a pair of distinct yellow to orange dorsolateral stripes, cream to yellowish-green mottling on its limbs, and black eyes. Males to 0.83 in (21 mm) in standard length, females to 0.94 in (24 mm).

Ecoregions
1 2 3 4 5

Phyllobates lugubris inhabits lowlands and foothills on the Atlantic slope, from extreme southeastern Nicaragua to central Panama, just west of the Panama Canal; sea level to 1,975 ft (600 m).

Natural history

A secretive but fairly common species. It occurs in the leaf litter of both primary and secondary forests, often on the vegetated banks of slow-moving waters. It is most active during dusk and dawn and on overcast humid days, and rarely ventures from its hiding places in shaded sections of the forest floor.

The male's territory is usually near a fallen log or in a dense thicket of vegetation. Their call is a constant low trill that lasts several seconds. Breeding takes places mostly during the wet season. No amplexus takes place; instead, the male and female position themselves vent to vent. The female lays her eggs, which the male then fertilizes. Eggs are deposited in the leaf litter, guarded and periodically moistened by the male during the 9–14 days it takes for them to hatch. The tadpoles will climb onto the male's back, who then transports them, 5–10 at a time, to a nearby forest stream, were they complete their development. Larval development in the aquatic phase may take another two months before metamorphosis.

The diet of *P. lugubris* consists of a wide variety of small invertebrates, mostly ants and mites. These dietary items provide chemical precursors to the complex toxins that the frogs sequester in their skin. Some frogs of the genus *Phyllobates* rank among the most toxic amphibians in the world, but

P. lugubris does not pose a threat to humans. In fact, some populations of this species showed no detectable levels of batrachotoxins, although other toxic components were present. Still, the bright dorsolateral stripes are thought to be an aposomatic (warning) coloration to deter predators.

In Costa Rica, *P. lugubris* appears more common in the southeast than in the north. Nevertheless, its populations appear stable and no major imminent threats are known to this species. Likely, deforestation and water pollution in its breeding habitat are among the bigger threats.

Male *P. lugubris* transporting tadpoles to a suitable water basin, where they will complete their development.

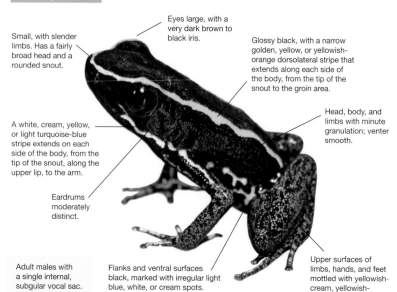

Eyes large, with a very dark brown to black iris.

Small, with slender limbs. Has a fairly broad head and a rounded snout.

Glossy black, with a narrow golden, yellow, or yellowish-orange dorsolateral stripe that extends along each side of the body, from the tip of the snout to the groin area.

A white, cream, yellow, or light turquoise-blue stripe extends on each side of the body, from the tip of the snout, along the upper lip, to the arm.

Head, body, and limbs with minute granulation; venter smooth.

Eardrums moderately distinct.

Adult males with a single internal, subgular vocal sac.

Flanks and ventral surfaces black, marked with irregular light blue, white, or cream spots.

Upper surfaces of limbs, hands, and feet mottled with yellowish-cream, yellowish-green, or turquoise.

Individuals from the south Atlantic coast of Costa Rica and nearby Bocas del Toro, Panama, often have more extensive yellow markings on their limbs and face.

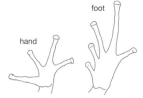

Digits on hands and feet long and skinny. Lacks webbing between fingers and toes. Finger and toe tips weakly expanded; each tip bears a pair of small scutes.

Pristimantis gaigei (p. 500)

Co-occurs with *P. lugubris*. It is similar in coloration but note copper (not black) iris. It is also larger.

Phyllobates vittatus (p. 307)

Very similar in appearance but does not co-occur.

Phyllobates vittatus
Golfo Dulce Poison Dart Frog

Endangered

A small frog. Glossy black, with a pair of distinct orange to reddish dorsolateral stripes, bluish-green mottling on its limbs, and black eyes. Males to 1.02 in (26 mm) in standard length, females to 1.22 in (31 mm).

Ecoregions
1 2 **3** 4 5

Phyllobates vittatus occurs in the Golfo Dulce region of southwestern Costa Rica and most likely also inhabits immediately adjacent Panama; sea level to 1,800 ft (550 m).

Natural history

This is a diurnal, terrestrial frog. It inhabits leaf litter of forested stream valleys in relatively undisturbed moist forest. It prefers dark areas in microhabitats with thick leaf litter, logs, and dense herbaceous vegetation. When approached, these shy frogs hide between tree roots, under logs, or in crevices.

Male *P. vittatus* usually call in the early morning and late afternoon, generally on cloudy days, hidden from view in leaf litter or underneath vegetation. Their call is a constant high-pitched trill produced in short bouts that last 4–6 seconds. Often, individual males respond to each other's vocalizations and a single call can set off a chain reaction of responses.

The courtship behavior of *P. vittatus* does not involve amplexus; instead, the male and female sit vent to vent. The female lays 7–21 eggs, generally in a shallow depression in the leaf litter, and the male fertilizes the eggs immediately after they are deposited. The male returns regularly to the egg clutch, emptying his bladder on the eggs about once a week to keep them from dehydrating.

When the eggs hatch, usually less than 18 days after deposition, the male transports the tadpoles on his back to small puddles, where they complete their development. Unlike species of *Dendrobates* that usually carry only one tadpole at a time, male *Phyllobates* have been seen with as many as 13 tadpoles on their back. Free-swimming tadpoles

have been observed in areas where rainwater has collected on the forest floor, but also in small pools formed in depressions in logs and in water-filled palm fronds that have fallen to the forest floor. Occasionally, tadpoles are deposited in pools or ponds near streams, and even in discarded water-filled cans and bottles.

During development, tadpoles feed on plant matter. Metamorphosis takes place approximately two months after hatching, and recent metamorphs already show the characteristic bright orange dorsolateral stripes at a size of about 0.5 in (13 mm) in standard length.

Like other dendrobatid frogs, *P. vittatus* is thought to acquire its skin toxins through its diet. In spite of the dangerous toxicity of related South American *Phyllobates*, the batrachotoxin levels in this species are relatively low and it poses no threat to humans. Nevertheless it is advisable to not touch your eyes or mouth after handling these frogs and to cover any cuts or scrapes to prevent toxins from entering the blood stream.

P. vittatus is fairly common but occupies only a small distribution range, making it vulnerable to habitat destruction. Its habitat in southwestern Costa Rica is threatened by encroaching agriculture (pineapple plantations and tree farms). Water contamination caused by runoff from agricultural lands and illegal gold mining may also be a problem.

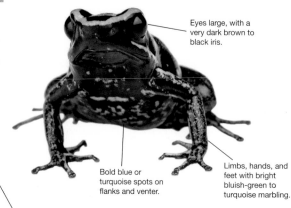

Adult males with a single internal, subgular vocal sac.

Eyes large, with a very dark brown to black iris.

A glossy-black frog, with a wide, reddish-orange or orange dorsolateral stripe that runs from the tip of the snout to the insertion of the hind limbs along each side of the body.

Bold blue or turquoise spots on flanks and venter.

Limbs, hands, and feet with bright bluish-green to turquoise marbling.

Small, with slender limbs; has a relatively robust head and a rounded snout.

A light blue or whitish lip stripe extends to the insertion of front limbs.

Eardrums fairly indistinct.

Head, body, and limbs finely granular; venter smooth.

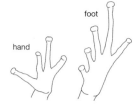

foot

hand

The width of the orange dorsolateral stripes, as well as the intensity and amount of the bluish spotting on body and limbs, varies between populations.

Hands and feet with long, skinny digits. Fingers and toe tips slightly expanded, topped with a pair of small dorsal scutes. No webbing between fingers and toes.

Similar species

Phyllobates lugubris (p. 305)

Similar in appearance but does not co-occur.

Genus *Silverstoneia*

This genus was created to accommodate eight cryptically colored species previously included in the genus *Colostethus*. They are characterized by having a light ventrolateral stripe and a light lateral diagonal stripe (which may be incomplete in some populations); absence of light dorsolateral stripes; basal toe webbing between toes III and IV; and, in adult males, the swollen appearance of the third (longest) finger. In addition, they lack the ability to sequester lipophilic alkaloid toxins in their skin and are therefore not toxic. Tadpoles of the frogs in this genus have unique umbrella-shaped mouthparts that serve to skim small food particles off the water surface.

These terrestrial frogs are usually found in leaf litter in or near streambeds, in the dry season, when water levels are low; in wetter months, they sometimes disperse into adjacent forest. Eggs are deposited in moist leaf litter; the male parent carries recently hatched tadpoles to a quiet section of a nearby stream, where they complete their development.

Two species of *Silverstoneia* occur in Costa Rica. *S. flotator* is widespread along both coasts and shows significant pattern variation throughout its range; *S. nubicola* just barely reaches Costa Rica and is known from a single locality in the extreme southeast, near the border with Panama. Both species are poorly defined and may contain several cryptic forms.

Silverstoneia flotator
Rainforest Rocket Frog

Least concern

A small terrestrial frog. Note brown dorsum, dark brown flanks, and two longitudinal white stripes, on each side of the body; the upper stripe begins in the groin and extends toward the upper eyelid (it is incomplete in some populations). Venter usually white or grayish; hind limbs uniform reddish-brown or tan, without dark transverse bands. Males to 0.67 in (17 mm) in standard length, females to 0.71 in (18 mm).

On the Atlantic slope, *Silverstoneia flotator* occurs locally in Costa Rica and in east-central Panama; on the Pacific slope, it is found in southwestern Costa Rica and adjacent western Panama; from sea level to 2,950 ft (900 m).

Ecoregions
1 2 3 4 5

Natural history

S. flotator is a diurnal, terrestrial species. It is found in leaf litter along forest streams, generally in relatively undisturbed forests, though it can also persist in altered habitats. Males are very vocal; they call at sunrise and during and immediately after daytime rainfall. The call is a slow series of whistling *peet-peet-peet* notes; it is primarily used to defend territories but also to attract potential mates. Males vigorously defend their territory against intruding males, and these tiny froglets can be seen wrestling one another; the dominant male will attempt to subdue any challengers by pinning them to the ground and sitting on top of their head.

Amplexus in *Silverstoneia* is typically cephalic (the male grasps the female around the head); the eggs are deposited in the leaf litter and then fertilized by the male. When the larvae hatch, the male collects the tadpoles and transports them on his back to a stream, where they continue their development; males have been observed to carry up to 12 tadpoles at a time. The tadpoles of this species have a large, funnel-shaped mouth that they use to filter floating food particles from the water surface. At night, feeding tadpoles swim just below the surface with their funnel-shaped mouth directed upward (see p. 312). When resting, they direct their mouthparts downward and use it as a suction device to cling to rocks in the current. The tadpoles have a powerful tail and are capable of navigating strong currents.

During the dry season, which is when their breeding activity peaks, these frogs can be abundant in the leaf litter and rocky substrate in drying streambeds. When the rainy season commences—and water levels rise, making the streambed habitat more treacherous—the frogs disperse into the adjacent forest. Especially during the wetter months of the year, it is not uncommon to encounter individual *S. flotator* far from streams.

S. flotator is a commonly observed species that is relatively tolerant of habitat alteration. Apparently it is not seriously affected by chytridiomycosis because it continues to persist in stream valleys where other amphibians have disappeared due to chytrid fungal infection.

Male *Silverstoneia flotator* routinely transport more than 10 tadpoles at a time.

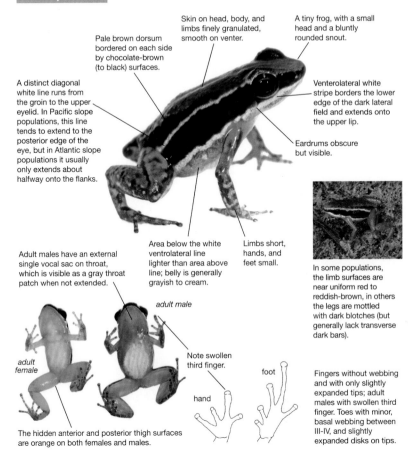

Skin on head, body, and limbs finely granulated, smooth on venter.

A tiny frog, with a small head and a bluntly rounded snout.

Pale brown dorsum bordered on each side by chocolate-brown (to black) surfaces.

A distinct diagonal white line runs from the groin to the upper eyelid. In Pacific slope populations, this line tends to extend to the posterior edge of the eye, but in Atlantic slope populations it usually only extends about halfway onto the flanks.

Venterolateral white stripe borders the lower edge of the dark lateral field and extends onto the upper lip.

Eardrums obscure but visible.

Adult males have an external single vocal sac on throat, which is visible as a gray throat patch when not extended.

Area below the white ventrolateral line lighter than area above line; belly is generally grayish to cream.

Limbs short, hands, and feet small.

In some populations, the limb surfaces are near uniform red to reddish-brown, in others the legs are mottled with dark blotches (but generally lack transverse dark bars).

adult male

adult female

Note swollen third finger.

foot

hand

The hidden anterior and posterior thigh surfaces are orange on both females and males.

Fingers without webbing and with only slightly expanded tips; adult males with swollen third finger. Toes with minor, basal webbing between III-IV, and slightly expanded disks on tips.

Similar species

Silverstoneia nubicola (p. 312)

Has a yellowish venter (white in *S. flotator*) and males have a black throat. Its hind legs are usually barred.

Allobates talamancae (p. 294)

Co-occurs locally with *S. flotator*. It is similar in appearance and habitat but it has a broad venterolateral stripe, a pale bronze or tan dorsolateral stripe on each side of the body, and lacks a diagonal white stripe.

311

Silverstoneia nubicola
Boquete Rocket Frog

Near threatened

A small terrestrial frog with a brown dorsum, dark brown flanks, and a pattern of two longitudinal white stripes on each side of the body; one of the stripes is a diagonal line that starts in the groin and extends toward the upper eyelid, but it may be incomplete in some populations. Venter usually yellowish; hind limbs marked with dark transverse bands. Males to 0.79 in (20 mm) in standard length, females to 0.91 in (23 mm).

Ecoregions
1 2 3 4 5

Silverstoneia nubicola occurs in extreme southwestern Costa Rica and adjacent western Panama; 3,450–5,250 ft (1,050–1,600 m). It also occurs in other regions of Panama, at lower elevations, and in western Colombia.

Natural history

This frog is diurnal. It is found in leaf litter along forested streams. Males call at dusk and dawn and, when it is overcast, at other times of the day; they make a rapid, whistling *peet-peet-peet* that lasts for about two seconds. These vocalizations serve as territorial markers as well as advertisement calls. Behavior and breeding biology is very similar to that of the much more common and widespread *S. flotator*. Amplexus is cephalic (the male grasps the female around the head); the eggs are laid in the leaf litter. The male transports hatched larvae to a nearby small stream, where they complete their development. The tadpoles of *S. nubicola* have modified, funnel-shaped mouthparts that they probably use to filter food particles from the water surface and to cling to rocks in the current, as has been observed in related *Silverstoneia* species.

In Costa Rica, *S. nubicola* is known from just a small area between Las Cruces and San Vito in the extreme southwestern border region with Panama. It is rarely observed in Costa Rica, but is widespread and common in Panama, where it can co-occur with *S. flotator*. No current data exists on the status of *S. nubicola* in Costa Rica, but other amphibians in the same area were severely impacted by chytridiomycosis. Nevertheless, in many areas *Silverstoneia* species seem to persist in the presence of chytrid fungal pathogens. This is a fairly adaptable species that inhabits undisturbed forests in parts of its range, but also occurs in somewhat disturbed settings.

Silverstoneia nubicola, as currently understood, is a confusingly variable species and future studies will likely show that it is in fact a complex of several cryptic taxa.

Silverstoneia tadpoles have unique funnel-shaped mouthparts that they probably use to filter food from the surface of the water in their breeding pool.

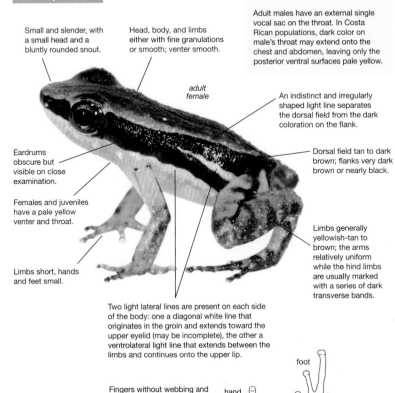

Small and slender, with a small head and a bluntly rounded snout.

Head, body, and limbs either with fine granulations or smooth; venter smooth.

Adult males have an external single vocal sac on the throat. In Costa Rican populations, dark color on male's throat may extend onto the chest and abdomen, leaving only the posterior ventral surfaces pale yellow.

adult female

An indistinct and irregularly shaped light line separates the dorsal field from the dark coloration on the flank.

Eardrums obscure but visible on close examination.

Dorsal field tan to dark brown; flanks very dark brown or nearly black.

Females and juveniles have a pale yellow venter and throat.

Limbs generally yellowish-tan to brown; the arms relatively uniform while the hind limbs are usually marked with a series of dark transverse bands.

Limbs short, hands and feet small.

Two light lateral lines are present on each side of the body: one a diagonal white line that originates in the groin and extends toward the upper eyelid (may be incomplete), the other a ventrolateral light line that extends between the limbs and continues onto the upper lip.

foot

Fingers without webbing and only slightly expanded tips. Adult males with a swollen third finger. Toes with minor, basal webbing between III-IV, and with slightly expanded disks on the tips.

hand

S. flotator (p. 310)

Lacks banded hind legs.

Allobates talamancae (p. 294)

Similar in appearance and habitat but has a broad venterolateral stripe and a pale brown dorsolateral stripe on each side of the body; it lacks a diagonal white stripe.

Family Eleutherodactylidae

Members of the family Eleutherodactylidae were
once contained in the huge catch-all genus
Eleutherodactylus, which was placed within the
subfamily Eleutherodactylinae (family Leptodactylidae).
But in 2008, an analysis of phylogenetic relationships
within the family Leptodactylidae led to its being
split into several groups that more accurately reflect
evolutionary relationships between species and species
groups. The Costa Rican species in the former family
Leptodactylidae are now divided over four families:
Craugastoridae, Eleutherodactylidae, Leptodactylidae
and Strabomantidae.

Frogs in the family Eleutherodactylidae are generally small, brown species that reproduce through direct development. Their evolutionary roots are in the West Indies, where they reach their highest diversity; some subgroups have also colonized regions in Central and South America.

The family Eleutherodactylidae is currently thought to contain 213 species. Of these, 192 form the genus *Eleutherodactylus*, whose members occur primarily on islands in the Caribbean Sea. The two species in this genus that occur in Costa Rica were recently introduced, and continue to expand their range as they colonize new territories.

Since all members of the family Eleutherodactylidae reproduce through direct development (and thus skip a free-living, aquatic tadpole stage), a clutch of eggs tucked away in a shipment of fruit or ornamental plants can result in hatched frogs establishing themselves in distant port cities. Undoubtedly, the species in this family that are native to Central America (the 11 species in the genus *Diasporus*) once reached their current home in a similar fashion. They most likely arrived on floating driftwood or other debris, a very long time ago.

Costa Rica is home to 7 species of Eleutherodactylidae: 5 native species of *Diasporus* and 2 species of introduced *Eleutherodactylus*. All are small, with a stocky body and small head, relatively short limbs, and unwebbed hands and feet. Identification of species is mostly based on the shape of their finger and toe disks, which can be exceedingly difficult to observe in the field.

Genus *Diasporus*

Frogs of the genus *Diasporus* are generally referred to as dink or tink frogs. These diminutive frogs produce a surprisingly loud *dink* call that is commonly heard at night, throughout much of the country. Finding the caller can prove to be tricky, however. These arboreal frogs often call from overhead, blocked from view by the leaf they are perched on. And, calling males tend to seek spots that are sandwiched between two leaves, so they can be hard to see even from above.

Identifying individual *Diasporus* can be challenging. Their very small size makes it difficult to discern diagnostic features under field conditions, and the extreme variability in their color and pattern (even a single individual looks very different at night than it does during the day) only complicates matters.

Eleven species of *Diasporus* are currently recognized, five of which occur in Costa Rica. They are tiny frogs (less than 1 in [25 mm]), with a narrow head, short limbs, and hands and feet that are free from webbing. Most species are identified based on the shape of the disk covers on their fingers and toes.

Diasporus diastema
Common Dink Frog

Least concern

A small, variably colored dink frog that lacks finger and toe webbing; it has oblong, moderately expanded finger and toe disks. It lacks conspicuously enlarged tubercles on its upper eyelids, but one or several small tubercles may be present. Males to 0.83 in (21 mm) in standard length, females to 0.95 in (24 mm).

Ecoregions
1 2 3 4 5

On the Atlantic slope, *Diasporus diastema* ranges from southeastern Honduras to the Canal Zone in Panama. On the Pacific slope, it occurs locally in the Guanacaste and Tilarán Mountain Ranges of Costa Rica; it is found more widely in the lowlands and foothills of the Talamanca Mountain Range in southwestern Costa Rica and adjacent Panama. From near sea level to 5,300 ft (1,615 m).

Natural history

D. diastema is a nocturnal, arboreal species. It is abundant throughout most lowland and foothill forests in the country. Its advertisement call, a loud clear *dink*, is often one of the most frequently heard late afternoon and nighttime sounds throughout its range. Males call from specific perches; these can be located in leaf piles on the forest floor, in dense shrubs, at sites at the top of the rainforest canopy, or on any arboreal perch in between. Call sites tend to be on top of a leaf, making these tiny frogs difficult to see when they perch somewhere above eye level. Generally, the call sites are relatively enclosed spaces such as the leaf axils of bromeliads or perches that are hidden within clusters of leaves.

The loud, explosive call serves as a territorial spacer but also to attract mates. A gravid female will approach a preferred calling male and nudge him, triggering the male to lead her to a nesting site, usually in a bromeliad axil, in a small tree hole, in a moss cushion, under bark, or within a leaf pile. The female deposits up to 20 eggs, but clutch sizes are usually smaller. The male fertilizes the freshly laid eggs, and the pair abandon their clutch; male *D. diastema* are not known to guard their eggs. In some cases multiple egg masses in various stages of development have been found in the same nest, indicating that multiple females may utilize a communal nest site. It is not known whether all clutches are fathered by a single male, but individual *D. diastema* have been observed to continue calling to attract additional females shortly after their original mate has laid her eggs. This species undergoes direct development,

skipping the free-swimming larval stage; young hatch as fully developed tiny froglets, not as tadpoles.

Although they usually call from concealed locations, on bright full moon nights these frogs often perch on exposed leaves. The reason for this behavior is not known.

Dink frogs have relatively short limbs and often walk across leaf surfaces in hand-over-hand fashion rather than jumping. Nevertheless, they are capable of powerful jumps when threatened and move quite nimbly through the dense vegetation of their arboreal habitat.

Due to their tiny size and the subtle nature of the diagnostic features that distinguish the various species of *Diasporus*, they are difficult to identify in the field. This can cause confusion in areas in which *D. diastema* overlaps with other *Diasporus*; it overlaps with *D. vocator* in the southern Pacific lowlands and foothills, and it may overlap with *D. hylaeformis* at middle elevations. It is also important to note that subtle variations in color, size, and calls suggest that different populations of this species in fact represent a complex of cryptic species. More research is needed in the taxonomy and population structure of *D. diastema* to clarify this issue.

Local populations of *D. diastema* appear to undergo irregular population fluctuations, possibly in response to weather patterns. Given its ability to persist in significantly altered habitats and adapt to changes in its environment, no major threat to this common species is currently apparent.

317

Often, a pair of light cream, tan, or pinkish dorsolateral stripes extends from the shoulder region to the pelvis. Usually a pinkish or orange-hued tubercle marks the posterior end of each stripe. Stripes are more easily discerned during the day, when the dorsal coloration is darkest.

Adult males lack a nuptial patch on the base of their thumbs.

A tiny arboreal frog. Has a small head and a rounded snout that is often tipped with a low, fleshy tubercle.

daytime coloration

A pattern of indistinct, alternating light and dark bands may be present on the hind limbs.

Eardrums indistinct.

Upper eyelids usually bear one or more low tubercles, but none are conspicuously enlarged or pointed.

Usually a light band is present across the top of the head, between the eyes; this band is bordered by a dark band or blotch on the back of the head.

Dorsum mostly smooth, but sometimes with scattered tubercles.

Posterior thigh surfaces marked with brown pigment.

daytime coloration

Limbs short.

hand

foot

Undersurfaces of hind limbs tan, orange, or brown; venter cream to yellow, marked with dark brown speckling.

Relatively small hands and feet lack interdigital webbing. Disk covers on fingers and toes distinctly expanded and clearly wider than digits; most disk covers rounded, some may be weakly pointed.

Color pattern strikingly different between day and night. During the day, dorsum mottled gray to dark brown, with various irregular dark markings. At night, colors fade and dorsum may be nearly uniform pink, tan, or salmon, with indistinct markings.

nighttime coloration

Throat yellowish, especially in adult males, whose yellow vocal sac is clearly visible as wrinkled skin on the throat when not in use.

Similar species

D. hylaeformis
(p. 320)

Has more sharply pointed finger and toe disks and often shows a significant amount of red in its color pattern. Generally found at higher elevations.

D. vocator
(p. 326)

Often darker in coloration, smaller in size, and has narrow, pointed finger and toe disks.

D. tigrillo
(p. 322)

NO PHOTO

Shows a distinctive spotted pattern on a tan-yellow background. Note its very restricted distribution range.

Pristimantis ridens
(p. 506)

Has distinctly raised nostrils; a fleshy, pointed tip on its snout; numerous enlarged tubercles on its upper eyelids; and greatly enlarged, rounded disks on the outer two (III and IV) fingers (the disks on the two inner fingers are significantly smaller).

Diasporus hylaeformis
Montane Dink Frog

Least concern

A small montane dink frog. Lacks finger and toe webbing; has moderately expanded and bluntly pointed finger and toe disks. It also lacks conspicuously enlarged tubercles on the upper eyelids, although some small tubercles may be present. Many individuals show a red or orange hue in their coloration. Males to 0.87 in (22 mm) in standard length, females to 1.06 in (27 mm).

Ecoregions
1 2 3 4 5

In Costa Rica, *Diasporus hylaeformis* inhabits the Tilarán and Central Mountain Ranges; it is also found in the Talamanca Mountain Range, in both Costa Rica and western Panama; 4,920–8,200 ft (1,500–2,500 m).

Natural history

In its preferred habitat, dense montane forests, *D. hylaeformis* is often heard though not easily seen. It calls at night from low vegetation; call sites are usually in dense leafy tangles and other concealed spots where a calling male is difficult to spot. The call is a loud, clearly modulated *dink*, often repeated two or three times at brief intervals. Calling intensity increases on humid nights or after a significant afternoon rain. Interestingly, on bright, full-moon nights, individual *D. hylaeformis* often perch exposed, on leaves in low vegetation; on such nights they usually do not call.

Calling males respond to the calls of other males; the vocalizations serve both as territorial spacers and to attract females. *D. hylaeformis* breeds through direct development; its eggs are deposited in humid, arboreal microhabitats, often in leaf axils of bromeliads or under loose tree bark. Tadpole development and metamorphosis is completed within the egg, and the larvae hatch as tiny frogs.

Like other species in its genus, these frogs often move through their arboreal habitat by walking rather than jumping. Especially when traversing leaf surfaces, they walk with a characteristically slow gait, although they may jump surprising distances if startled. This means of locomotion is perhaps an adaptation to life in treetops, where a poorly aimed jump could place these frogs far from their preferred location.

This fairly adaptable frog can often be heard calling from secondary vegetation in areas with moderately altered habitats, including in areas close to human settlements. *D. hylaeformis* is still common in some middle elevation and highland areas that have been severely impacted by chytrid fungal infections, and where many of the local amphibian species have disappeared.

Description

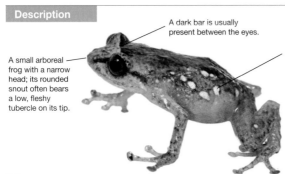

A dark bar is usually present between the eyes.

A small arboreal frog with a narrow head; its rounded snout often bears a low, fleshy tubercle on its tip.

At night, dorsal coloration can lighten significantly; in some individuals, nighttime coloration may reveal light blotches that are not visible, or are much less pronounced, during the day.

nighttime coloration

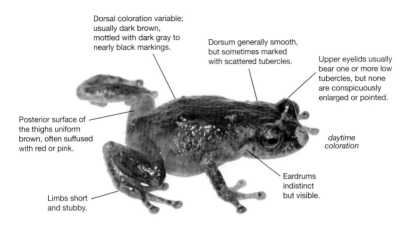

Dorsal coloration variable; usually dark brown, mottled with dark gray to nearly black markings.

Dorsum generally smooth, but sometimes marked with scattered tubercles.

Upper eyelids usually bear one or more low tubercles, but none are conspicuously enlarged or pointed.

Posterior surface of the thighs uniform brown, often suffused with red or pink.

daytime coloration

Eardrums indistinct but visible.

Limbs short and stubby.

Many individuals with reddish or pink hues on the dorsum. Particularly pronounced in adult females, which sometimes are nearly uniform red.

Adult males lack a nuptial patch on the base of their thumbs, but have a large, extendable vocal sac on the throat. Expanded vocal sac is yellow to pink.

Some individuals show a pale yellow or pink middorsal pinstripe.

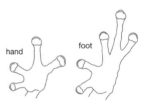

hand   foot

Hands and feet relatively small; without interdigital webbing. Disk covers on fingers and toes moderately expanded and clearly wider than digits; some disk covers are bluntly pointed or have a teardrop shape, while others are round to oblong.

Ventral surfaces are yellowish to cream, with discrete dark speckling.

Similar species

D. diastema (p. 317)

Disk covers on the fingers and toes are more rounded. Generally does not have red suffusion in its dorsal coloration. Occurs at lower elevations.

D. vocator (p. 326)

Slightly smaller and has narrowly elongated and pointed disk covers on its digits. Generally occurs at lower elevations.

D. ventrimaculatus (p. 324)

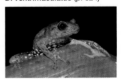

Has bold white spots on its ventral surfaces. Note very restricted distribution range.

Diasporus tigrillo
Spotted Dink Frog

Data deficient

A bright yellow-orange dink frog with a pattern of scattered brown spots on its dorsum, reminiscent of the pattern on spotted cats (hence *tigrillo*). Males to 0.71 in (18 mm) in standard length, females not known but probably slightly larger.

Ecoregions
1 2 3 4 5

Diasporus tigrillo has only been found in two localities in the Río Lari Valley, in extreme southeastern Costa Rica (Limón Province); 985–1,500 ft (300–460 m). Based on known records, this is a Costa Rican endemic, although it may eventually be found in adjacent Panama too.

Natural history

Until recently, only two individuals of *D. tigrillo* were known, collected in 1964. A survey of the original collection site, a premontane rainforest in an indigenous reserve, resulted in the rediscovery of this species in 2015. All individuals seen were adult males, collected in late afternoon or early evening from low vegetation in areas with disturbed secondary growth. Even though only few individuals of this species have ever been recorded, its presence in disturbed habitats indicates that *D. tigrillo* is likely adaptable and may be more widespread than is presently assumed. This species inhabits areas rarely visited by biologists and is therefore likely underreported.

Limited natural history observations indicate that males call from relatively low call sites, between 1.5–3 ft (40–100 cm) above the ground, usually deep inside dense brushy tangles. It is possible that *D. tigrillo* is more terrestrial than some other species in the genus and climbs out of the leaf litter at dusk to an elevated perch to call. Its call is similar to that of other dink frogs, but has been described as more closely resembling a short whistle rather than an explosive *dink*.

The type locality for *D. tigrillo* coincides with that of an enigmatic species that has not been observed in decades in Costa Rica, the Panamanian brook tree frog (*Duellmanohyla lythrodes*).

Description

- A small frog, with a short, rounded snout.
- Skin on head, body, and limbs smooth, with widely scattered low tubercles.
- Ventral surfaces coarsely granular.
- Eardrums indistinct.
- Upper eyelids usually bear one or more low tubercles, but none are conspicuously enlarged or pointed.
- Limbs short.
- Adult males lack nuptial patches on inner fingers, but have a single, extendable vocal sac on the throat.
- Dorsum yellow-orange; each scattered tubercle marked with brown pigment, creating a pattern similar to that of spotted jungle cats.
- Venter yellow, with small dark markings.
- Posterior thigh surfaces uniform yellow-orange.
- Vocal sac yellow.

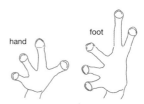

Hands and feet lack webbing. Disks on finger and toe tips weakly pointed and slightly wider than the digits themselves.

D. diastema (p. 317)

Lacks orange coloration and brown spots. Has rounded disk covers on its fingers and toes. Also note brown posterior thigh surfaces.

Pristimantis ridens (p. 506)

Has distinctly raised nostrils and a fleshy, pointed tip on its snout. Also has numerous enlarged tubercles on its upper eyelids. The two outer fingers have greatly enlarged, rounded disks, while the disks on the two inner fingers are significantly smaller.

Diasporus ventrimaculatus
Spot-bellied Dink Frog   **Not assessed**

A small montane dink frog. It lacks finger and toe webbing; has distinctly expanded, rounded to weakly pointed finger and toe disks. Dorsal coloration variable, ranging from nearly black to bright red; the venter is typically marked with large white spots. Has a very small distribution range. Males to 0.91 in (23 mm) in standard length, females to 0.98 in (25 mm).

Ecoregions
1 2 3 4 5

A Costa Rican endemic, only known from a very small area on the west side of the Río Terbi, in the southeastern section of the Valle del Silencio, Talamanca Mountain Range; 8,200–8,370 ft (2,500–2,550 m).

Natural history

This species was discovered in 2008 in a very small area within the Valle del Silencio, a high elevation valley, which is the only flat area in an otherwise steeply mountainous section of the Talamanca Mountain Range. It contains a large river and flooded areas with mature oak forest, bamboo, giant terrestrial bromeliads, ferns, and mosses. Although *D. ventrimaculatus* is reportedly abundant in the section of the valley where it was discovered, its total area of occurrence is tiny and estimated to be less than 1 mi² (about 2.5 km²).

D. ventrimaculatus is the only species of Costa Rican *Diasporus* in which males, females, and juveniles each have a distinct coloration.

Calling males produce a *dink* call from concealed locations, including the leaf axils of terrestrial bromeliads and perches in the tree canopy. They call with great frequency on humid nights; males are sometimes found in close proximity to one another. The loud call serves both to mark off territory and to attract mates. Presumably, the mating pair deposits its eggs in a concealed nest

location, where the larvae undergo direct development; however, these aspects of this species' breeding behavior have yet to be confirmed. Limited observations indicate that males occupy higher perches than do females, and that both sexes are often found between the leaves of the large terrestrial bromeliad *Greigia sylvicola* (which is prevalent in the Valle del Silencio) or in leaf litter accumulations associated with these bromeliads.

Based on limited surveys, this species' populations appear robust and stable. Given its tiny distribution range, however, it is potentially susceptible to any events that impact the habitat in its home range. In addition, many highland amphibians in the general vicinity of the Valle del Silencio have succumbed to chytrid fungal infections in recent history; such pathogens may also pose a risk to the survival of *D. ventrimaculatus*. Additional research is urgently needed to ascertain the full extent of the distribution range and population status of this species.

Dorsal coloration varies with age and sex. Adult males generally pinkish to red, while adult females are heavily marked with black spots, sometimes appearing nearly solid black. Juveniles more cryptic, mottled with brown and gray.

Adult males lack nuptial patches on inner fingers; but have a single, extendable vocal sac on the throat. Vocal sac orange or red.

female

Venter red to pink in males, black in females; marked with diagnostic large white blotches.

Ventral surfaces coarsely granular.

A small, robust frog with a small head and short, truncated snout. Skin on head, body, and limbs ranges from smooth to distinctly granular.

Upper eyelids lack distinctly enlarged tubercles, but may have scattered, small tubercles.

Usually shows a dark bar across the top of the head, between upper eyelids; sometimes difficult to discern in very dark individuals.

hand

foot

Color of posterior thigh surfaces matches dorsal coloration.

Eardrums indistinct.

Limbs short; hands and feet relatively large and lack webbing. Long, slender digits bear narrow lateral skin fringes. Disks on finger and toe tips rounded to weakly pointed, and distinctly wider than digits.

Diasporus hylaeformis (p. 320)

Lacks bold white spots on its venter. Occurs in mountainous areas close to small range of *D. ventrimaculatus*.

Diasporus vocator
Pacific Dink Frog   **Least concern**

A very small dink frog. Most disk covers on the finger and toe tips are distinctly pointed. Dorsum usually shows dark coloration. Males to 0.63 in (16 mm) in standard length, females to 0.71 in (18 mm).

Ecoregions
1 2 3 4 5

Diasporus vocator occurs in the Pacific lowlands and foothills from southwestern Costa Rica to central Panama, and crosses over onto the Atlantic slope in Panama; from near sea level to 5,415 ft (1,650 m).

Natural history

This tiny frog produces a loud, distinctive call but is more often heard than seen. Mostly terrestrial, it hides in leaf litter on the forest floor during the day. At dusk and after dark, especially on rainy nights or after strong afternoon rains, males call from hidden perches on the ground or from low vegetation. Its call is similar to that of other species in the genus *Diasporus* and consists of a surprisingly loud, clear *dink*. This call is often repeated two or three times at brief intervals, and neighboring males respond to each other's calls, making it difficult to hone in on a single calling individual.

This species reproduces through direct development. Small clutches of 4–6 eggs are deposited in leaf litter or in leaf axils of bromeliads or other suitable plants; its larvae develop entirely inside the eggs. Calling and breeding activity takes place mainly during the rainy season, roughly from June through November.

Throughout its range, *D. vocator* is a common species in appropriate habitat. It prefers dense rainforests, but tolerates some habitat alteration; it persists in the interior of disturbed areas that have sufficient canopy cover to maintain adequate humidity levels and a sufficiently thick leaf litter layer on the forest floor. Known populations of *D. vocator* appear to be secure, and this species is often seen.

Proper identification under field conditions can be difficult due to this species' tiny size and correspondingly diminutive diagnostic features. There is considerable variation between populations in coloration and in the overall shape and size of these frogs; it is possible that multiple cryptic forms are contained in what we now consider *D. vocator*, and additional research is needed to clarify this.

Description

Often has a dark bar between the eyes, across the top of the head; the area in front of the bar—on top of the snout—is usually paler than the dorsum.

Usually has an indistinctly defined dark band that extends backward from the rear margin of each eye toward the groin.

Dorsum usually dark brown, olive, or gray.

nighttime coloration

Adult males lack nuptial patches on inner fingers; they have a single, extendable vocal sac on the throat.

Eyes very large; upper eyelids generally smooth, although some individuals have a small, low tubercle on the top of each eyelid.

A tiny frog. Has a fairly long snout and only slightly raised nostrils, giving the snout a pointed appearance when seen in profile.

Skin on head, body, and limbs very finely granular; ventral surfaces coarsely granular.

Posterior thigh surfaces mottled dark brown.

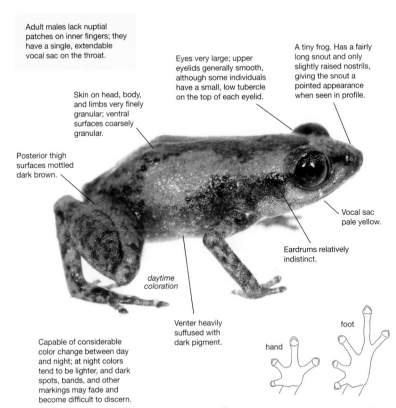

Vocal sac pale yellow.

Eardrums relatively indistinct.

daytime coloration

Venter heavily suffused with dark pigment.

Capable of considerable color change between day and night; at night colors tend to be lighter, and dark spots, bands, and other markings may fade and become difficult to discern.

hand

foot

Limbs short. Hands and feet lack webbing. Disks on finger and toe tips narrow, barely wider than digits. Most disks elongated and sharply pointed.

Similar species

D. diastema (p. 317)

Finger and toe disks are more prominently expanded and less pointed. Usually has at least one enlarged tubercle on its upper eyelid and often a fleshy tubercle on the tip of its snout.

Pristimantis ridens (p. 506)

Has distinctly raised nostrils, a fleshy, pointed tip on its snout, and numerous enlarged tubercles on its upper eyelids. The two outer fingers (fingers III and IV) have greatly enlarged, rounded disks, while the disks on the two inner fingers are significantly smaller.

Genus *Eleutherodactylus*

The genus *Eleutherodactylus* includes at least 192 species, most of which inhabit the islands of the West Indies. Some species, however, are native to northern Central America, Mexico, and the United States.

Most Caribbean species have a limited distribution range, and often occur on just a single island. Many of the these island species are threatened with extinction, while a few have rapidly expanded their range as they, or their eggs, are transported on cargo ships to other islands and port cities on the continent, or even as far away as the Hawaiian Islands. In some areas, the newcomers are so successful that they threaten native frog species—and may require population control or eradication.

Two species of introduced *Eleutherodactylus* occur in Costa Rica. One, *E. johnstonei*, has been in the country for several decades and occurs mainly in areas that produce ornamental plants. The other, *E. coqui*, is a recent arrival and appears to be restricted to an urban area in Turrialba. Introduction of the latter species has caused problems on the Hawaiian Islands, where it is now spreading rapidly due to a lack of native predators. The situation in Costa Rica should be carefully monitored.

Eleutherodactylus coqui
Puerto Rican Coquí

Least concern

A small, robust frog. It lacks webbing on hands and feet, but note tree-frog-like expanded disks on its fingers and toes. This species lacks conspicuously enlarged tubercles on its upper eyelids and limbs and has distinctly visible eardrums that are surrounded by a pronounced tympanic ring. Snout has a sharp-edged canthal ridge. Males to 1.42 in (36 mm) in standard length, females to 2.05 in (52 mm).

Ecoregions

1 2 3 4 5

Native to Puerto Rico, but introduced and expanding rapidly into areas of southeastern United States, the Hawaiian Islands, and several islands in the Caribbean. In Costa Rica, there is an introduced population in a residential area in the city of Turrialba, at approximately 2,100 ft (650 m).

Natural history

E. coqui is best known for its call, a loud two-noted *co-qui*. During the day these frogs seek shelter in the leaf litter; at night, adults ascend vegetation to forage for insect prey, while juveniles tend to remain in the leaf litter.

To attract a mate, adult males call from arboreal perches, usually fairly close to the ground but occasionally from the tree canopy. Females select a mate based on its call; once formed, the pair selects a suitable location to deposit their eggs. They seem to prefer enclosed areas such as rolled up leaves or leaf axils of bromeliads. Interestingly, fertilization of the eggs is internal. The pair will clasp their hind legs together, with their vents touching; the sperm enters the female and fertilizes her eggs as they are inside her body (in most frogs, the eggs are fertilized after they leave the female's body). The larvae skip a free-swimming tadpole stage and instead complete their development and metamorphosis while inside the egg. Male *E. coqui* guard the nest and defend it against intruding predators (including other coqui frogs, who are known to eat the eggs) until the young have hatched.

These frogs have proven themselves adept at colonizing new habitats, including human dominated landscapes. A single population of this species is currently found in the city of Turrialba, but there is a risk that *E. coqui* will expand its range in Costa Rica at the expense of native frogs or other native animals.

Skin on head, body, and limbs mostly smooth, without significantly enlarged or pointed tubercles (some small, inconspicuous tubercles may be present on the upper eyelids).

Head broad, with a relatively pointed snout that has a sharply defined canthal ridge between the eye and the nostril.

A small to medium-sized species; has a robust body and short, stocky limbs.

Dorsal coloration variable, usually olive-green, brown, or grayish-brown. Color pattern may be uniform or patterned. Commonly observed markings include an inconspicuous W-shaped mark in the shoulder region, paired light dorsolateral stripes, and a pale middorsal pinstripe.

Ventral surfaces coarsely granular.

A broad cream or tan band may be present on top of the head, between the eyes.

Eardrums distinctly visible, surrounded by a pronounced tympanic ring and bordered above by a glandular supratympanic skin fold.

foot

hand

Venter white or pale yellow with brown dusting.

Hands and feet lack interdigital webbing. Tips of digits with greatly expanded rounded or truncated disks.

Pristimantis ridens (p. 506)

Similar in overall appearance but smaller and with distinctly enlarged tubercles on the upper eyelids and greatly enlarged disks on its outer two fingers (finger III and IV) only. It generally has red markings in the groin area and on its hind limbs.

Eleutherodactylus johnstonei
Johnstone's Whistling Frog   **Least concern**

A small dink-frog-like species. Has a narrow head, large eyes, short limbs, and unwebbed hands and feet. Its digits are tipped with small round disks. The dorsum is distinctly tuberculate. Males to 0.98 in (25 mm) in standard length, females to 1.38 in (35 mm).

Ecoregions
1 2 3 **4** 5

Eleutherodactylus johnstonei is native to the Lesser Antilles but has been introduced to many Caribbean islands. In Central America, it can be found in Costa Rica (in San José) and in Panama (in Panama City). It has established populations in various locations in South America.

Natural history

In recent years this species has colonized many new areas after being inadvertently introduced on several Caribbean islands and the Central and South American mainland. These frogs and their eggs travel as stowaways in crates of fruits and ornamental plants; they tend to establish themselves near the ports that they are carried to.

E. johnstonei is mostly found in altered habitat or man-made structures, including greenhouses, plantations, farms, gardens, and even people's homes. It tends to shy away from forested areas but may occur in tree fall areas, clearings, and open areas along the forest edge.

In its native habitat, males call during the rainy season, attracting mates with an oft-repeated, two-note whistle. *E. johnstonei* reproduces through direct development. Small clutches of 10–30 eggs are deposited in moist terrestrial or arboreal microhabitats, where their larvae develop and metamorphose inside the egg and hatch as tiny froglets.

E. johnstonei is an invasive species in Costa Rica. Although its current distribution is limited, and it mostly occurs in areas that do not support many native frogs, its populations should be carefully monitored. It could potentially cause problems for native herpetofauna and attempts should therefore be made to halt the spread of this species.

Upper eyelids with several indistinct, low tubercles but never with distinctly projecting or enlarged tubercles.

Snout appears rounded when viewed from the side, but is relatively pointed when seen from above; canthal ridge sharply defined.

Dorsum with dense concentrations of low tubercles.

Tympanum indistinct but visible.

Heel and limbs with linear series of projecting tubercles.

A small, narrow-headed frog, with large eyes.

Dorsum generally marbled with shades of gray, tan, or brown; the dorsal tubercles are often marked with a darker coloration. Although variable in its color pattern, commonly seen markings include dark chevrons, a thin light middorsal stripe, or paired light dorsolateral stripes.

foot

hand

Ventral surfaces are grayish-cream to white.

Hands and feet lack webbing; digits with small but clearly defined round disks.

Diasporus diastema (p. 317)

Lacks a tuberculate dorsum and does not have round finger and toe disks.

Diasporus hylaeformis (p. 320)

Has relatively smooth dorsal skin and bluntly pointed disk covers on its digits.

Pristimantis ridens (p. 506)

Has conspicuously enlarged tubercles on its upper eyelids, smooth dorsal skin, and greatly enlarged disks on its outer two fingers (finger III and IV) only.

Family Hemiphractidae

These frogs mainly occur in South America. Of the more than 100 species in this family, three range north into Central America; only *Gastrotheca cornuta* occurs in Costa Rica.

The members of the family Hemiphractidae display a wide variety of morphologies, habitat preferences, and other natural history traits. There are several distinct subgroups within this family, and its taxonomy is uncertain. Historically, these frogs were thought to constitute a subfamily of the family Hylidae (tree frogs), undoubtedly because of the arboreal habits of many species and the presence of distinct disks on their fingers and toes. However, not all species share these traits, and many are terrestrial frogs without enlarged finger and toe disks.

One of the few traits that is common to all species currently included in the family Hemiphractidae is their unique breeding behavior: females carry the fertilized eggs on their back. Depending on the species, these eggs are carried to a body of water where the larvae complete their development, or the eggs hatch on the female's back and she carries the tadpoles to a body of water. In several species in the genera *Gastrotheca* and *Flectonotus*, the female carries her eggs around in a specialized brood pouch on her back, where the eggs undergo direct development, skipping the free-swimming tadpole stage.

Genus *Gastrotheca*

Sixty-five species of this predominantly South American genus are currently known; only one species, *Gastrotheca cornuta*, ranges into Costa Rica. Female *Gastrotheca* have a modified skin fold on their back that acts as a brood pouch. Fertilized eggs are placed in this brood pouch, which opens immediately above the vent, sometimes with help from the male; the female carries the eggs with her as she conducts her daily activities. Eggs undergo direct development; on hatching, tiny fully developed frogs emerge from the dorsal opening of the brood pouch.

Gastrotheca cornuta
Horned Marsupial Frog

Endangered

This is the only frog in Costa Rica with pointed, flaplike projections on its upper eyelids and expanded disks on the fingers and toes. Standard length to 3.19 in (81 mm).

Ecoregions
1 2 3 4 5

On the Atlantic slope, *Gastrotheca cornuta* occurs from central Costa Rica to central Panama; within Costa Rica it has been found from 650–1,650 ft (200–550 m). It also occurs on the Pacific slope, from eastern Panama through the Pacific lowlands of Colombia to western Ecuador.

Natural history

Although rarely observed, this species may not be as rare as is generally assumed. Its nocturnal habits and high canopy habitat make it difficult to observe. Its call—a loud, explosive *bop* not unlike the sound of a champagne bottle being uncorked—can be heard over a considerable distance. One or two calls are emitted at a time; the calls are separated by long intervals that may last over 10 minutes, which makes it difficult to locate these frogs.

The highly specialized breeding biology of this frog, in which the female carries fertilized eggs in a skin pouch on her back until fully developed frogs hatch from the eggs, makes this species completely independent of bodies of water for reproduction. Female *G. cornuta* have been observed with up to nine eggs in the brood pouch. Their eggs rank among the largest in the amphibian world and can measure up to 0.47 in (12 mm) in diameter.

Although *G. cornuta* does not rely on bodies of water for its reproduction, most of these frogs are observed in undisturbed wet lowland and foothill forests in the vicinity of sizeable rivers and creeks. As a canopy dweller, it does not tolerate the removal of canopy trees, which causes the understory habitat to dry out. The biggest threat to this species' survival is deforestation and other forms of habitat alteration, although in some parts of its range chytridiomycosis may have negatively impacted it as well.

G. cornuta is very rare in Costa Rica, and only a few individuals have ever been recorded, from scattered locations in the forested foothills of Limón Province. No marsupial frogs had been reported in Costa Rica for almost 20 years after they were intitially discovered in the country. But a few recent sightings, in 2013 and 2014, indicate that *G. cornuta* still persists in remote forested areas. Although habitat alteration has affected much of the area, suitable forests still exists on indigenous lands that are rarely surveyed, and this spectacular frog likely persists there.

G. cornuta has a highly unusual breeding strategy; females carry their fertilized eggs in a brood pouch on their back until the eggs hatch into miniature froglets. The opening of this brood pouch is visible above the vent as an inverted V.

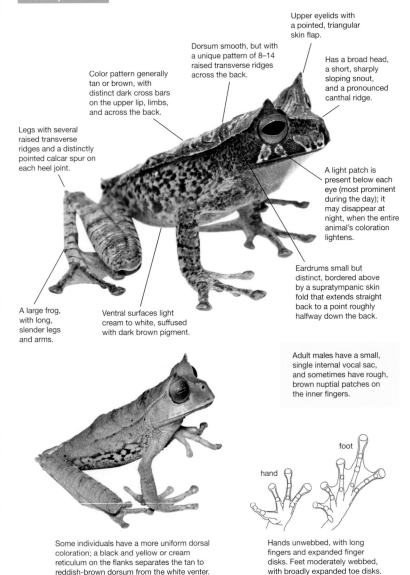

Upper eyelids with a pointed, triangular skin flap.

Dorsum smooth, but with a unique pattern of 8–14 raised transverse ridges across the back.

Has a broad head, a short, sharply sloping snout, and a pronounced canthal ridge.

Color pattern generally tan or brown, with distinct dark cross bars on the upper lip, limbs, and across the back.

Legs with several raised transverse ridges and a distinctly pointed calcar spur on each heel joint.

A light patch is present below each eye (most prominent during the day); it may disappear at night, when the entire animal's coloration lightens.

A large frog, with long, slender legs and arms.

Ventral surfaces light cream to white, suffused with dark brown pigment.

Eardrums small but distinct, bordered above by a supratympanic skin fold that extends straight back to a point roughly halfway down the back.

Adult males have a small, single internal vocal sac, and sometimes have rough, brown nuptial patches on the inner fingers.

foot

hand

Some individuals have a more uniform dorsal coloration; a black and yellow or cream reticulum on the flanks separates the tan to reddish-brown dorsum from the white venter.

Hands unwebbed, with long fingers and expanded finger disks. Feet moderately webbed, with broadly expanded toe disks.

Similar species

This species is unmistakable—there is no other frog remotely like it in Costa Rica.

Family Hylidae
(Tree Frogs)

This large and extremely diverse family includes over 950 species worldwide. Species occur in many regions of the world—including Eurasia, northern Africa, and Australia—but the family Hylidae reaches its greatest diversity in the New World tropics. This is the largest family of frogs in Costa Rica, which is home to 44 species, distributed among 15 genera.

This family's common name is something of a misnomer, as not all species dwell in trees; some members of this family live on the forest floor or even burrow below its surface. Nevertheless, all Costa Rican tree frogs do spend a significant portion of their life in trees. On the tips of their toes and fingers, they have expanded, roughly equal-sized, round disks that help them stick to vertical surfaces; a specialized joint structure in the toes and fingers provides maximum adhesion by allowing a frog to keep its adhesive disks flat against the climbing surface, regardless of the angle formed between that and the frog's limbs. In addition, almost all Costa Rican tree frogs have some amount of webbing between their fingers and toes (generally, the feet are more extensively webbed than the hands). Like the disks on their digits, this webbing can improve adhesion to leaf surfaces; and, for species that breed in ponds or streams, it may also aid in swimming. Some species of tree frog are even known to use their extensively webbed hands and feet as an airfoil; as they glide down from the canopy, their spread hands and feet help them slow their descent and steer towards a suitable landing site.

The tree frogs of Costa Rica are extremely variable in shape and size. On one end of the spectrum is the tiny *Isthmohyla zeteki*, whose adult males reach approximately 0.94 in (24 mm); on the other are giants like *Trachycephalus typhonius*, whose adult females reach at least 4.49 in (114 mm).

Hylids live in a wide variety of habitats and climates, including, at one extreme, the hot, dry lowlands of Guanacaste, and, at the other, the sometimes cool, wet summits of Costa Rica's mountains. In many parts of Costa Rica, particularly in the lowlands, hylids are among the more conspicuous amphibians, especially on warm, rainy nights, when large numbers of breeding males fill the air with their advertisement calls.

All Costa Rican hylid tadpoles pass through a free-swimming stage. Depending on the species, the watery habitats parents may select for the development of their tadpoles include rapid mountain streams, temporary ponds, treetop bromeliads, and water-filled cavities in living trees or fallen logs.

Many hylid species are easily identified based on their specific, and often showy, color pattern. However, others can be quite difficult to identify in the field. Though some small tree frogs resemble glass frogs, note that the eyes of tree frogs are generally directed sideways, while the eyes of glass frogs are directed forward. In addition, glass frogs have transparent skin on the belly; tree frogs do not.

Costa Rican frogs in a few other families, including Craugastoridae and Pristimantidae, may superficially resemble tree frogs because they have expanded disks on their toes and fingers, and they are sometimes found in trees or perched on low vegetation. Close examination of such frogs, however, often reveals that the disks on the hands and feet are neither round nor of equal size. Also, members of the family Pristimantidae lack webbing between the toes and fingers.

But even distinguishing among Costa Rican members of the family Hylidae can be a challenge. Some species show great variation in appearance; and, some species are very similar in appearance to other species. In these cases, geographic location or elevational distribution may help in making a correct identification.

Since the late 1980s, many tree frog species, especially those populations found at middle and high elevations, have seen dramatic declines. Many species that were once abundant went missing several decades ago. A few species have recently started to show signs of recovery. In addition, some enigmatic species have always been exceedingly rare, and the chances of encountering them are marginal at best. *Isthmohyla xanthosticta*, for example, is only known from a single individual.

Revisions of tree frog taxonomy have led to significant changes, and starting in 2005 many of Costa Rica's Hylidae have been reassigned and renamed. And, several new species of tree frog have recently been added to the herpetofauna of Costa Rica, including one introduced species, the Cuban tree frog (*Osteopilus septentrionalis*). Some of the most notable changes affecting Costa Rican species involve splitting the old genus *Hyla* into eight genera. In addition, the horned marsupial frog (*Gastrotheca cornuta*) has been removed from Hylidae and placed in the family Hemiphractidae (formerly considered a subfamily of Hylidae).

One noteworthy phylogenetic grouping in the family Hylidae is the subgroup of so-called leaf frogs, assigned to the subfamily Phyllomedusinae. The frogs in this subgroup all are relatively large, have long, spindly limbs, and are identified by their vertically elliptical pupils. This subgroup has two genera, *Agalychnis* and *Cruziohyla*, and includes such well-known examples as the red-eyed leaf frog (*Agalychnis callidryas*). All other Costa Rican tree frogs belong to the subfamily Hylinae and have horizontally elliptical pupils.

Costa Rican tree frogs vary in shape, color, and size but are easily recognized by the rounded disks on their fingers and toes and the presence of webbing on the hands and feet, as seen in this masked tree frog (*Smilisca phaeota*).

Genus *Agalychnis*

This genus contains eight species, five of which occur in Costa Rica. One Costa Rican species formerly in *Agalychnis* was recently removed and is now placed in the genus *Cruziohyla*.

These are relatively large, skinny, long-legged tree frogs, with webbed hands and feet and with large finger and toe disks. *Agalychnis* are typically green, with contrastingly colored flanks, thighs, upper arms, and hands and feet. This group includes several species famous for their red eyes; others have equally striking yellow, gold, or silver irises. All have vertical pupils.

Frogs in the genus *Agalychnis* generally breed in temporary or permanent pools of water, preferably those without fish that can prey on their larvae. Eggs are laid in clumps and attached to vegetation or rocks overhanging the water. On hatching, the tadpoles drop into the water, where they complete their development. Larvae of the various *Agalychnis* species are usually seen near, or just below, the water surface, swimming at a 45 degree angle, and moving only the filamentous, whiplike tip of their tail to maintain a suitable position in the water column. When disturbed, the tadpoles rapidly dive to the bottom.

Agalychnis annae
Golden-eyed Leaf Frog

Endangered

This large green frog has yellow irises, vertical pupils, and blue flanks. Males to 2.91 in (74 mm) in standard length, females to 3.31 in (84 mm).

Ecoregions
1 2 3 4 5

Agalychnis annae is found at middle elevations in the Tilarán, Central, and northern Talamanca Mountain Ranges, in both Costa Rica and adjacent Panama; 2,550–5,400 ft (780–1,650 m).

Natural history

Like other tree frogs, *A. annae* is arboreal and nocturnal; during the day, it sleeps on the underside of vegetation, with its limbs folded tightly against its sides to mask the outline of the body and prevent dehydration.

A. annae mainly breeds during the rainy season (May–November). It lays its eggs on the upper surfaces of large leaves that overhang pools. Males generally vocalize from relatively high perches overhanging a potential breeding pond; their advertisement call is a single *chac* note, repeated at intervals of several minutes. Amplexus takes place near the male's calling site; the pair descends to the pool, where the female absorbs water into her bladder before ascending to the oviposition site. Eggs have been found 1–10 ft (35 cm–3 m) above the water surface. On a rainy night, the tadpoles wriggle out of their gelatinous egg case and drop into the water below, where they complete their development. Like other *Agalychnis* tadpoles, the larvae have a threadlike tail tip that they move vigorously to maintain a diagonal, head-up position in the water column.

Long thought to be endemic to Costa Rica, *A. annae* was recently found in adjacent Panama. It was once quite common, even in urban areas, but has declined dramatically throughout its entire range. Indeed, since the early 1990s, it disappeared from most known sites, including protected areas such as the Monteverde Cloud Forest Preserve and Tapantí National Park. In recent years it has begun to recover, primarily in the metropolitan areas of the Central Valley, which are generally polluted

and unforested, while it remains largely absent from its natural rainforest habitat.

In the Central Valley, it can be found in landscaped gardens, city parks, coffee plantations, and other places where there is sufficient vegetation within which to hide and a nearby source of standing water in which to breed. It has been suggested that these generally contaminated sites may be even less hospitable to chytrid fungal pathogens than they are to the frogs themselves. Continued monitoring of recovering populations is needed to better understand the reasons for this species' decline and apparent recovery, and searches for additional populations, especially in undisturbed habitat within its historic range, are necessary to better evaluate the current conservation status of *A. annae*.

Calling male with inflated vocal sac. Note how at night their colors darken, showing a dark green dorsum and brownish-purple flanks.

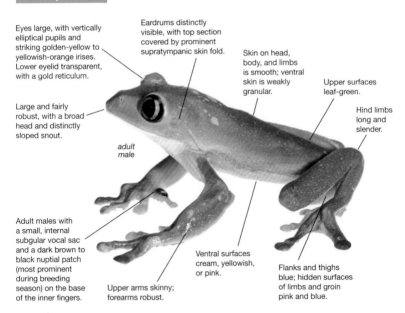

Eyes large, with vertically elliptical pupils and striking golden-yellow to yellowish-orange irises. Lower eyelid transparent, with a gold reticulum.

Eardrums distinctly visible, with top section covered by prominent supratympanic skin fold.

Skin on head, body, and limbs is smooth; ventral skin is weakly granular.

Upper surfaces leaf-green.

Large and fairly robust, with a broad head and distinctly sloped snout.

Hind limbs long and slender.

adult male

Adult males with a small, internal subgular vocal sac and a dark brown to black nuptial patch (most prominent during breeding season) on the base of the inner fingers.

Upper arms skinny; forearms robust.

Ventral surfaces cream, yellowish, or pink.

Flanks and thighs blue; hidden surfaces of limbs and groin pink and blue.

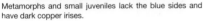
Metamorphs and small juveniles lack the blue sides and have dark copper irises.

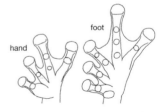

Fingers short and stocky, with greatly enlarged rounded disks; hands with significant webbing. Toes with large rounded disks that are only slightly smaller than the finger disks; feet with extensive webbing.

In adults, the combination of yellow irises and blue flanks is unique. Note that *A. annae* occurs at higher elevations than other *Agalychnis* species (with the sole exception of *A. lemur*).

Agalychnis callidryas
Red-eyed Leaf Frog

Least concern

A green tree frog with red eyes, vertical pupils, and dark blue to purplish-brown sides that are marked with light vertical bars. Males to 2.91 in (74 mm) in standard length, females to 3.31 in (84 mm).

Ecoregions
1 2 3 4 5

Agalychnis callidryas is widespread in Central America. On the Atlantic slope, it occurs from Mexico to eastern Panama; it also occurs throughout the Pacific slope of Costa Rica, except for some of the drier areas of the northwest; from near sea level to 4,100 ft (1,250 m).

Natural history

A. callidryas is nocturnal and arboreal; during the day it sleeps on the underside of a broad leaf, with limbs folded tightly against its sides to mask the outline of the body and to prevent dehydration. In suitable habitat, it is abundant. This frog occurs in lowlands and foothills and prefers humid forests with an intact canopy; it is usually found near temporary or permanent pools bordered by vegetation. But because it is fairly tolerant of altered habitat, it can also be found in gardens or agricultural zones, as long as there are trees and breeding ponds surrounded by broad-leaved plants.

Individual *A. callidryas* remain in the tree canopy during the day. On rainy nights, these frogs move back and forth between the tree canopy and vegetation surrounding their breeding pond, generally making slow hand-over-hand movements but sometimes jumping—then gliding down—from elevated positions.

Breeding activity happens near a body of standing water. It usually begins in late May or early June, triggered by the first rains of the rainy season; a second peak in breeding activity usually takes place toward the end of the rainy season, although breeding pairs can be seen throughout the year as long as there is sufficient rainfall. At dusk, males start to emit their advertisement call, a short single or double *chuck*, made as they descend to a perch near the surface of the water. Calling males do not remain in one position for very long. They move from perch to perch, often turning to call in different directions in an attempt to increase the chances of attracting a potential mate. Females attracted by the vocalizations of a particular male approach the calling individual, and axillary amplexus will ensue. Females tend to select the largest males, which are capable of producing the loudest calls.

Developing eggs on a palm leaf. Note the ant on the clutch; many invertebrate and vertebrate predators feed on red-eyed leaf frog eggs.

Agalychnis tadpoles have a characteristic whiplike filament on their tail tip; they typically maintain a relatively fixed position in the water column, floating at a 45 degree angle, only wiggling the tail tip to stay in a desirable spot.

343

Prior to egg-laying, the pair descends into a pond and remains there for several minutes, during which time the female fills her bladder with water. After the the pair emerges from the pond and selects a suitable site, the female deposits a clutch of 20–50 sticky eggs, which are instantly fertilized by the male. It is not uncommon for competing males to try to dislodge an amplecting male during oviposition, or even fertilize an egg clutch at the same time as an amplecting male. Molecular studies have shown that eggs in a single clutch can have multiple paternity. When the eggs are deposited, the female then empties her recently filled bladder onto the eggs, whose surrounding jelly coat then absorbs the water and swells. After laying her eggs the female, with the male still on her back, may return to a pond in order to refill her bladder and begin the process anew; females sometimes lay up to five clutches in the course of a single night. Before dawn, the frogs disengage and climb back to the forest canopy.

A. callidryas generally deposits its eggs on a large leaf overhanging a body of water, although a variety of other oviposition sites have been observed. Occasionally, the leaf holding an egg clutch is folded over the top of the eggs and held in place with the gluelike jelly coating of the eggs. Although related leaf frogs in the genus *Phyllomedusa* are known to actively fold leaves to protect their ergs, this behavior is not generally seen in frogs in the genus *Agalychnis*; perhaps *Agalychnis* merely deposit their eggs on leaves that are prone to curl.

Hatching tadpoles free themselves by wriggling rigorously until the egg capsule breaks—and then the tadpole drops into the water below. Sometimes newly hatched tadpoles land at the edge of the pond rather than in it, or they are not able to roll off the leaf because they have become stuck to the jelly of other eggs. In both cases, these tadpoles generally survive; they can live out of water for up to 20 hours and may wash into the pond with subsequent rain, or they can move to reach the water by using tail flips. Even inside the egg, *A. callidryas* tadpoles are able to respond to environmental cues. Experiments have shown that individual tadpoles are able to hatch and abandon the clutch prematurely when vibrations of the leaf that the clutch is located on signal that a predator may be present.

This species is very common throughout its range; prospects for its continued survival seem very positive. *A. callidryas* is tolerant of habitat alteration and can live in secondary forest as long as there is an intact canopy. Large numbers of these frogs were collected annually for the commercial pet trade, especially in Nicaragua. The commercial trade in all species of *Agalychnis* is now regulated by the Convention on International Trade of Endangered Species (CITES).

Description

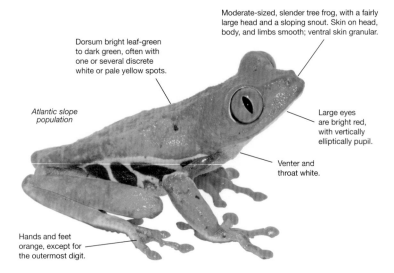

Moderate-sized, slender tree frog, with a fairly large head and a sloping snout. Skin on head, body, and limbs smooth; ventral skin granular.

Dorsum bright leaf-green to dark green, often with one or several discrete white or pale yellow spots.

Atlantic slope population

Large eyes are bright red, with vertically elliptically pupil.

Venter and throat white.

Hands and feet orange, except for the outermost digit.

Body slender and somewhat flattened; limbs long and skinny.

Pacific slope population

Eardrums distinct; bordered by a weak supratympanic skin fold.

Thighs and upper arms blue in Atlantic populations and orange in Pacific slope populations.

Flanks dark blue (in most Atlantic slope populations) to purplish-brown (in Pacific slope populations), with cream-colored or yellow vertical bars.

Lower eyelid transparent, with a gold reticulum.

hand   foot

Fingers short but robust, with moderate webbing and large, rounded disks. Feet extensively webbed; toes bear enlarged rounded disks.

juvenile

adult male

Metamorphs and very young juveniles can change color; they are green by day and reddish brown at night; note yellowish-bronze eyes and absence of light vertical bars on the flanks.

Adult male with its single, internal vocal sac expanded; note the rough, brown nuptial patches on its inner fingers.

Similar species

Other red-eyed tree frogs in the genera *Duellmanohyla* and *Ptychohyla* have horizontal pupils.

Agalychnis saltator (p. 348)

Also has red eyes, vertical pupils, and blue flanks, but lacks pale vertical bars on the sides.

Agalychnis spurrelli (p. 350)

Has red eyes and vertical pupils, but its flanks are orange (not blue).

Agalychnis lemur
Lemur Leaf Frog

Critically endangered

An extremely skinny, long-legged tree frog. Also note brown or green dorsum and huge, striking silver-white eyes with vertical pupils. Males to 1.61 in (41 mm) in standard length, females to 2.09 in (53 mm).

Ecoregions
1 2 3 4 5

Agalychnis lemur occurs on the Atlantic slope, throughout middle elevation forests, from northwestern Costa Rica (Tilarán Mountain Range) to western Panama. On the Pacific slope, there is an isolated population in northwestern Costa Rica; there are also Pacific slope populations in Panama and Colombia. From 1,440 to 5,250 ft (440 to 1,600 m).

Natural history

This arboreal, nocturnal frog occurs at low and middle elevations in pristine humid forests. At night, it is seen walking or climbing on low vegetation, in a deliberate hand-over-hand fashion. During the day, these frogs sleep attached to the underside of leaves. Breeding activity peaks during the wet season (April–August), when males emit a widely spaced, single *chick* note to attract females. Eggs are deposited on vegetation, attached to a leaf or twig that overhangs a shallow pond, marsh, or temporary rain pool. Egg clutches are small and contain 15–30 pale green or cream-colored eggs, encased in clear jelly; hatching takes place 1–2 weeks after oviposition, and the larvae drop into the water below, where they complete their development.

This species was always fairly uncommon throughout its range, but in recent decades it has undergone a drastic decline. Most known populations in Costa Rica have disappeared altogether and several Panamanian populations have declined significantly. As a species that resides in the forest interior, *A. lemur* is very susceptible to habitat alteration, though chytridiomycosis seems to be the most significant threat in undisturbed areas. Currently, this species is only known to survive in a handful of sites in Costa Rica—in the Caribbean lowlands near Limón—and in scattered locations in western Panama. While populations in eastern Panama are regularly encountered still, members of these populations differ in size and appearance from Costa Rican animals and may represent a distinct species.

A. lemur has seen some changes in its taxonomic placement in recent years and can be found in older literature as *Phyllomedusa lemur* and *Hylomantis lemur*.

Recent biochemical research on *A. lemur* skin peptides have shown them to have potential anti-bacterial, anti-cancer, and anti-diabetes functions, underscoring the importance of preserving this rapidly disappearing species.

Tadpoles about to hatch. The larvae will drop into a pool below and complete their development there.

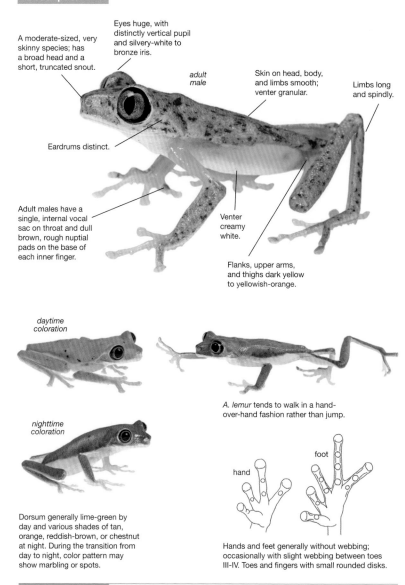

A moderate-sized, very skinny species; has a broad head and a short, truncated snout.

Eyes huge, with distinctly vertical pupil and silvery-white to bronze iris.

adult male

Skin on head, body, and limbs smooth; venter granular.

Limbs long and spindly.

Eardrums distinct.

Adult males have a single, internal vocal sac on throat and dull brown, rough nuptial pads on the base of each inner finger.

Venter creamy white.

Flanks, upper arms, and thighs dark yellow to yellowish-orange.

daytime coloration

nighttime coloration

Dorsum generally lime-green by day and various shades of tan, orange, reddish-brown, or chestnut at night. During the transition from day to night, color pattern may show marbling or spots.

A. lemur tends to walk in a hand-over-hand fashion rather than jump.

foot

hand

Hands and feet generally without webbing; occasionally with slight webbing between toes III-IV. Toes and fingers with small rounded disks.

Other *Agalychnis* species have webbed hands and feet and a different eye color.

Agalychnis saltator
Blue-sided Gliding Leaf Frog   **Least concern**

A red-eyed tree frog with vertical pupils, a green dorsum (can be tan to brown at night), and uniform blue flanks that lack a pattern of light vertical bars. Males to 2.13 (54 mm) in standard length, females to 2.60 in (66 mm).

Agalychnis saltator occurs on the Atlantic slope, in isolated populations, from northeastern Honduras to southeastern Costa Rica; from near sea level to 2,950 ft (900 m).

Ecoregions
1 2 3 4 5

Natural history

This is a relatively uncommon species that is found at low and middle elevations. It prefers pristine, humid and wet forests, where it inhabits the forest canopy. *A. saltator* only descends from the treetops to breed in temporary pools. It is an explosive breeder, and large breeding aggregations may gather when conditions are right; after periods of exceptionally heavy rain, large, wriggling clumps of frogs converge during the early morning hours to mate and lay eggs. These dense aggregations—there are reports of 25–400 frogs covering just a few meters of vertical vine surface—take place on vegetation overhanging temporary pools in forested areas.

Amplexing pairs deposit their egg clutches amid the mosses that cover vines, sometimes until the entire vine surface is enveloped with a layer of their dark eggs. Tadpoles will hatch and drop into the pool below, where they complete their development. Because of the large number of frogs in these breeding aggregations and because fertilization takes place externally, it is often impossible to assess which males actually fertilize which eggs, and some clutches may be fertilized by more than one male.

These large concentrations of frogs and eggs are a potential feast for predators, and adults and eggs fall prey to a variety of animals. Long columns of ants have been seen on egg-covered vines, carrying small amounts of egg yolk to their nests. Snakes such as the northern cat-eyed snake (*Leptodeira septentrionalis*), a well-known nocturnal predator of frogs and their eggs, have been observed in these breeding aggregations. The blunt-headed vine snakes (genus *Imantodes*) and parrot snakes (genus *Leptophis*) prey on adults at night or during the day, respectively. Other predators reported to feed on adult frogs in such breeding aggregations include hawks, coatis, and white-faced capuchin monkeys.

Communal breeding activity in *A. saltator* is generally triggered by rainfall and other environmental cues; when the rainfall begins, adult frogs move rapidly from their forest canopy retreat toward the breeding site. Individual frogs leap from considerable heights. During the descent, the frog extends its limbs and spreads out its fingers and toes, using the interdigital webbing as a parafoil to increase its surface area. This "parachuting" behavior slows down the fall and allows the frog to steer toward a suitable landing area such as an understory palm or other large-leaved plant. The large swaying leaves absorb much of the impact of the frog's landing.

When ascending trees, *A. saltator* climbs aerial roots, vines, and lianas with a rapid hand-over-hand motion that, along with parachuting, is an adaptation to life in trees. Parachuting has also been observed in some other Costa Rican tree frogs, including two other species of *Agalychnis* (*A. callidryas* and *A. spurrelli*), *Ecnomiohyla miliaria*, *Trachycephalus typhonius*, and *Scinax boulengeri*.

Because it spends most of its time in the canopy, *A. saltator* is rarely seen outside its breeding season. And, because it occurs in isolated populations—its preferred primary forest habitat is patchy—populations are likely underreported. That said, it is a widespread species that seems to maintain stable populations in the areas from which it is known, several of which are in protected sites. Also, commercial trade in this species for the international pet trade is now regulated by CITES.

348

Description

Iris bright red, with a vertically elliptical pupil. Lower eyelid transparent, with a gold reticulum.

Has a relatively large head and a bluntly rounded snout.

Dorsal skin smooth; venter granular.

Daytime dorsal coloration bright leaf-green, sometimes with one or a few raised cream (to pale yellow) dorsal spots. At night, the dorsal base color can change to tan, brown, or reddish-brown.

daytime coloration

A pattern of faint, irregular dark bands usually present across the dorsum; most visible at night.

A moderate-sized, slender species; limbs long and skinny.

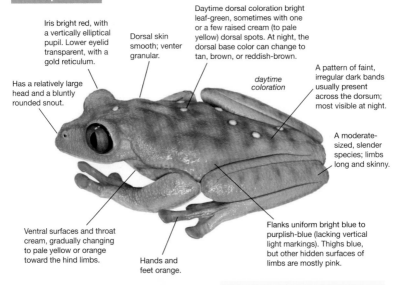

Ventral surfaces and throat cream, gradually changing to pale yellow or orange toward the hind limbs.

Hands and feet orange.

Flanks uniform bright blue to purplish-blue (lacking vertical light markings). Thighs blue, but other hidden surfaces of limbs are mostly pink.

Adult males with a single, internal subgular vocal sac; also note a rough brown nuptial patch (most prominent during breeding season) on the base of each thumb.

The absence of cream vertical bars on the blue sides distinguishes *A. saltator* from the otherwise similar *A. callidryas*.

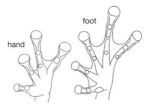

foot

hand

Hands with minimal webbing and greatly enlarged, round disks on all fingers. Feet with moderate webbing and large rounded toe disks.

Similar species

Tree frogs in the genera *Duellmanohyla* and *Ptychohyla* also have red eyes, but note horizontal pupils and the absence of blue on the flanks.

A. callidryas (p. 343)

Larger. Has vertical white, yellow, or cream-colored bars on its flanks.

A. spurrelli (p. 350)

Much larger. Has bright orange flanks.

349

Agalychnis spurrelli
Orange-sided Gliding Leaf Frog

Least concern

A large, green tree frog with dark red eyes, vertical pupils, and bright orange flanks, thighs, and upper arms; its orange hands and feet are huge and extensively webbed. Males to 2.20 in (56 mm) in standard length, females to 2.83 in (72 mm).

Ecoregions

1 2 3 4 5

Agalychnis spurrelli occurs in isolated populations in southeastern and southwestern Costa Rica; from near sea level to 2,460 ft (750 m). It is also found in the Atlantic lowlands of Panama, the Pacific lowlands of Colombia, and in northwestern Ecuador (these populations were formerly considered a different species, *A. litodryas*).

Natural history

A. spurrelli is an elusive inhabitant of lowland forest canopies. It frequently escapes detection for long periods of time, but heavy noctural rains sometimes trigger breeding activity in which aggregations of hundreds of individuals may congregate around the temporary pools that are used for breeding. Egg masses are deposited on vegetation overhanging the breeding pond, on the upper surfaces of leaves, vines, and branches, 3–25 ft (1.5–8 m) above the water surface. On hatching, tadpoles drop into the water below.

Breeding generally takes place early in the morning. Immediately after oviposition, the amplecting pairs separate and disappear back into the canopy, climbing in slow hand-over-hand fashion. *A. spurrelli* is known to use its extensively webbed hands and feet to "parachute" from the canopy, performing a rapid and controlled descent from the treetops by using its spread extremities as a parafoil to slow down and direct its fall. Gliding frogs aim for large-leaved plants such as small understory palms and heliconias, whose foliage can absorb the impact of their rapid descent.

The elusive habits of *A. spurrelli* make it difficult to accurately assess its population and conservation status. However, based on the number of frogs gathering at known breeding aggregations, it is likely a much more common species than is generally assumed. Since this is a canopy species, it is probably most heavily affected by destruction of its forest habitat.

During explosive breeding events, hundreds of *A. spurrelli* may gather simultaneously to breed.

Pupils vertically elliptical. Eyes usually a darker red than those of other Costa Rican "red-eyed tree frogs." Lower eyelid transparent, with a gold reticulum.

Skin on head, body, and limbs smooth; venter granular.

On some individuals, the dorsum shows isolated, raised cream or white spots that are sometimes outlined in black.

Large and robust, with a broad head and sloping snout.

Throat and lower lip white.

Eardrums prominent.

Flanks, upper arms, and thighs with bold yellow to orange coloration; most intense orange is on the hands and feet.

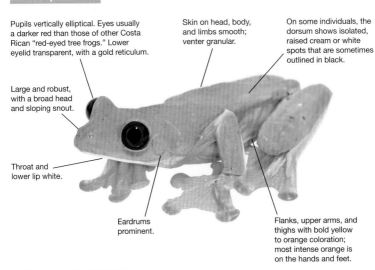

juvenile

Small juveniles are sometimes reddish-brown at night.

Adult males with a single, internal subgular vocal sac; during the breeding season, they show rough, brown nuptial patches on the base of the thumbs.

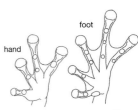

foot

hand

adult

Dorsum uniform leaf green; at night, adult's dorsum usually turns a darker shade of green.

Orange hands and feet huge, with extensive webbing; fingers and toes with greatly expanded, rounded disks.

A. callidryas
(p. 343)

A. saltator
(p. 348)

A. lemur
(p. 346)

Cruziohyla calcarifer
(p. 357)

Has a pattern of vertical light bars on otherwise blue or brown flanks.

Has solid blue flanks.

Sometimes has orange flanks, but it never has red eyes.

Has yellow-bordered purplish eyes and a pattern of vertical bars on the flanks.

Genus *Anotheca*

This genus contains a single, unique species. *Anotheca spinosa* is a large tree frog with a crownlike series of knobs and spines on top of its head. It is strikingly patterned, with distinctive dark brown or black bars with pale outlines, spots, and blotches on a silver-gray to light brown background. This species displays a unique breeding strategy perfectly adapted to a life in the treetops. Its eggs are laid in water-filled tree holes, and the tadpoles develop on a diet of unfertilized, nutritive eggs provided by the female parent.

Anotheca spinosa
Crowned Tree Frog

Least concern

This is the only frog in Costa Rica with a series of projecting, bony spines on the back of its head. Males to 2.72 in (69 mm) in standard length, females to 3.31 in (84 mm).

Ecoregions
1 2 3 4 5

Anotheca spinosa occurs in isolated populations over a large geographic area. It is known from the Atlantic slope of Mexico, Honduras, Costa Rica, and Panama (there are no records for this species in Belize, Guatemala, El Salvador, or Nicaragua). In Costa Rica, it occurs in scattered sites both on the Atlantic slope (in the northern half of the country) and in the extreme southwest; 1,150–4,425 ft (350–1,350 m).

Natural history

A. spinosa occurs at low and middle elevations, in humid and wet forests; it can also be found in areas with mature second growth. Little information exists on the population size of this species because it lives high in the forest canopy. Nevertheless, its call, a loud, far-carrying *boop-boop-boop*, can be heard in appropriate habitat.

A. spinosa is able to complete its entire life cycle within the confines of the treetops. It reproduces in water-filled cavities in the canopy, although breeding sites much closer to the ground have also been observed. It lays its eggs just above the water surface, glued to the inside wall of a tree hollow or bamboo internode, or to the underside of tank bromeliad leaves. *Anotheca* tadpoles are able to breathe atmospheric oxygen immediately after hatching, unlike most early stage tadpoles, which predominantly absorb oxygen from the water through external gills.

The tadpoles of this species are fed with unfertilized food eggs delivered on repeated visits by the mother at intervals of about five days. The female lays these nutritive eggs only when stimulated by her tadpoles; they trigger egg-laying by rapidly swimming around the female and touching her skin with their beaks. The nutritive eggs—along with the ability of the newly hatched larvae to take air at the water surface—are both adaptations to life in tiny bodies of water that offer little food and oxygen.

Although *A. spinosa* is rarely seen, it is probably more common than the limited observations suggest. The biggest threat to this species is likely the degradation and loss of closed canopy forest. It is not know whether chytridiomycosis is a major threat.

Over the past decade, monitoring of a population in Rara Avis Rainforest Preserve suggests that this species is both increasing significantly in population size and also extending its elevational range, expanding from primary forest to mature secondary forest at lower elevations. Long-term studies (started in 1992) in that preserve did not detect *A. spinosa* until 2005. Since then, sightings have been increasing. Different individuals are regularly encountered in the same area, suggesting that specific reproductive sites are utilized repeatedly over many years. In addition, adults and juveniles are now found in a wide range around the original detection site, suggesting that *A. spinosa* is either re-colonizing the area after a prior decline (before the start of the surveys), or that perhaps habitat or climate conditions in the area are changing gradually to increase its suitabilty. Research on this phenomenon is ongoing.

Note the white eggs visible in the digestive tract of these *A. spinosa* tadpoles; they are fed solely with infertile eggs provided by their mother.

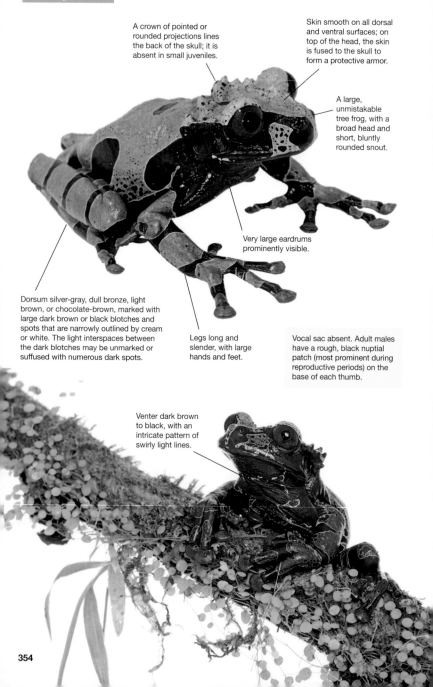

A crown of pointed or rounded projections lines the back of the skull; it is absent in small juveniles.

Skin smooth on all dorsal and ventral surfaces; on top of the head, the skin is fused to the skull to form a protective armor.

A large, unmistakable tree frog, with a broad head and short, bluntly rounded snout.

Very large eardrums prominently visible.

Dorsum silver-gray, dull bronze, light brown, or chocolate-brown, marked with large dark brown or black blotches and spots that are narrowly outlined by cream or white. The light interspaces between the dark blotches may be unmarked or suffused with numerous dark spots.

Legs long and slender, with large hands and feet.

Vocal sac absent. Adult males have a rough, black nuptial patch (most prominent during reproductive periods) on the base of each thumb.

Venter dark brown to black, with an intricate pattern of swirly light lines.

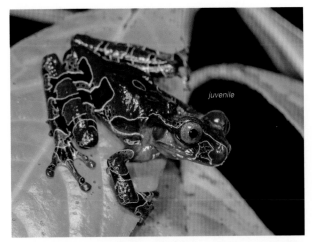

Juveniles have only weakly developed spiny projections on their head.

A. spinosa shows considerable variation in its coloration and amount of spotting.

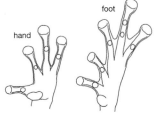

Hands without webbing; fingers with enlarged, round disks on the tips. Feet moderately webbed. Toes with enlarged, round disks.

Similar species

No other Costa Rican frog has spiny projections on its head.

Genus *Cruziohyla*

A recently created genus that contains two species of spectacular tree frog, one of which, *Cruziohyla calcarifer,* occurs in Costa Rica. This species was split from the genus *Agalychnis* based on biochemical, morphological, and ecological characteristics. Defining morphological characteristics include a bicolored iris and extensively webbed hands and feet. It also has a unique breeding strategy, in which tadpoles live in water-filled depressions in fallen trees or that lie between the buttressed roots of standing trees.

Cruziohyla calcarifer
Splendid Leaf Frog

Least concern

A spectacular tree frog. Note dark vertical bars on yellow to orange flanks; eyes with purplish-gray centers and yellow borders; and a prominent flaplike calcar tubercle present on each heel joint. Males to 3.19 in (81 mm) in standard length, females to 3.43 in (87 mm).

Ecoregions
1 2 3 4 5

Cruziohyla calcarifer occurs in isolated localities. On the Atlantic slope, it ranges from eastern Honduras to central Panama; on the Pacific slope, it can be found in eastern Panama, Colombia, and northwestern Ecuador. From near sea level to 2,300 ft (700 m).

Natural history

Spotting this spectacular frog is a Holy Grail for frog lovers. This elusive, nocturnal frog occurs at low elevations, where it is found in the canopy of rainforests. Although infrequently seen—and known from just a few scattered localities—its secretive lifestyle undoubtedly causes it to be under-reported. Recent sightings greatly expanded the known elevational range of *C. calcarifer*, indicating that it is likely more widespread and common than currently believed.

These frogs generally only descend to the understory to breed. Males initiate calling to potential mates (they produce a single *whuunk* note) from high in the canopy, then descend to breed near ground level. This large frog generally climbs using hand-over-hand motions, but it is known to occasionally jump and glide from the treetops, using its large hands and feet as an airfoil to slow down and direct its fall.

The breeding biology of *C. calcarifer* is unique among Costa Rican frogs in that its tadpoles develop in pools of water that are situated on top of, or inside, fallen logs, or sometimes in a pool contained within the buttressed roots of a large tree—never in a water-filled depression in the ground. Its eggs are attached to vegetation, roots, or vines that overhang the pool of water. Amplecting pairs have been seen throughout most of the year; reproduction can take place as long as rainfall and humidity sustain the water level in the breeding basin. Often, the same oviposition site is used during consecutive breeding bouts, until the water reservoir becomes unsuitable—usually when the log that it is located in has rotted to the point where it no longer holds water. The female lays her eggs early in the morning; generally clutches contain 10–54 eggs. The eggs appear moldy—indeed barely look like frog eggs—likely an adaptation to dissuade predators. Larval development is very slow; when reared under captive conditions, tadpoles have taken up to seven months to reach metamorphosis. Presumably it takes about as long in the wild.

C. calcarifer is only known from undisturbed forest habitat, and thus the biggest threats to its continued existence are deforestation and other forms of habitat alteration, encroaching agriculture, and pollution from crop spraying. Augmentation of its breeding habitat by placing water-filled tanks in known habitats, thus creating additional breeding pools, has been a succesful technique to boost local populations of *C. calcarifer*.

Egg clutches look moldy and are thus well camouflaged. They are generally attached to vines or twigs suspended over a water-filled cavity in a fallen log.

Adult males with a small, internal subgular vocal sac and a rough, brown nuptial pad (prominent during reproductive periods) on each inner finger.

Eyes with a vertically elliptical pupil and a beautiful pale purple to gray iris, bordered by yellow. Eyelids are not reticulated.

Eardrums prominent.

Skin on head, body, and limbs mostly smooth to minutely granular; ventral surfaces granular.

A large, robust tree frog, with a broad head and acutely sloping snout.

White, cream, or yellowish stripes mark the upper lip, trailing edges of limbs, and the area around the vent.

Dorsum dark green, with bright yellow to orange flanks, upper arms, and thighs. Flanks and thighs marked with a series of vertical black bars.

Limbs robust.

Ventral surfaces deep orange.

Low, white skin folds line the trailing edges of the limbs. Folds on the legs connect to a flaplike, triangular calcar spur on each heel joint.

Like the other leaf frogs in the genus *Agalychnis*, *C. calcarifer* tends to walk hand-over-hand along branches, rather than jump, as it moves through its arboreal habitat.

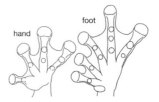

Hands and feet with significant webbing. Broadly expanded disks on all fingers and toes.

Agalychnis annae (p. 341)

The only other tree frog in Costa Rica with a vertical pupil and yellow iris, but lacks orange flanks with dark barring. It does not overlap in distribution range.

Genus *Dendropsophus*

The genus *Dendropsophus* contains 98 species of small tree frogs that were relatively recently split off from the genus *Hyla*. The majority of these frogs occur in South America, but several range into Central America, including three Costa Rican species. All species in this genus have cells that contain 30 chromosomes, not a terribly useful field mark.

Costa Rican *Dendropsophus* are cricketlike frogs, in size, shape, and call. These tan or brown frogs are common inhabitants of lowland forests and areas with secondary vegetation. They breed in temporary rain pools, marshes, and flooded fields, generally in areas without a closed-canopy cover. Eggs are attached to vegetation, either above the water or directly at the water surface. The tadpoles of these frogs have a unique, high tail fin that extends onto the body and ends in a thin filamentous tip, which is used to maintain a relatively stationary position in the water column.

Dendropsophus ebraccatus
Hourglass Tree Frog

Least concern

A beautifully patterned tree frog, with a tan to yellow dorsum, marked with a variable, vivid pattern of dark brown blotches that are often outlined by cream or yellow. Typically with a cream to pale yellow lip stripe that is broadly expanded below each eye and a brown eye mask that extends posteriorly, covering the eardrums and continuing onto the body. It has significant webbing between its fingers. Note bright yellow to orange thighs. Males to 1.06 in (27 mm) in standard length, females to 1.38 in (35 mm).

Ecoregions

1 2 3 4 5

On the Atlantic slope, *Dendropsophus ebraccatus* occurs in isolated populations in Chiapas, Mexico, northern Guatemala, Belize, Honduras, and northern Nicaragua; starting in southern Nicaragua, its populations become more or less continuous, and it is found throughout Costa Rica, Panama, Colombia, and northwestern Ecuador. On the Pacific slope, it occurs locally in Costa Rica and Colombia. From near sea level to 5,250 ft (1,600 m).

Natural history

This is a common and widespread species of humid lowland forests; it is quite adaptable and persists in disturbed areas, including parks, gardens, and agricultural areas. Like other tree frogs it is nocturnal and its activity peaks during periods of rain, although individuals can be found nearly year-round in appropriate habitat.

Breeding activity in *D. ebraccatus* commences with the first heavy rains of the wet season. Males call from reeds and grasses emerging from the breeding pond or from shrubby vegetation along the edge of the water. The call is a loud, insectlike *wreek* that is sometimes repeated a few times in rapid succession. Its call is similar to that of the two other members of this genus that occur in Costa Rica; it sometimes triggers aggressive responses from other *Dendropsophus* species.

Male *D. ebraccatus* are aggressive; besides producing mating calls to attract females, they intersperse their mating call with aggressive clicking calls meant to establish their territory. Females choose their partner based on call characteristics. A female will seek out her chosen mate and, when approached closely enough, the male will climb onto her back and engage in axillary amplexus. Often, non-calling satellite males will wait near a calling male and attempt to amplex with an aproaching female, who in turn may attempt to dislodge the interloper from her back by squeezing through tight spaces or under low branches. Once paired up, the male and female move toward the breeding pool, generally a temporary or permanent

pond, and lay their eggs in a single layer on the leaf surface of emergent vegetation, a short distance above the water line. Upon hatching, tadpoles wriggle free from their jelly mass and drop into the pond, where they complete their development.

D. ebraccatus is quite similar in appearance and biology to the two other Costa Rican members of its genus, and throughout its range it generally shares its breeding ponds with at least one other species of *Dendropsophus*. In a few low-lying locations near the continental divide in northwestern Costa Rica, all three species may co-occur.

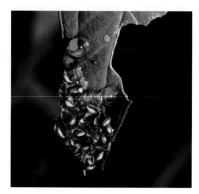

Eggs are invariably attached to leaves that overhang standing water. Tadpoles will hatch and drop into the water below.

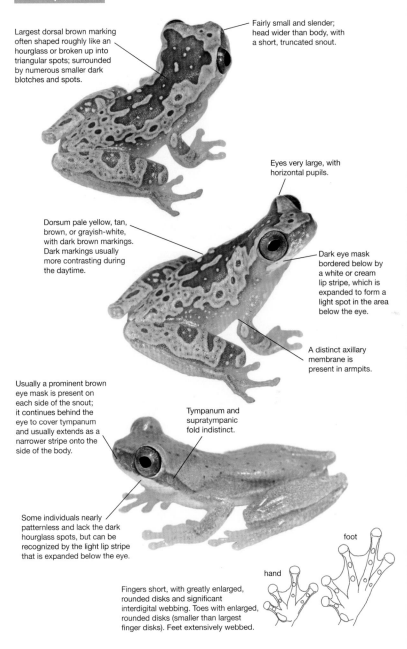

Largest dorsal brown marking often shaped roughly like an hourglass or broken up into triangular spots; surrounded by numerous smaller dark blotches and spots.

Fairly small and slender; head wider than body, with a short, truncated snout.

Eyes very large, with horizontal pupils.

Dorsum pale yellow, tan, brown, or grayish-white, with dark brown markings. Dark markings usually more contrasting during the daytime.

Dark eye mask bordered below by a white or cream lip stripe, which is expanded to form a light spot in the area below the eye.

A distinct axillary membrane is present in armpits.

Usually a prominent brown eye mask is present on each side of the snout; it continues behind the eye to cover tympanum and usually extends as a narrower stripe onto the side of the body.

Tympanum and supratympanic fold indistinct.

Some individuals nearly patternless and lack the dark hourglass spots, but can be recognized by the light lip stripe that is expanded below the eye.

foot

hand

Fingers short, with greatly enlarged, rounded disks and significant interdigital webbing. Toes with enlarged, rounded disks (smaller than largest finger disks). Feet extensively webbed.

Thighs uniform bright yellow or orange.

Patternless individuals that lack the characteristic hourglass markings may be relatively common in some populations.

Adult males with vocal slits and large, extendable vocal sac. Nuptial patches absent.

Pair of *D. ebraccatus* in amplexus. Note that males are significantly smaller than females.

Similar species

D. microcephalus (p. 363)

Lacks the boldly spotted hourglass pattern, dark eye mask, and expanded white spot below the eye.

D. phlebodes (p. 365)

Has an irregular brown reticulum on its dorsum and a pair of short, brown dorsolateral stripes. It lacks the white lip stripe and light spot below the eyes.

Dendropsophus microcephalus
Small-headed Tree Frog

Least concern

A small yellow to tan tree frog, typically with a thin, dark brown dorsolateral line (on each side of the body) that is bordered above by a narrow white line. Fingers with minimal webbing. Also note uniform yellow thighs. Males to 1.06 in (27 mm) in standard length, females to 1.22 in (31 mm).

Ecoregions

1 2 **3 4 5**

Dendropsophus microcephalus has a huge distribution range. It occurs along the Atlantic slope in southern Veracruz and northern Oaxaca states (Mexico), central Guatemala, Belize, Honduras, and northern Nicaragua. It switches to the Pacific slope in Nicaragua and continues south through Costa Rica and Panama, reaching the Amazon basin of South America. In Costa Rica it ranges from near sea level to 2,650 ft (810 m); it sometimes occurs at higher elevations in other countries.

Natural history

This widespread and abundant frog is generally found in disturbed habitats, where it is usually associated with permanent or temporary wetlands, flooded pastures, cattle ponds, roadside ditches, or marshy situations in or near forest. *D. microcephalus* is never found in primary forest, and is one of the very few amphibian species that may be increasing in numbers as it benefits from the conversion of forest into open scrubby habitat for human settlement and agriculture.

This nocturnal frog can be seen—and heard—year-round, although it becomes significantly more active during the wet season, when breeding takes place. Successive periods of rainfall trigger large numbers of these frogs to gather in the vegetation surrounding suitable breeding ponds.

Males call from elevated perches on emergent vegetation in, or along the edge of, the breeding pond; they seem to prefer small herbaceous plants or grasses and sedges in the shallow margins of the pond. The advertisement call of the male is similar to that of other species in the genus; he produces a loud, insectlike *creek-eek-eek-eek*, interspersed with aggressive clicking calls made to ward off rival males. Males generally arrive at the breeding pool at dusk and commence calling. If weather conditions are favorable (usually dictated by persistent rain), females will follow the sound of the calls later in the evening. If not, the females stay put and the males will eventually cease calling for the night and retreat. Amplexus in *D. microcephalus* is axillary and takes place in shallow sections of the breeding pond. Small clumps of eggs are attached to vegetation near the water surface and hatch into tadpoles that complete their larval development in the breeding pool.

Studies on the calling behavior in this species show that males respond to other calling males present at the breeding pool, and space their own calls so that rival males are able to detect the territorial function of their call. In addition, isolating their individual calls from those of others in a frog chorus allows approaching females to identify a preferred mate. Calling males have been noted to increase their call frequency when females are nearby. Not only do individual frogs adjust their calling activity in response to other males in a chorus, entire frog choruses have been observed to align their calls with those in distant, but audible, breeding choruses. Interestingly, one recent study showed that human-generated noise pollution also influenced calling activity in male *D. microcephalus*. When exposed to anthropogenic ambient noise, males shortened their calling period at the breeding pool. Since these calls are the trigger to guide females down to the breeding pool, shortening the calling period may reduce the frequency of successful breeding in a noisy environment.

Some authors recognize different subspecies of *D. microcephalus*. Populations from northwestern Costa Rica differ slightly in details of their finger webbing and coloration from those found in southwestern Costa Rica. Northern populations are sometimes attributed to the subspecies *D. m. cherrei*, which has less patterning on the dorsum and lacks a dark bar between the eyes; southern Costa Rican populations perhaps represent the subspecies *D. m. microcephalus*. Regardless, both forms possibly intergrade in the central Pacific lowlands of Costa Rica, and the variation displayed in both potential subspecies may make recognition of such formal taxa irrelevant.

A delicate little frog. Its head is only slightly wider than its body; also note short, truncated snout.

Dorsum yellow to tan, with a network of darker lines and markings (often shaped like an X) or a pair of longitudinal bars connected with a crossbar.

Tympanum fairly indistinct; supratympanic fold only faintly indicated.

Eyes very large, with horizontal pupils.

Thighs uniform yellow.

Ventral surfaces white.

Thin, dark brown dorsolateral line (bordered above by a narrow white line) on each side of the body; dorsolateral line starts at the tip of the snout, follows the curve of the upper eyelid, and extends backward toward the insertion of the hind limbs.

daytime coloration

Adult males with vocal slits and large, extendable vocal sac; when not inflated, the sac appears as yellowish wrinkly skin on the male's throat. During the breeding season, adult males have a small, whitish nuptial patch at the base of each thumb.

nighttime coloration

Capable of considerable color change between day and night. At night, the dorsum is pale yellow, with brown or tan markings. During the day, the dorsum is dark yellow or tan, or light brown, with darker brown or reddish markings.

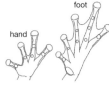

Fingers short, with very little webbing and moderately enlarged, rounded disks. Feet with extensive webbing; toes with enlarged, rounded disks (smaller than largest finger disks).

D. ebraccatus (p. 360)

Has extensive finger webbing, and (usually) a dorsal pattern of bold dark brown blotches; unmarked *D. ebraccatus* are similar to *D. microcephalus* but have a pronounced light lip stripe.

D. phlebodes (p. 365)

Occurs mainly on the Atlantic slope but may co-occur locally in northwestern Costa Rica; on each side of the body, it has only a short, incomplete brown dorsolateral line that is never bordered by a white line.

Dendropsophus phlebodes
Veined Tree Frog   **Least concern**

A yellowish-tan to light brown tree frog, usually with a short, faint, brown dorsolateral stripe that extends only to the shoulder region. Dorsum shows a dark reticulum. It has minimal finger webbing. Also note yellow thighs. Males to 0.94 in (24 mm) in standard length, females to 1.10 in (28 mm).

Ecoregions
1 2 3 4 5

On the Atlantic slope, *Dendropsophus phlebodes* occurs in southeastern Nicaragua, Costa Rica, and eastern Panama. It also occurs on the Pacific slope, from central Panama to northwestern Colombia. From near sea level to 2,035 ft (620 m).

Natural history

D. phlebodes is a nocturnal species that inhabits vegetation near breeding pools. This explosive breeder reproduces during the rainy season in shallow pools. At night, calling males produce a loud, insectlike *creek-eek-eek-eek* while perched on grasses or other emergent vegetation on the edge of, or in, the breeding pool. Amplexus is axillary. Eggs are laid in small clumps of up to 400 that float on the water surface and are usually attached to vegetation. The eggs hatch into tadpoles that complete their larval development in the breeding pool.

D. phlebodes is very similar to *D. microcephalus* (p. 363) in morphology, biology, and behavior. In Costa Rica, *D. phlebodes* generally occurs on the Atlantic slope, while *D. microcephalus* occurs on the Pacific slope, although in a few locations near the continental divide in the northwestern parts of the country both species co-occur. In addition, the closely-related *D. ebraccatus* is found on

both slopes and shares breeding ponds with both species. Due to the very similar breeding biology of these species, competition over scarce resources at a breeding pool leads not only to aggressive interactions between individuals of the same species, but also at times between individuals of different species. Each of the three *Dendropsophus* species reacts to the similar calls of the two other species. In general, *D. phlebodes* tends to be the least aggressive of the three species and occurs in the lowest density.

D. phlebodes is adaptable and somewhat tolerant of habitat alteration; it is found in flooded pastures at the edge of forest, in secondary forests, in plantations, and even in the immediate vicinity of humans, provided enough vegetation cover remains. Pollution, spraying of pesticides, and contaminated runoff into breeding pools appear to be key threats to local populations of this fairly common species.

Dorsum is a relatively uniformly yellowish-tan to light brown, usually with an indistinct, short dorsolateral line on each side of the body that extends back from the eye onto the shoulder region. The dorsolateral line is never bordered above by a white line.

Eyes large, with horizontal pupils.

Thighs uniform yellow.

daytime coloration

Capable of considerable color change between day and night. At night, the dorsal reticulum tends to be more visible; during the day, when the background color of the dorsum becomes darker (tan or light brown), the reticulum is less visible.

Ventral surfaces white.

The dorsum is generally marked with a reticulum of irregular brown lines.

Adult males with a small, whitish nuptial patch (prominent during the breeding season) at the base of each thumb.

A small, inconspicuous tree frog with a head that is slightly wider than its body and a short, truncated snout.

Tympanum fairly indistinct; lacks a conspicuous supratympanic fold.

foot

hand

Adult males with vocal slits and a large, extendable vocal sac; when not inflated, it appears as yellowish wrinkly skin on the male's throat.

Fingers short, with moderately enlarged, rounded disks and basal webbing. Toes with slightly enlarged, rounded disks that are similar in size to the largest finger disks. Feet extensively webbed.

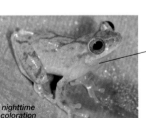

nighttime coloration

D. microcephalus
(p. 363)

Invariably has a pair of dark brown dorsolateral lines, each bordered above by a thin white stripe.

D. ebraccatus
(p. 360)

Has a light lip stripe that is expanded below each eye, and usually a bold pattern of large brown blotches on its back and limbs.

Tlalocohyla loquax
(p. 443)

Larger and more robust, with bright red or orange coloration in its webbing, flanks, and thighs.

Smilisca puma (p. 436)

Larger, lacks webbing between its fingers, and usually has a light lip stripe.

366

Genus *Duellmanohyla*

The genus *Duellmanohyla* consists of eight stream-breeding tree frogs previously included in the genus *Hyla*. The distribution of this genus is disjunct throughout Central America; three species occur in Costa Rica and adjacent Panama. The Costa Rican members of this genus are easily recognized by having large eyes with a red iris and horizontal pupils. All share a more or less prominent white stripe on the upper lip that usualy expands below the eye (*Ptychohyla legeri* shares the red eyes and horizontal pupil seen in *Duelllmanohyla*, but usually lacks the white labial stripe).

Tadpoles of these frogs have a ventrally oriented funnel-shaped oral disc that helps them adhere to rocks in their stream habitat.

Duellmanohyla lythrodes
Panamanian Brook Tree Frog **Endangered**

A red-eyed tree frog with horizontal pupils. A distinct light stripe runs along the upper lip and extends to the groin; a white mark below the eye connects with the lip stripe. Dorsum green to brown; has uniform yellow posterior thigh surfaces and a yellow venter. Males to 1.30 in (33 mm) in standard length; no females have ever been recorded but they are likely somewhat larger than males.

Ecoregions
1 2 3 4 5

Duellmanohyla lythrodes is an enigmatic species, known only from a few localities on the Atlantic slope of southeastern Costa Rica and adjacent northwestern Panama, between 550 and 1,445 ft (170 to 440 m).

Natural history

This rare species is only known from a handful of specimens. A single male was discovered in southern Limón Province in 1964, and three additional males were later found in nearby western Panama in the 1970s; no additional sightings of *D. lythrodes* have been reported since. All individuals were collected in lowland rainforest, as they perched or called near streams. One indiviudal was in a tree about three meters above the ground; the other individuals were on low vegetation. The call of this species has been described as a soft, bell-like sound, repeated 5–7 times. Like other species in its genus, *D. lythrodes* likely breeds in streams, but it is not known whether eggs are deposited on vegetation, on rocks, or in the water, and its tadpole remains unknown.

Due to its perceived rarity and the very subtle difference between the few known individuals of *D. lythrodes* and the more common and inherently variable species *D. rufioculis*, opinions differ on whether *D. lythrodes* should be treated as a separate species or a regional variant of *D. rufioculis*. Only future research into this species and discovery of additional populations will ultimately provide the answer to that question.

- A small to medium-sized tree frog, with a broad head and a short snout that appears truncated when seen from above.
- Eyes large, with bright red eyes and horizontal pupils.
- Skin on head, dorsum, and limbs smooth; venter granular.
- Tympanum fairly distinct, moderate in size (about half the size of the diameter of the eye).
- Limbs moderate in length.
- Adult males with a brown nuptial patch on the thumb base, vocal slits, and an extendable vocal sac on the throat.
- Dorsum green to light brown; capable of considerable color change and may show varied patterns of brown and green mottling.
- A white lip stripe that extends posteriorly onto the sides of the body to the groin is broadly expanded below each eye.
- White stripes present along the margins of the forearms and feet and around the vent.
- Posterior thigh surfaces yellow, without dark mottling.
- Venter yellow; throat and vocal sac white.

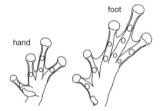

Fingers short, with minimal webbing and only moderately expanded, rounded disks. Feet with moderate webbing and fairly small, rounded disks on the toes.

D. rufioculis (p. 370)

Very similar in appearance, but has a white venter and a slightly smaller tympanum (less than 45% of the diameter of the eye).

D. uranochroa (p. 372)

Has a white lip stripe that is not expanded to connect to the eye. It has a brighter green dorsum and a larger tympanum (measuring more than two-thirds of the diameter of the eye).

Duellmanohyla rufioculis
Rufous-eyed Brook Tree Frog **Least concern**

A red-eyed tree frog with horizontal pupils. A distinct light stripe runs along the upper lip and extends to the groin; a white mark below the eye connects with the lip stripe. Its dorsum is brown to olive-green, while the venter is white; posterior thigh surfaces are tan to yellow, usually suffused with dark pigment. Males to 1.18 in (30 mm) in standard length, females to 1.57 in (40 mm).

Ecoregions
1 2 **3** 4 5

Duellmanohyla rufioculis ranges along the Atlantic and Pacific slopes of the Costa Rican mountain ranges; 2,300–5,185 ft (700–1,580 m).

Natural history

D. rufioculis occurs at low and middle elevations in humid forests. It is most frequently seen at night on vegetation along small forest streams. Males produce a soft, rasping call. In spite of the fact that this species is relatively commonly observed, little information exists on its breeding biology. Eggs are deposited in small pools in rocky or gravel-bottomed streams. Their tadpoles are adapted to life in streams and have a long, muscular tail with a low fin to navigate tight spaces between rocks and swim in rapid currents; their oral disk is well-developed and faces downward, allowing the tadpoles to cling to the lee side of rocks in the current. Tadpoles have been observed in seeps and very small trickles that barely contained water.

Peak activity for *D. rufioculis* is during the wet season, roughly from July to December; but with

sufficient rainfall, calling males can be heard almost year-round.

D. rufioculis is quite variable in size and coloration, and some authors suggest that this species may consist of a complex of multiple cryptic species. The exact taxonomic position of the similar *D. lythrodes* is uncertain.

D. rufioculis all but disapeared from several foothill sites during the 1990s, but has since made a significant recovery and is currently found reliably in areas where it was previously undetectable. It persists in areas that are known to host the chytrid fungal pathogen *Bd*, and appears to successfully co-exist with it. Due to its preference for densely vegetated streams, loss of riparian corridors (through habitat destruction) and water pollution as a result of contaminated runoff are likely the biggest threats to this species.

Description

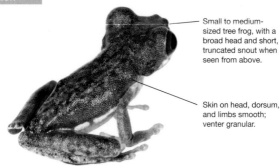

Small to medium-sized tree frog, with a broad head and short, truncated snout when seen from above.

Skin on head, dorsum, and limbs smooth; venter granular.

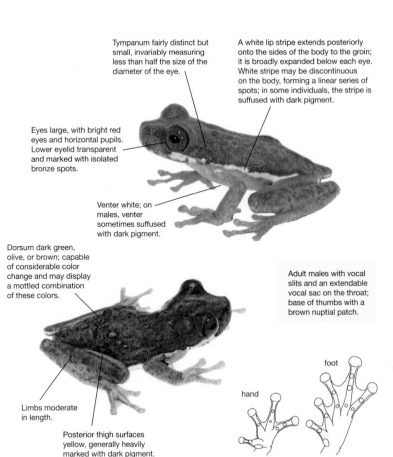

Tympanum fairly distinct but small, invariably measuring less than half the size of the diameter of the eye.

A white lip stripe extends posteriorly onto the sides of the body to the groin; it is broadly expanded below each eye. White stripe may be discontinuous on the body, forming a linear series of spots; in some individuals, the stripe is suffused with dark pigment.

Eyes large, with bright red eyes and horizontal pupils. Lower eyelid transparent and marked with isolated bronze spots.

Venter white; on males, venter sometimes suffused with dark pigment.

Dorsum dark green, olive, or brown; capable of considerable color change and may display a mottled combination of these colors.

Adult males with vocal slits and an extendable vocal sac on the throat; base of thumbs with a brown nuptial patch.

Limbs moderate in length.

Posterior thigh surfaces yellow, generally heavily marked with dark pigment.

foot

hand

Fingers slender, with minimal webbing and only slightly expanded round disks. Toes moderately webbed, with fairly small rounded disks.

Similar species

D. lythrodes (p. 368)

NO PHOTO

Has a yellow venter. Its tympanum is slightly larger, measuring a little more than half the diameter of the eye.

D. uranochroa (p. 372)

Has a brighter green dorsum and lacks an expanded white blotch below the eyes. It has much larger eardrums that measure two-thirds of the diameter of the eye.

Ptychohyla legleri (p. 420)

Lacks an expanded white spot below the eye, and often lacks a white lip stripe (although a light lateral stripe is generally present between the front and hind limbs).

371

Duellmanohyla uranochroa
Red-eyed Brook Tree Frog   **Critically endangered**

A red-eyed tree frog with horizontal pupils. Note a distinct light stripe that runs from the upper lip to the groin but lacks a white mark below the eye that connects with the lip stripe. This frog has a leaf-green dorsum and yellow venter; posterior thigh surfaces are yellow. Males to 1.46 in (37 mm) in standard length, females to 1.57 in (40 mm).

Ecoregions
1 2 3 4 5

Duellmanohyla uranochroa inhabits foothill and middle elevation forests in Costa Rica and adjacent western Panama; 985–5,740 ft (300–1,750 m).

Natural history

D. uranochroa occurs in foothills and at middle elevations, in humid and wet evergreen forests. It is usually found in the vicinity of rapidly flowing forest streams. During the day, individuals have been found clinging to small, moss-covered trees or leaves; they are also found in the leaf axils of bromeliads and in aroid plants.

Males call at night from low vegetation, often at some distance from a stream. The advertisement call is a rapid, melodic, bell-like *ting-ting-ting*, which is repeated at irregular intervals. Breeding activity reaches its peak during the onset of the wet season, usually in May or June, and mostly takes place on rainy nights. During humid nights in the dry season, adult males can be heard calling from treetops in the forest interior, suggesting that these frogs move away from their streamside breeding habitat during dry periods.

Amplexus and courtship behavior in this species are unknown. Eggs are deposited in small pools in rocky streams. The tadpoles dwell on the bottom of the stream, where they develop in the sandy or gravel substrate; they have a powerful physique, with a long, muscular tail to position themselves in the current. Tadpoles also have a large, downward-facing oral disk that is used for suction to adhere to rocks. Large tadpoles develop the characteristic red eyes before their legs emerge, and are therefore easily recognized.

D. uranochroa declined precipitously during the late 1980s and early 1990s, and for a period of more than 10 years no individuals were found at sites with former, well-known populations. Since many of these populations inhabited undisturbed forest habitat in protected areas, habitat alteration was not a likely explanation for the declines. Instead, climate change, airborne pollution, and chytridiomycosis have been suggested as possible factors.

In 2007, this species was observed yet again in the Monteverde area. Today several populations of *D. uranochroa* have been identified in isolated areas in the Tilarán, Central, and Talamanca Mountain Ranges, and the species appears to be recovering. Nevertheless, this beautiful frog has a long way to go before it recolonizes its historic range. Interestingly, some of the recovering populations are located in areas that are known to host the chytrid fungal pathogen *Bd*, yet the frogs seem to be able to survive there. Perhaps the recovering frogs have developed an increased level of resistance against chytridiomycosis, or the pathogen has become less virulent over time. At least one recovering population is located in disturbed edge habitat; this open habitat is possibly less suited for *Bd* than the forest interior habitat generally occupied by *D. uranochroa*. Additional research is urgently needed to explore these questions.

Small to medium-sized tree frog, with a broad head and relatively short snout. Eyes large, bright red, with horizontal pupils.

Skin on head, dorsum, and limbs smooth; venter granular.

Adult males have vocal slits, an extendable vocal sac on the throat, and a brown nuptial patch at the base of each thumb.

Dorsal coloration bright leaf-green.

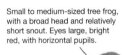
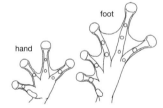

A white lip stripe extends along the sides of the body to the groin. Stripe generally not expanded to form a white blotch below each eye.

Tympanum fairly distinct and large, measuring two-thirds or more of the diameter of the eye.

Venter, hands, feet, and posterior thigh surfaces are bright yellow to yellowish-orange.

Thin white lines adorn the outer margins of the forearms and legs and the area above the vent.

Limbs moderate in length.

foot

hand

Even at an early stage of development, tadpoles of this species can be identified by their red eyes.

Fingers short, with minimal webbing and moderately expanded, round disks. Feet with moderate webbing; round disks are fairly small.

Similar species

D. rufioculis
(p. 370)

Has a white venter and significantly smaller eardrums. Also note white blotch below eye that connects with lip stripe.

D. lythrodes
(p. 368)

NO PHOTO

Has smaller eardrums and an expanded white blotch below each eye that connects with the lip stripe.

Ptychohyla legleri
(p. 420)

Has a lateral white stripe that is interrupted in the shoulder region.

Agalychnis callidryas
(p. 343)

Red-eyed leaf frogs in the genus *Agalychnis* have vertically elliptical pupils.

Genus *Ecnomiohyla*

Very little is known about the natural history of these nocturnal inhabitants of the high canopy.

These distinctive frogs were previously placed in the genus *Hyla*. The phylogenetic basis of the new genus *Ecnomiohyla* is strictly molecular, although these tree frogs do share easily recognizable traits: they are large to very large; have conspicuous skin fringes along the outer margin of the forearms and tarsus; and possess huge, extensively webbed hands and feet. In addition, all male *Ecnomiohyla* have a protuberant prepollex with a projecting spine (sometimes covered by a retractable sheath of skin) at the base of the thumb. In most species, the skin on the top of their head is fused (co-ossified) with the skull, or contains embedded small, bony plates (osteoderms).

Fourteen species are currently recognized in this genus; most are known from just a few specimens. Four species of *Ecnomiohyla* occur in Costa Rica.

Ecnomiohyla bailarina
Golden-eyed Fringe-limbed Tree Frog   **Not assessed**

A moderate-sized tree frog, with large, fully webbed hands and feet. All known individuals have striking golden-yellow eyes. Skin fringes along the outer margin of the lower limbs consist of widely spaced, triangular projections. Skin on the head and dorsum is extremely granular and contains small, embedded bony plates; skin on the flanks and undersides lacks spine-tipped tubercles. The prepollex at the base of each thumb has a bluntly pointed, bony projection, and, in adult males, also bears two patches of small black spines. Males to 2.68 in (68 mm) in standard length; only known adult female 2.28 in (58 mm).

Ecoregions
1 2 3 4 5

Ecnomiohyla bailarina was recently described from a single individual in the Darién Province in Panama, near the border with Colombia. In 2015, a second locality was documented at a single site in the foothills of the Talamanca Mountains in southeastern Costa Rica, at an elevation of approximately 985 ft (300 m).

Natural history

This species was discovered in 2011 on Cerro Bailarín, Darién Province, Panama, but was not formally described until 2014, based on the only known specimen, an adult male. By 2013, three additional individuals had been collected in southern Costa Rica, some 375 mi (600 km) from the collection site in Panama, which were assigned to *E. bailarina* in 2015. At this point, no other populations of *E. bailarina* are known from intermediate areas.

E. bailarina is nocturnal and spends much of its time in the canopy, hidden from view; little is known about its biology. Individuals have been detected by calls emanating from the treetops. The only three Costa Rican individuals that have been collected were found at the same time, as they were approaching one another. One of the three was a male found calling from a tree, at a height of 55 ft (17 m); the other two individuals, an adult male and an adult female, were approaching this tree from lower vegetation.

The call of *E. bailarina* consists of a rapid series of pulsed notes that lasts about two seconds and is repeated at sporadic intervals, sometimes hours apart. A second, weaker two-toned call has been documented from an adult male interacting with the approaching female and may represent courtship vocalization.

E. bailarina inhabits areas of humid mature forest interspersed with patches of older secondary vegetation; the Costa Rican locality was near a large clearing, caused by a blowdown of several trees a few years prior. It is assumed that these frogs breed in water-filled tree cavities, as has been documented in other species of *Ecnomiohyla*, but additional surveys are needed to confirm this.

The paucity of observations—and the large distance between the two sets of observations—make it impossible to assess the conservation and population status of this species.

Extremely well camouflaged. Coloration variable. Generally some shade of green; individuals may change color in response to changes in temperature or light levels, becoming brown or olive-gray.

Adult males with a relatively small internal, extendable vocal sac on the throat. Adult males with two discrete dense clusters of black, keratinized hard spines on the prepollex; absent in females.

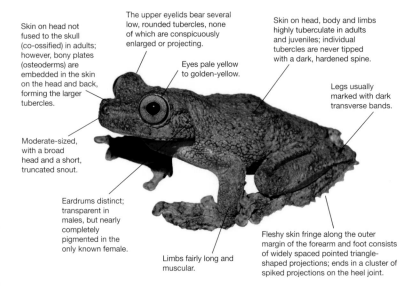

Skin on head not fused to the skull (co-ossified) in adults; however, bony plates (osteoderms) are embedded in the skin on the head and back, forming the larger tubercles.

The upper eyelids bear several low, rounded tubercles, none of which are conspicuously enlarged or projecting.

Eyes pale yellow to golden-yellow.

Skin on head, body and limbs highly tuberculate in adults and juveniles; individual tubercles are never tipped with a dark, hardened spine.

Legs usually marked with dark transverse bands.

Moderate-sized, with a broad head and a short, truncated snout.

Eardrums distinct; transparent in males, but nearly completely pigmented in the only known female.

Limbs fairly long and muscular.

Fleshy skin fringe along the outer margin of the forearm and foot consists of widely spaced pointed triangle-shaped projections; ends in a cluster of spiked projections on the heel joint.

Has huge hands and feet. Fingers and toes tipped with very large round disks. Interdigital webbing is very extensive and reaches the base of the digital disks on at least one or two fingers and most toes. Base of each thumb with a well-developed prepollex that ends in a hard, bluntly pointed projection.

E. sukia (p. 381)

Adult males lack patches of black keratinized spines on the prepollex. Dermal limb fringes are scalloped, and do not consist of isolated triangle-shaped projections.

E. miliaria (p. 379)

Has a sharp, recurved prepollical spine and lacks patches of black keratinized spines on the prepollex. It also has spine-tipped tubercles scattered over its flanks, venter, and legs.

Ecnomiohyla fimbrimembra
Highland Fringe-limbed Tree Frog   **Endangered**

A very large tree frog, with huge, nearly fully webbed, hands and feet; fleshy fringes along the trailing edges of its limbs; and a distinct prepollex with a projecting spine at the base of each thumb. Adults brown, with relatively smooth skin; juveniles show a mottled gray-green dorsum, with coarsely tuberculate skin. Males to 3.11 in (79 mm) in standard length, females to 3.62 in (92 mm).

Ecoregions
1 2 3 4 5

Ecnomiohyla fimbrimembra is known from only few individuals. It has been observed in undisturbed forests on the slopes of the Tilarán, Central, and Talamanca Mountain Ranges of Costa Rica; it also occurs in western Panama; 2,460–6,235 ft (750–1,900 m).

Natural history

E. fimbrimembra is known from just a few individuals and information on this species is sparse. Individual *E. fimbrimembra* have been discovered at night perched on vegetation relatively close to the ground. One individual was located in the spray zone of a waterfall. It has been found in relatively pristine humid forests with large trees and is presumed to be a canopy specialist. The powerful limbs and huge, webbed hands and feet may be adaptations to a an arboreal lifestyle. Other species in the genus are known to glide from the treetops, using the skin between their spread fingers and toes as an airfoil to control the speed and direction of their descent. This behavior has not been observed in *E. fimbrimembra*, but its morphology certainly hints at this ability.

A single, large tadpole (3.03 in [77 mm]) ascribed to this species was discovered in a water-filled, man-made hole in the ground. This breeding site is likely an anomaly, however; based on the breeding biology of related species, *E. fimbrimembra* more likely breeds in water-filled cavities in tree branches and hollowed out sections of tree trunks. Observations on captive bred individuals indicate that these frogs attach their eggs to the wall of a tree cavity, above the water level, and that the larvae drop into the water when they hatch. Given the large size of the tadpole, larval development may last a long time.

Due to its secretive nocturnal habits, preference for tree canopies in densely forested regions, and the fact that its call remains undescribed, *E. fimbrimembra* is likely underreported; it may be more common than is currently assumed. It is difficult to adequately assess its population and conservation status. Most of the known sites for this species are located in protected areas, but logging and removal of large canopy trees are likely among its biggest threats.

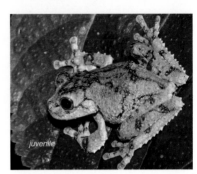

juvenile

Juveniles have a mottled gray-green dorsum and coarsely tuberculate skin. Both juveniles and adults have fleshy fringes along their limbs, but they are more prominent in younger frogs.

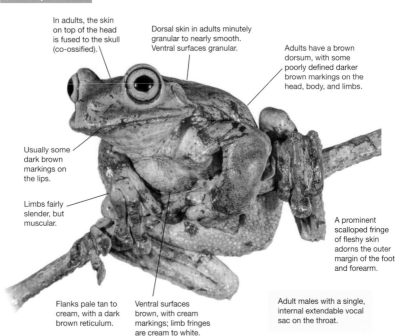

In adults, the skin on top of the head is fused to the skull (co-ossified).

Dorsal skin in adults minutely granular to nearly smooth. Ventral surfaces granular.

Adults have a brown dorsum, with some poorly defined darker brown markings on the head, body, and limbs.

Usually some dark brown markings on the lips.

Limbs fairly slender, but muscular.

A prominent scalloped fringe of fleshy skin adorns the outer margin of the foot and forearm.

Flanks pale tan to cream, with a dark brown reticulum.

Ventral surfaces brown, with cream markings; limb fringes are cream to white.

Adult males with a single, internal extendable vocal sac on the throat.

A very large tree frog, with a broad head and a bluntly truncated snout.

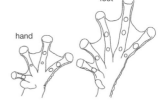

Hands and feet extremely large, with extensive webbing; the interdigital webbing almost reaches the base of the large round finger and toe disks. Has a well-developed prepollex at the base of each thumb. Adult males with small black nuptial spines on the base of the thumb and prepollex.

E. miliaria (p. 379)

Is quite similar to juvenile *E. fimbrimembra* but has more extensively webbed hands and feet. Not known to co-occur.

Ecnomiohyla miliaria
Lowland Fringe-limbed Tree frog **Vulnerable**

A very large tree frog with huge, fully webbed, hands and feet, rugose skin, and a scalloped fringe of skin along the trailing edges of its forearm and tarsus. At the base of each thumb is a distinct prepollex (tipped with a sharp, projecting spine in adult males). This is one of the largest tree frogs in Central America; males to 4.33 in (110 mm) in standard length; largest known female 3.39 in (86 mm), but females potentially reach an even larger size than males.

Ecoregions
1 2 3 4 5

On the Atlantic slope, *Ecnomiohyla miliaria* has been found in scattered localities in extreme southeastern Nicaragua and eastern Costa Rica; from near sea level to 2,950 ft (900 m). On the Pacific slope, it occurs in southwestern Costa Rica and, locally, in western and central Panama and in Colombia; 3,280–4,365 ft (1,000–1,330 m).

Natural history

Little is known about the biology of this rarely seen species. It is a nocturnal canopy inhabitant of undisturbed lowland and foothill humid forests. Individuals have been reported to jump and glide from elevated perches, using the expansive webbing between their spread fingers and toes to direct and slow down their descent. Thick rugose skin and the embedded osteoderms (bony plates) may help prevent dehydration and are thought to be adaptations to a harsh canopy environment.

Limited observations of the breeding biology of *E. miliaria* indicate that it breeds in arboreal, water-filled cavities. Males call from within or near a suitable breeding cavity to attract a female. The call is a series of slurred short barks, lasting several seconds; calls are repeated very infrequently, sometimes spaced hours apart. Eggs are likely deposited in the water or attached to the wall of the

water-filled cavity; tadpoles develop in their small arboreal pool. Tadpoles of a related species have been found with frog eggs in their digestive tract, suggesting that the female possibly returns regularly to feed her offspring with infertile, nutritive eggs, a behavior also seen in another tree-cavity breeding tree frog, *Anotheca spinosa*, and some smaller tree frogs in the genus *Isthmohyla*. However, this behavior has yet to be confirmed in *E. miliaria*.

Additional surveys are needed to more accurately assess the population and conservation status of *E. miliaria*. Its secretive lifestyle undoubtedly causes it to be underreported, and *E. miliaria* may be more common and widespread than is currently understood. Deforestation of its primary forest habitat is likely to be the biggest threat to this spectacular species.

Description

Skin on head not fused to the skull (co-ossified) in adults; however, bony plates (osteoderms) are embedded in the skin on the head and down the center of the back.

A huge tree frog with a broad head and truncated snout.

The upper eyelids bear one or more enlarged, projecting tubercles and several low, rounded tubercles.

adult

Dorsal skin in adults and juveniles rugose, densely marked with numerous tubercles. Ventral surfaces also tuberculate.

Legs marked with dark brown or dark olive-green transverse bands.

Ventral surfaces generally tan or light brown, with a pattern of small whitish markings.

Interdigital webbing usually dark brown.

Adult males with a relatively small internal, extendable vocal sac on the throat.

Limbs fairly short and robust.

Adults have hardened (keratinized) black spiny tips on the tubercles covering the venter and sections of the flanks and heels.

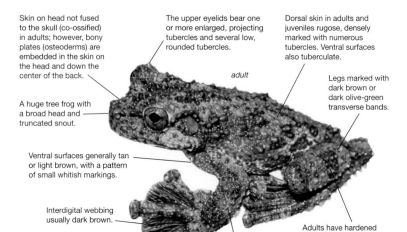

Dorsum with variable coloration; usually light brown, marked with irregular green spots and blotches. Capable of significant color change between day and night, generally appearing lighter at night than during the day.

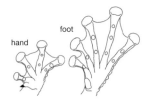

Huge hands and feet. Fingers and toes tipped with large disks; the interdigital webbing is very extensive, usually reaching the base of the digital disks on one or two fingers. The base of each thumb has a well-developed prepollex, tipped with a recurved, sharp spine in adult males. No clusters of small, black keratinized spines present on the prepollex of males and females.

Similar species

E. sukia (p. 381)

Lacks enlarged, projecting tubercles on the upper eyelids.

E. bailarina (p. 375)

Smaller and lacks spine-tipped tubercles on its undersurfaces.

E. fimbrimembra (p. 377)

Has relatively smooth skin as an adult, and lacks spine-tipped tubercles on its undersides.

Ecnomiohyla sukia
Shaman Fringe-limbed Tree Frog

Not assessed

A moderate-sized tree frog, with very large, fully webbed, hands and feet. Its limbs bear scalloped skin fringes along their outer margins; on the hind limbs, these fringes continue across the heel joint. The skin on the head and dorsum has enlarged tubercles that contain small bony plates (osteoderms); the skin on its flanks and undersurfaces lacks spine-tipped tubercles. The prepollex at the base of each thumb has a pointed, bony projection in adult males, but lacks patches of small, keratinized spines. Males to 2.48 in (63 mm) in standard length, females to 2.72 in (69 mm).

Ecoregions

1 2 3 4 5

A Costa Rican endemic known to occur on the Atlantic slope, near Siquirres and Limón, in the humid foothills of the Talamanca Mountain Range; 1,310–3,280 ft (400–1,000 m). It is likely considerably more widespread than current observations suggest.

Natural history

E. sukia is a nocturnal canopy dweller. It generally occurs in relatively undisturbed humid forests, although it has occasionally been found in mature secondary forest. Because it breeds in water-filled arboreal cavities, the presence of large, mature trees is critical.

These frogs are more commonly heard than seen, as they call from high overhead. At times, however, individuals have been observed at lower levels in the forest, perched on leaves or branches. Activity seems to increase during the dry season (February–April), although calling individuals can sometimes be heard sporadically throughout the year. The call of *E. sukia* has been described as a series of 13–20 short, staccato barks, generally completed under 10 seconds. Pauses between vocalizations may exceed one hour. Even during the peak reproductive season, adult males sometimes don't call more than a few times each night. There is some evidence that both males and females call.

This species breeds in water-filled cavities in living trees. Males likely call from the cavity or from its vicinity to attract females to it. The eggs are most likely laid in the water or attached to the wall of the water-filled cavity; the tadpoles develop in their arboreal pool. A tree cavity located approximately 45 ft (15 m) above the ground was confirmed as a breeding site for *E. sukia*. It reportedly contained a single adult and 10 early stage tadpoles. Due to the lack of food in such small, isolated breeding pools, biologists speculate that the larvae feed on infertile eggs provided at regular intervals by the female parent.

Very little information is available on the biology, distribution, or conservation status of this species. Given the availability of seemingly suitable habitat in several protected areas to the north and south of the known localities for *E. sukia*, it seems likely that this species is significantly more widespread than is currently thought.

In adults, skin on head not fused to the skull (co-ossified); however, bony plates (osteoderms) are embedded in the skin on the head and back and underlie the larger tubercles.

No conspicuously enlarged or projecting tubercles on the upper eyelids, although several low, rounded tubercles are present.

Head, body, and limbs with scattered tubercles, none of which are tipped with a dark, hardened spine.

Limbs fairly long and robust.

Moderate-sized, with a broad head and a short, truncated snout.

Ventral surfaces white or cream, usually with small brown flecking; undersides of limbs generally with larger, more diffuse markings than head or body.

daytime coloration

Legs without dark transverse bands.

A scalloped fleshy skin fringe lines the outer margin of forearm and foot, and continues across the heel as a series of fleshy flaps.

Adult males have a relatively small internal, extendable vocal sac on the throat.

Dorsum uniform reddish-brown, brown, tan, or gray, marked to varying degrees with irregular dark markings or green blotches. Changes color in response to changes in temperature and light levels; often lighter at night than during the day.

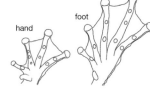

hand   foot

Very large hands and feet. Fingers and toes tipped with large round disks. Interdigital webbing is very extensive and reaches the base of the digital disks on at least one or two fingers and most toes. Base of each thumb with a well-developed prepollex; in adult males the prepollex ends in a hard, bluntly pointed projection (prepollex lacks clusters of black, keratinized spines).

E. bailarina (p. 375)

Has limb fringes that consist of a series of triangle-shaped (rather than rounded) projections; in adult males, note distinct clusters of black keratinized spines on the prepollex.

E. miliaria (p. 379)

Adult males have a sharp, recurved prepollical spine. Also note spine-tipped tubercles scattered over the flanks, ventral skin, and the underside of the legs.

Genus *Hyloscirtus*

The tree frogs in this recently redefined genus were previously included in the genus *Hyla*. It contains 34 species of mostly South American tree frogs.

These stream-breeding tree frogs are primarily grouped together based on molecular characteristics, although all species in the genus have fleshy skin fringes on their digits. They are green or brown, occasionally marked with middorsal or dorsolateral light stripes. Males, and sometimes females, have a prominently raised, round glandular structure on the chin (mental gland). Costa Rica is home to two species in this genus.

Hyloscirtus colymba
Stream Tree Frog   **Critically endangered**

A smallish tree frog. Green overall, with green bones, a whitish mental gland on the chin; and a thin white stripe on each side of the body that starts at the tip of the snout, passes above the eye, and ends in the shoulder region. Males to 1.46 in (37 mm) in standard length, females to 1.54 in (39 mm).

Ecoregions
1 2 3 4 5

On the Atlantic slope, *Hyloscirtus colymba* occurs in isolated populations, from southeastern Costa Rica to central Panama; 2,000–3,600 ft (610–1,100 m). On the Pacific slope, it occurs in a few areas in eastern Panama; it may range into Colombia and Ecuador.

Natural history

This rare—and rarely seen—frog is very secretive in its habits. Eggs are deposited in small, fast moving streams, likely during the dry season, when water levels are lower and flow rates less rapid. Males call from underneath large boulders; the mating call is a very cricketlike series of high pitched chirps. Individuals tend to stop calling for a long period of time at the least sign of danger, making it very difficult to pinpoint the location of a calling male. Eggs are possibly deposited in a foam nest under rocks. The tadpoles of this species develop in the stream and have a large oral disk that they use to cling to rocks in the current.

Due to its elusiveness, no reliable information exists on the population status of this species. *H. colymba* is only known from a few isolated localities and populations appear to have declined significantly in recent years, possibly due to chytridiomycosis. It continues to be found in very low numbers in a few Panamanian localities but no recent records for Costa Rica exist.

Description

Dorsum generally enamel-green, but may be yellow-green to bluish-green (very rarely brown). Usually uniform, but sometimes marked with small, dark green spots; on close examination, a pattern of minute white specks is often visible.

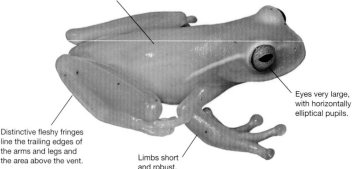

Eyes very large, with horizontally elliptical pupils.

Distinctive fleshy fringes line the trailing edges of the arms and legs and the area above the vent.

Limbs short and robust.

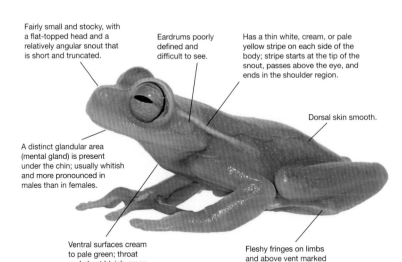

Fairly small and stocky, with a flat-topped head and a relatively angular snout that is short and truncated.

Eardrums poorly defined and difficult to see.

Has a thin white, cream, or pale yellow stripe on each side of the body; stripe starts at the tip of the snout, passes above the eye, and ends in the shoulder region.

Dorsal skin smooth.

A distinct glandular area (mental gland) is present under the chin; usually whitish and more pronounced in males than in females.

Ventral surfaces cream to pale green; throat and chest bluish-green.

Fleshy fringes on limbs and above vent marked with thin light lines.

Adult males with a single internal vocal sac on throat and a distinctly developed prepollical protuberance on the base of each thumb.

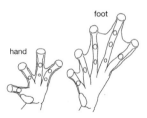

foot

hand

Moderate-sized hands and feet, and robust digits. Finger and toe disks relatively small and not much wider than digits; basal webbing between fingers; extensive webbing between toes.

H. palmeri
(p. 386)

Larger and uniform green; it lacks dorsolateral light stripes behind the eye.

Hypsiboas rufitelus
(p. 391)

Has a distinct prepollex and is generally green. Sometimes has light dorsolateral stripes, but has distinctive red webbing between its fingers and toes and pointed tubercles on its heels.

Isthmohyla angustilineata
(p. 394)

Generally brown, with narrow greenish or white dorsolateral lines that continue to the insertion of the hind limbs.

Hyloscirtus palmeri
Palmer's Tree Frog

Least concern

A moderate-sized tree frog. Uniform dark green, with green bones, a whitish mental gland, and a small triangular calcar spur on the heel. Males to 1.77 in (45 mm) in standard length, females to 1.97 in (50 mm).

Ecoregions
1 2 3 4 5

On the Atlantic slope, *Hyloscirtus palmeri* occurs in scattered localities, in humid forests, from central Costa Rica (1,965–2,460 ft [600–750 m]) to central Panama. On the Pacific slope, it occurs in central Panama, western Colombia, and northwestern Ecuador.

Natural history

Not much is known about the biology of *H. palmeri* in Costa Rica, where it is rarely encountered. It is known from scattered locations, most notably in the Río Sucio and Guayacan areas. It is likely more widespread than currently thought.

H. palmeri occurs in lowland and foothill locations, usually near cascading streams in forests. It breeds in streams, and likely reproduces in the dry season, when water levels and flow rates decrease.

Males call from rocks along the stream bank or from underneath rocks in the stream, making them difficult to see. Rarely do they call from low vegetation overhanging the water. The advertisement call has been described as a long series of repeated, high-pitched whistles. Its aquatic larvae can be seen in pools in small streams; they have a large oral disk that they use to adhere to rocks in the current.

Description

The peritoneum is visible through the translucent ventral skin, causing this species to occasionally be mistaken for an oversized glass frog.

Underside of the limbs green with a yellowish suffusion; throat and chest bluish-green; and the remaining ventral surfaces translucent green.

Males and females have a round mental gland, visible as a greenish-white circular structure on the chin.

Medium-sized and robust, with a broad head. Its relatively short snout is nearly truncate when seen in profile.

Dorsum uniform dark green.

Eyes a distinctive silver-gray.

A small triangular calcar spur present on each heel.

Adult males with a single subgular vocal sac and a distinctive prepollical protuberance on the base of each thumb; they do not have nuptial pads.

adult

Toe webbing pale orange to tan.

Trailing edges of the arms, legs, and the area above the vent marked with an indistinct, thin white stripe.

This species has pale green bones.

Eardrums indistinct.

metamorph

foot

hand

Limbs relatively short.

Hands and feet with short, stocky digits. Hands with basal webbing between fingers. Feet extensively webbed. Finger and toe disks round; only slightly wider than digits.

H. palmeri is larger and more robust than any Costa Rican glass frog. It also differs in having a calcar spur and having pale green bones (Costa Rican glass frogs have either white or dark green bones).

H. colymba (p. 384)

Has short, dorsolateral light stripes that extend from the tip of the snout to the shoulder region.

Hypsiboas rufitelus (p. 391)

Hypsiboas rufitelus has a green dorsum and occasionally light dorsolateral stripes, but note bright red (not brown or tan) webbing between its fingers and toes.

387

Genus *Hypsiboas*

The members of this genus were formerly included in the genus *Hyla*. The genus *Hypsiboas* currently has 91 species of tree frog. Most occur in South America, but 5 species in this genus occur in lower Central America, two of which reach Costa Rica.

These are medium-sized to very large tree frogs. In all species, adult males have a well-developed prepollex on the base of the inner fingers, which tends to bear a sharp spine. These structures are used during amplexus to give the male a better grip on the female. In some species, these spines are also used in territorial male-male aggression, hence the common name gladiator frog. Some species in this genus, including the Costa Rican species *Hypsiboas rosenbergi*, construct a breeding basin on the edge of an existing body of water to lay their eggs in.

Hypsiboas rosenbergi
Gladiator Tree Frog

Least concern

A large tree frog with a relatively flat head and long snout. Adults usually show a series of vertical bars on the flanks and a thin black line that runs from the tip of the snout toward the posterior, along the middle of the dorsum. Also has a prominent prepollex on the base of its thumbs and extensively webbed hands and feet. Males to 3.11 in (79 mm) in standard length, females to at least 3.74 in (95 mm).

Ecoregions
1 2 **3** 4 5

Hypsiboas rosenbergi is found in isolated populations throughout the Pacific lowlands, from central and southwestern Costa Rica, through Panama and Colombia, to western Ecuador; from near sea level to 2,950 ft (900 m).

Natural history

This species can breed year-round, but it displays heightened reproductive activity at the end of dry spells. Male *H. rosenbergi* scoop out a bowl-shaped depression near the edge of a stream or pond. These depressions fill up with water that seeps through from the adjacent body of water. A male enters this artificial pond and starts calling to attract females. His mating call can be heard from a great distance. It is a percussive series of low, short *tonk-tonk* notes that resemble the sound of a stick striking a hollow log.

Amplexus in *H. rosenbergi* is axillary; the eggs are deposited in a floating film on the surface of the artificial pond. Males tend to their eggs, and aggressively defend the nest area against intruders.

They are especially on their guard for competing male *H. rosenbergi*, who will try to invade their nest. If the surface tension of the water within the basin is broken, the eggs will sink to the bottom and die from lack of oxygen. For this reason, males actively patrol the perimeter of their nest and fight off rival male trespassers, sometimes using their prepollical spines as weapons. Males are occasionally injured during combat.

H. rosenbergi is relatively common thoughout its range and appears to be adaptable. It is frequently found in areas of secondary vegetation, flooded pastures, agricultural areas, and other kinds of disturbed habitat. There does not appear to be a significant threat to the continued survival of this species.

Description

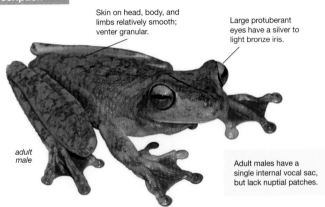

Skin on head, body, and limbs relatively smooth; venter granular.

Large protuberant eyes have a silver to light bronze iris.

adult male

Adult males have a single internal vocal sac, but lack nuptial patches.

One of the largest tree frogs in Costa Rica; has a large head with a long, flat snout that is bluntly rounded in profile.

Eardrums prominently visible.

Generally, with fairly indistinct dark vertical bars on the flanks and a diagnostic thin black line along the middle of the back, which starts at the tip of the snout and continues backward onto the dorsum. Both pattern elements may be obscured in some individuals.

Legs very long, sometimes with a calcar protuberance on the heel joints.

adult male

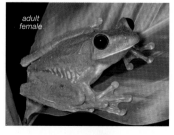

Ventral surfaces cream or white.

Dorsum varies in color; sometimes mottled, cream, tan, brown, or reddish-brown, with or without obscure markings on the back, sides, and limbs. Males usually have a darker dorsal color than do females.

adult female

Adult females with mostly uniform dorsal coloration

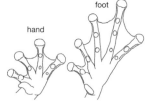

foot

hand

juvenile

Juveniles do not resemble adults. Their dorsum is nearly uniform green, often marked with black polka dots. Green juveniles somewhat resemble glass frogs.

The base of each inner finger shows a characteristic fleshy prepollex (which covers a sharp prepollical spine in adult males). Large hands; fingers with extensive webbing; greatly enlarged, round disks on each finger. Feet large and extensively webbed, with large rounded toe disks.

Similar species

H. rufitelus (p. 391)

No other large tree frog in Costa Rica has a thin, black line along the middle of its back. Juvenile *H. rosenbergi* resemble some glass frogs (family *Centrolenidae*), but have very large eyes and rounded disks on the fingers and toes.

Resembles juvenile *H. rosenbergi* but has red webbing between its fingers and toes. No range overlap in Costa Rica.

Hypsiboas rufitelus
Red-webbed Tree Frog

Least concern

Unmistakable; the only frog in Costa Rica with combination of beautiful lime-green to bluish-green dorsum and bright red finger and toe webs. Males to 1.93 in (49 mm) in standard length, females to 2.17 in (55 mm).

Ecoregions
1 2 3 4 5

Hypsiboas rufitelus occurs on the Caribbean slope, from eastern Nicaragua to central Panama; from near sea level to 2,130 ft (650 m).

Natural history

H. rufitelus is infrequently observed though not uncommon. Its call, a series of high pitched clucks, is regularly heard throughout the year, though these frogs are mainly observed in the wet season (August–November), when they emerge on rainy nights from their arboreal retreats and descend to mate.

Calling males are generally found in dense vegetation surrounding standing water in forested areas. Males perhaps use their prepollical spine as a weapon during aggressive male-male encounters, but this behavior is poorly documented. The distinct prepollex and spine may aid the male in grasping the female during amplexus. Mated pairs display axillary amplexus and lay their eggs in a floating raft on the water surface, usually in shallow sections of the breeding pool. *H. rufitelus* prefers murky swamps with lush aquatic and emergent vegetation. Once breeding activity subsides, the frogs disperse into the surrounding forest habitat.

The distribution of *H. rufitelus* is very localized, but if appropriate swampy breeding habitat is available in its forested lowland range then this species can be quite common. It appears to be tolerant of some habitat disturbance and is not considered to be under threat.

daytime
coloration

nighttime
coloration

Capable of considerable color change; the two photos above are of the same individual.

Adult males have a single external vocal sac, but lack nuptial pads.

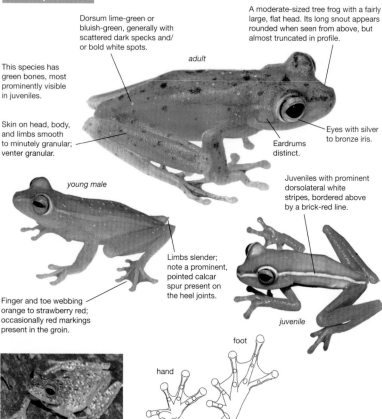

A moderate-sized tree frog with a fairly large, flat head. Its long snout appears rounded when seen from above, but almost truncated in profile.

Dorsum lime-green or bluish-green, generally with scattered dark specks and/or bold white spots.

adult

This species has green bones, most prominently visible in juveniles.

Skin on head, body, and limbs smooth to minutely granular; venter granular.

Eyes with silver to bronze iris.

Eardrums distinct.

young male

Juveniles with prominent dorsolateral white stripes, bordered above by a brick-red line.

Limbs slender; note a prominent, pointed calcar spur present on the heel joints.

Finger and toe webbing orange to strawberry red; occasionally red markings present in the groin.

juvenile

foot

hand

Some individuals, generally adult females, with dense concentrations of white spots that may obscure the green dorsal coloration.

Males and females have a well-developed prepollex (smaller in females) at the base of their thumbs. In males, this prepollical protuberance contains a bony spine that is covered by a retractible sheath. Hands moderately webbed, with only basal webbing between the inner two fingers; fingers have expanded, round disks. Feet with extensive webbing and with rounded disks on all toes.

Hyloscirtus palmeri (p. 386)

Juvenile *H. rosenbergi* are occasionally misidentified as *H. rufitelus* but lack the orange-red webbing between the toes and fingers characteristic of the latter species. No range overlap in Costa Rica.

An overall darker green tree frog with brown webbing (not red).

Genus *Isthmohyla*

All of the 14 species of tree frog that make up the genus *Isthmohyla* are restricted to the isthmus of Central America, hence the scientific name of this new genus. These species were previously included in the genus *Hyla* but were recently removed based on molecular and genetic characteristics. These 14 species are quite variable in their morphology and biology, and no readily identifiable field characteristic unites them all. All 11 Costa Rican *Isthmohyla* inhabit middle to high elevations; several are exceedingly rare.

Isthmohyla angustilineata
Narrow-lined Tree Frog

Critically endangered

A fairly small montane tree frog. Dorsum brown or green, with a thin white dorsolateral line that runs from behind the eye to the groin area, along each side of the body. Venter with dark spots. Lacks webbing between the fingers. Males to 1.34 in (34 mm) in standard length, females to 1.46 in (37 mm).

Ecoregions
1 2 3 4 5

In Costa Rica, *Isthmohyla angustilineata* occurs in the Tilarán, Central, and northern Talamanca Mountain Ranges; 1,640–6,690 ft (500–2,040 m). It is also found locally in the mountains of adjacent western Panama.

Natural history

Little is known about this rare species. On rainy nights, males call from twigs and vegetation close to the ground. The mating call has been described as a pattern of two short, pulsed notes that are repeated roughly every minute to minute and a half. Reproduction takes place in small water-filled depressions in the forest floor. Tadpoles have been found in rain pools in March and April, suggesting that breeding activity takes place during January or February.

I. angustilineata inhabits relatively undisturbed forests, at middle to high elevations. It was apparently never a common species anywhere, but individuals were once regularly seen in Tapantí National Park and on Cerro Chompipe, in Braulio Carrillo National Park. By the late 1980s, *I. angustilineata* appeared to have vanished. A 2005 sighting of a single individual in a regenerating pasture adjacent to oak forest on the slopes of Barva Volcano (6,725 ft [2,050m]) was the first in nearly two decades. Since then, additional sporadic observations on Cerro Chompipe and in the Monteverde area (Tilarán Mountain Range), as well as in western Panama, indicate that *I. angustilineata* still persists locally and may be recovering somewhat. However not enough data is available to fully understand the reasons for its decline, its current population status, and the potential threats it still faces.

Although juveniles often have a bright green dorsum, some individuals—like this metamorph—already have the copper to tan dorsum normally seen in adults.

A small to moderate-sized tree frog, with a short, rounded snout and a sharply angled canthal ridge.

Dorsal skin smooth, ventral skin granular.

Anterior and posterior surfaces of the thighs, as well as the underside of the legs and groin, pale yellow.

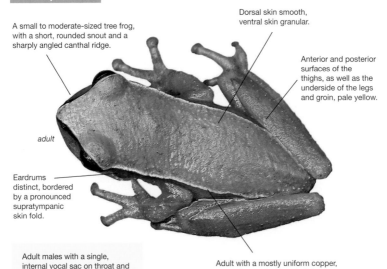

adult

Eardrums distinct, bordered by a pronounced supratympanic skin fold.

Adult males with a single, internal vocal sac on throat and brown nuptial patches on inner fingers during breeding season.

Adult with a mostly uniform copper, tan, or brown dorsum, dark brown flanks, and a very thin white dorsolateral line separating the lateral coloration from the dorsal field.

Venter and lower flanks white to cream; invariably heavily marked with dark brown specks.

hand   foot

juvenile

Some juveniles with a bright green dorsum and head; this color is gradually replaced with the brown adult coloration.

Hands relatively large, with robust digits, large disks, and no webbing between fingers. Feet with large toe disks and moderate amounts of webbing.

Hyloscirtus colymba (p. 384)

Has a green dorsum and incomplete dorsolateral white stripes that end in the shoulder region.

Isthmohyla calypsa
Spiny Short-snouted Tree Frog   **Critically endangered**

A medium-sized tree frog with large eyes, a very short, bluntly truncated snout, and distinct spiny tubercles covering the upper surfaces of its head, body, and limbs. Males to 1.42 in (36 mm) in standard length, females to 1.61 in (41 mm).

Isthmohyla calypsa is found in the extreme southern Talamanca Mountain Range of Costa Rica and adjacent western Panama. In Costa Rica, it is known only from the Pacific slope of Cerro Pando; 5,940–6,300 ft (1,810–1,920 m).

Ecoregions
1 2 3 4 5

Natural history

This species can hardly be confused with any other tree frog in the region (some populations of *I. lancasteri* [p. 400] have relatively tuberculate dorsal skin but never spine-like tubercles). For decades, *I. calypsa* was thought to be a spiny highland form of the variable and closely related *I. lancasteri*, but differences in breeding biology and morphology warrant species status for *I. calypsa*.

Within its limited range, *I. calypsa* is found at night on the vegetation lining fast flowing streams. It breeds on leaves, generally depositing clutches of 10–36 yellowish eggs on the underside of leaves that overhang the streams or their banks.

Reproductive activity takes place primarily April–December and peaks around the start of the rainy season (October–November). The male's mating call is a single ascending note, generally made from a favorite calling site that he uses repeatedly. Mating and egg-laying don't take place until after the first rains. The first round of breeding in the rainy season is usually the most prolific, but females may return later in the season to breed 1–2 more

times to deposit additional egg masses. Some 23–56 days after ovipostion, hatching larvae wriggle free from the egg jelly during heavy rains, and drop into the water below to complete their development there. In spite of living in torrential streams, these tadpoles do not have specialized or enlarged oral disks to adhere to rocks in the current.

This species used to be relatively common in the early 1990s in montane rainforest habitat on Cerro Pando, but it started to decline in 1992 and has since disappeared entirely. It now appears to be extinct in Costa Rica, but may still persist in Panamanian sections of its small range. *I. calypsa* was only known from undisturbed, pristine areas; habitat alteration—or destruction—and pollution were therefore not likely reasons for its decline. Chytrid fungal infections have been confirmed in several species in the area, including in dead individual *I. calypsa* in 1993. A predictive modeling study to determine the potential preferred habitat conditions for the amphibian chytrid pathogen, *Bd*, included the entire distribution range of *I. calypsa*.

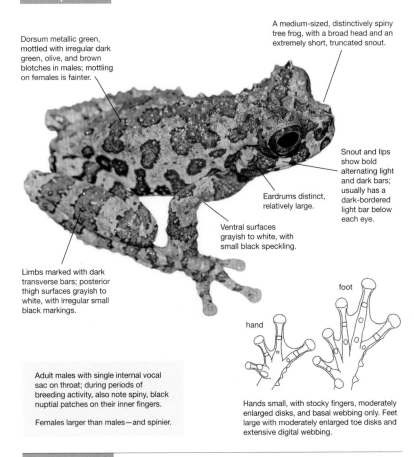

A medium-sized, distinctively spiny tree frog, with a broad head and an extremely short, truncated snout.

Dorsum metallic green, mottled with irregular dark green, olive, and brown blotches in males; mottling on females is fainter.

Snout and lips show bold alternating light and dark bars; usually has a dark-bordered light bar below each eye.

Eardrums distinct, relatively large.

Ventral surfaces grayish to white, with small black speckling.

Limbs marked with dark transverse bars; posterior thigh surfaces grayish to white, with irregular small black markings.

foot

hand

Adult males with single internal vocal sac on throat; during periods of breeding activity, also note spiny, black nuptial patches on their inner fingers.

Females larger than males—and spinier.

Hands small, with stocky fingers, moderately enlarged disks, and basal webbing only. Feet large with moderately enlarged toe disks and extensive digital webbing.

Similar species

I. lancasteri (p. 400)

Similar in appearance to *I. calypsa*, but is less spiny and has a boldly contrasting pattern of alternating black and yellow bands on the posterior surfaces of its thighs.

Isthmohyla debilis
Weak-voiced Tree Frog
Critically endangered

A small montane tree frog. Note green dorsum, a white or cream spot below each eye, and a complete (or partial) lateral light stripe (white, cream, or yellow) between the front and hind limbs. Males to 1.18 in (30 mm) in standard length, females to 1.26 in (32 mm).

Ecoregions
1 2 3 4 5

On the Atlantic slope, *Isthmohyla debilis* occurs in scattered localities in the Central and Talamanca Mountain Ranges; it also occurs in adjacent western Panama; 2,985–4,750 ft (910–1,450 m). On the Pacific slope it is also found in the El Copé area in southwestern Panama.

Natural history

I. debilis is a rare species known from just a few localities in middle elevation and in cloud forest habitat. It is nocturnal and arboreal and generally found in low vegetation overhanging streams, or on rocks in the streambed. Males produce a very soft cricketlike call from dense vegetation just above the water surface. The tadpoles develop in these streams; they have a large oral suction disk that helps them cling to rocks in the current.

This species declined dramatically in the 1990s, and recent surveys in Costa Rica have not resulted in any sightings. Populations in neigboring western Panama collapsed in the late 1990s and were last detected in the Bocas del Toro region in 1998. Recent surveys (2008–2009) in the Comarca Ngöbe-Buglé, in the central mountains of Panama, detected several individual *I. debilis*. Although not extinct, the future of *I. debilis* seems precarious at best and additional surveys are urgently needed to

assess the status of Costa Rican populations. Habitat alteration and chytridiomycosis are the most likely threats to this species.

I. debilis is an exceedingly rare species in Costa Rica and any observations of this species should be reported.

Adult males with a small, internal vocal sac on throat; when breeding, males have rough, brown nuptial pads on inner fingers.

Capable of considerable color change. At night, dorsal coloration darker and duller; during the day, dorsum is bright leaf-green.

Dorsal skin smooth.

A small tree frog with a small head and a short, truncated snout.

Dorsum uniform grayish-green, leaf-green, bluish-green, or olive; sometimes with dark green or brown mottling.

An irregular dark brown or bronze stripe runs along the canthal ridge, between nostril and eye.

A light spot adorns the upper lip, below each eye.

There are no dark transverse bands on the legs; posterior thigh surfaces and groin yellow, occasionally with dark brown motling.

Limbs short and stocky.

Has a pronounced supratympanic skin fold bordering its small eardrums.

foot

hand

Has a complete or partial white, cream, or yellow stripe between the front and hind limbs; if partial, generally only the posterior part of the stripe is present.

Hands small, with robust digits, basal webbing only, and small disks. Feet with moderate interdigital webbing and small toe disks.

I. xanthosticta
(p. 412)

Most similar morphologically to *I. debilis*, but has a broad dark lateral stripe marked with yellow spots on its flanks.

I. angustilineata
(p. 394)

Has a thin white dorsolateral line on each side of the body between the eye and groin (not a light lateral line between the limbs).

I. pictipes
(p. 404)

Some individuals may have similar coloration but invariably lack a light lateral stripe.

Duellmanohyla rufioculis
(p. 370)

Members of the genus *Duellmanohyla* (p. 367) and also *Plectrohyla legleri* (p.420) have a lateral light stripe and a suborbital light spot, but are distinguished by their red eyes.

Isthmohyla lancasteri
Mottled Short-snouted Tree Frog

Least concern

A small to medium-sized tree frog with a characteristically short, truncate snout. Dorsum is smooth to moderately tuberculate. Note distinctive alternating black and yellow transverse bars on the hidden surfaces of the thighs. Males to 1.34 in (34 mm) in standard length, females to 1.50 in (38 mm).

Isthmohyla lancasteri occurs in Costa Rica and western Panama, on the Atlantic slopes of the Talamanca Mountain Range; 1,150–3,935 ft (350–1,200 m).

Ecoregions
1 2 3 4 5

Natural history

This is a fairly common species. It occurs at middle elevations in humid forests, where it is most often found in low vegetation near streams. Breeding takes place predominantly during the dry season, when water levels are low; males call from rocks in quiet parts of the stream or from nearby low bushes. The eggs are deposited in single long strands containing 70–80 chocolate-brown eggs. The egg strands are usually attached to emergent vegetation, sinking to the bottom of the stream after several days. Tadpoles develop amid the gravel and debris on the streambed and use their oral disk to cling to rocks and pebbles.

The taxonomic status of *I. lancasteri* has long been confusing, in part due to the variation in color pattern and skin structure found in various populations of this species. The extremely spiky highland forms differ in breeding biology and are considered a separate species, *I. calypsa* (p. 396).

I. lancasteri is relatively tolerant of habitat alteration and it persists in open areas, roadsides, and pastures, as long as the humidity remains high and some tree cover is present.

I. lancasteri shows considerable variation in pattern and skin texture from population to population.

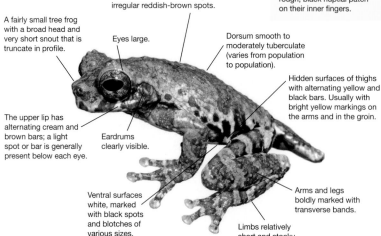

A fairly small tree frog with a broad head and very short snout that is truncate in profile.

Dorsum tan to cream, with a metallic green wash on some parts. Dorsum and head show a pattern of irregular reddish-brown spots.

A single internal vocal sac is present in males; sexually active males also have a rough, black nuptial patch on their inner fingers.

Eyes large.

Dorsum smooth to moderately tuberculate (varies from population to population).

The upper lip has alternating cream and brown bars; a light spot or bar is generally present below each eye.

Eardrums clearly visible.

Hidden surfaces of thighs with alternating yellow and black bars. Usually with bright yellow markings on the arms and in the groin.

Ventral surfaces white, marked with black spots and blotches of various sizes.

Arms and legs boldly marked with transverse bands.

Limbs relatively short and stocky.

The combination of a short, truncated snout and alternating black and yellow bars on the thighs is unique among Costa Rican frogs.

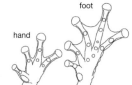

hand foot

Hands and feet of moderate size, with robust digits and moderately enlarged finger and toe disks. Hands with basal webbing only, feet extensively webbed.

I. calypsa (p. 396)

Has many large tuberculate spines on the dorsal surfaces of the body and head; its posterior thigh surfaces are white to gray with black markings.

Scinax boulengeri (p. 423)

The only other tree frog in Costa Rica with a bold pattern of dark and light bars on the hidden surfaces of its thighs, but it has a very long and flattened snout profile.

Isthmohyla picadoi
Picado's Bromeliad Tree Frog   **Near threatened**

A fairly small but robust frog. Has a distinctive pair of toothlike odontoids in the front of the lower jaw (not readily visible unless mouth is open). Dorsum is nearly uniform grayish-brown, yellowish-tan, or reddish-orange. Often has a narrow dark line encircling the wrist. Typically hands have only a fringe of basal webbing. Males to 1.34 in (34 mm) in standard length, females to 1.50 in (38 mm).

Ecoregions
1 2 3 4 5

In Costa Rica, *Isthmohyla picadoi* occurs on both slopes of the Central and Talamanca Mountain Ranges; it also ranges into the Panamanian highlands, as far as the Serranía de Tabasará; 6,300–9,515 ft (1,920–2,900 m).

Natural history

I. picadoi is only known from montane forests, where it inhabits arboreal bromeliads. These frogs can complete their entire life cycle inside these epiphytic plants; adults, juveniles, eggs, and tadpoles of this species have all been found in water-filled bromeliad leaf axils. Although there is insufficient data to make a definitive conclusion, it is likely that the tadpoles are fed nutritive eggs by the parents, a behavior seen in other frogs in the region (e.g., *I. zeteki*, *Anotheca spinosa*, *Oophaga granulifera*, and *O. pumilio*).

Originally thought to be a species of pristine forests, recent findings indicate that population density is often highest in secondary edge habitat contiguous with mature forest, probably because bromeliad density is higher in such open areas. The majority of these frogs have been extracted from arboreal bromeliads, sometimes more than 65 ft (20 m) above the ground. Size of the plant appears critical; frogs have been found only in bromeliad rosettes measuring more than 10 in (25 cm) in diameter, with the larger plants holding the most frogs. There appears

to be a trend towards vertical stratification, with males generally occurring lower in the canopy than females or juveniles.

Males call at night, producing a single slightly drawn-out note, followed by a series of two- and three-note chirps that lasts several seconds; it has been described as: *creeeeeek crik crik crikek crikek crikek crikek crikekek.* Although exact timing of the reproductive season in this species is not known, calling males and metamorphosing juveniles have been reported in January, hinting at a protracted breeding season that may peak during the wet season.

Little information exists on the population size and status of *I. picadoi* due to the difficulty of surveying these canopy specialists. However, recent findings indicate that it is more adaptable and likely more common than previously assumed. Populations persist in areas that have suffered amphibian declines due to chytrid fungal pathogens, suggesting that this species is relatively stable. Habitat loss in areas outside of protected zones is likely the biggest threat to *I. picadoi*.

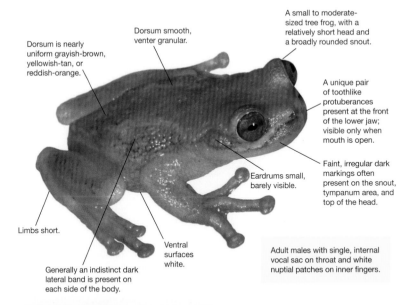

Dorsum is nearly uniform grayish-brown, yellowish-tan, or reddish-orange.

Dorsum smooth, venter granular.

A small to moderate-sized tree frog, with a relatively short head and a broadly rounded snout.

A unique pair of toothlike protuberances present at the front of the lower jaw; visible only when mouth is open.

Faint, irregular dark markings often present on the snout, tympanum area, and top of the head.

Eardrums small, barely visible.

Limbs short.

Ventral surfaces white.

Generally an indistinct dark lateral band is present on each side of the body.

Adult males with single, internal vocal sac on throat and white nuptial patches on inner fingers.

Although generally fairly uniform in coloration, some individuals are marked with dark mottling on the head and back. Also note the distinctive forward-facing dark red eyes.

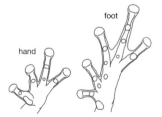

foot

hand

Has short stocky digits. Hands with only vestigial interdigital webbing and moderately expanded, round disks. Feet with reduced webbing, especially between inner three toes; toes with moderately expanded round disks.

I. zeteki (p. 414)

Smaller. It has more extensive finger webbing; distinctly visible eardrums; and only a single toothlike odontoid on the lower jaw.

Isthmohyla pictipes
Dark-footed Tree Frog

Endangered

A moderate-sized montane tree frog. Posterior thigh surfaces dark brown, with discrete white, cream, or bright yellow spots. Dorsum uniform green, olive, brown, or purplish-black. Venter dark. Males to 1.53 in (39 mm) in standard length, females to 1.77 in (45 mm).

Ecoregions
1 2 3 4 5

Isthmohyla pictipes occurs in Costa Rica, in the Central and Talamanca Mountain Ranges; 6,330–9,185 ft (1,930–2,800 m). As currently understood, this species is endemic to Costa Rica, but may also occur in neighboring western Panama.

Natural history

A stream-breeding species of high-elevation, moist forests. *I. pictipes* breeds in fast-flowing mountain streams in its cool, wet climate. Calling males have been observed on rocks in or near the stream, or on low vegetation overhanging the water. Tadpoles cling to rocks using a well-developed oral disk. It is not known where this species lays its eggs, but it is thought that oviposition takes place on leaves overhanging the water.

Not much is known about the population structure or biology of this species, but it is reported to be declining locally. Recent surveys at the locations of historic populations in the Central Mountain Range did not result in any sightings, but *I. pictipes* reportedly persists in other areas such as Cerro de la Muerte, in the northern Talamanca Mountain Range. Chytridiomycosis is a definite threat to this species, as are the release of trout in mountain streams and habitat loss. More research is needed to assess the conservation status of this species.

Description

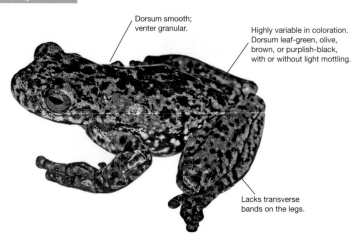

Dorsum smooth; venter granular.

Highly variable in coloration. Dorsum leaf-green, olive, brown, or purplish-black, with or without light mottling.

Lacks transverse bands on the legs.

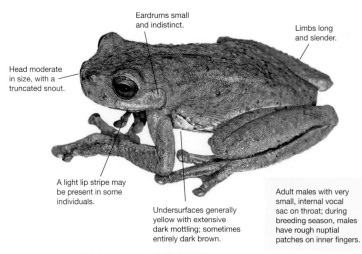

Eardrums small and indistinct.

Limbs long and slender.

Head moderate in size, with a truncated snout.

A light lip stripe may be present in some individuals.

Undersurfaces generally yellow with extensive dark mottling; sometimes entirely dark brown.

Adult males with very small, internal vocal sac on throat; during breeding season, males have rough nuptial patches on inner fingers.

Hidden posterior surfaces of thighs dark brown, with distinct white, cream, or bright yellow spots; pattern may extend onto anterior thigh surfaces and groin.

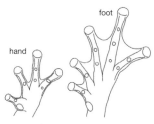

foot

hand

Hands large, with robust, long fingers; fingers with moderately enlarged disks and basal webbing only. Feet with long toes, large disks, and moderate amounts of webbing.

Similar species

I. xanthosticta (p. 412)

The extremely rare *I. xanthosticta* has a bronze stripe between its nostril and eye, and a dark stripe marked with yellow spots on the flanks.

I. pseudopuma (p. 406)

Shows a highly variable color pattern but almost always has a pale venter (not yellow or dark brown). Posterior thigh surfaces lack spots.

Isthmohyla pseudopuma
Meadow Tree Frog   **Least concern**

A highly variable, moderate-sized tree frog, with no easy field marks. Usually tan, brown, or yellow; has a pattern of yellowish-cream spots in the otherwise dark brown groin area. Tubercles on the soles of the feet are dark. Calling males have a bilobed vocal sac. Males to 1.77 in (45 mm) in standard length, females to 2.05 in (52 mm).

Isthmohyla pseudopuma occurs in Costa Rica in the Tilarán, Central, and Talamanca Mountain Ranges; it is also found on the Pacific slope of adjacent mountains in western Panama; 3,600–7,545 ft (1,100–2,300 m).

Ecoregions
1 2 3 **4** 5

Natural history

I. pseudopuma is a common species of middle elevations. It occurs in both undisturbed forests and in secondary growth, including plantations and agricultural areas. It is nocturnal; during the day, these frogs hide in vegetation, preferably bromeliads and other plants with protective leaf axils. Outside the breeding season, foraging individuals can be observed low in trees and on bushes at night.

This species breeds in large numbers after the first heavy rains of the wet season (April–May) flood pastures and fill up road side ditches and temporary rain pools. It is known to breed in very small bodies of water, as small as a cattle hoof print, but usually breeds in larger ponds. Males call from the edge of the pool, producing a low, single-note call that is repeated at intervals of just over one second. Males often outnumber females at breeding sites, and heavy competition ensues. Sometimes, multiple males attempt to dislodge an amplectant male, resulting in a clump of males surrounding a female. Eggs are laid in the water, where they are attached to emergent vegetation (usually grass) in multiple small clutches. Tadpole development is rapid, an adaptation to life in a small ephemeral pond. In breeding ponds where not enough plant matter is available to feed the tapoles, larval *I. pseudopuma* may become cannibalistic and feed on tadpoles of other and their own species. There is some indication that tadpoles respond to drying breeding pools

by accelerating their development and metamorphosing earlier, but with a smaller body size.

Quite adaptable to habitat alteration, *I. pseudopuma* can be very common locally; it is often found in close proximity to human settlements, though it is likely adversely affected by intensive land use practices and pollution. The well-studied populations in Monteverde, Costa Rica, declined during the late 1980s and early 1990s, when many other species of amphibians in the area declined or disappeared, but *I. pseudopuma* has since recovered. It is unclear what caused the initial decline or what allowed *I. pseudopuma* to recover while other co-occurring species have not.

Pair of *I. pseudopuma* in amplexus. Note the bright yellow coloration of the male during the breeding season.

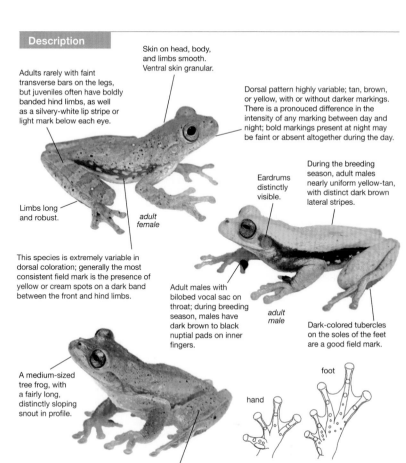

Description

Adults rarely with faint transverse bars on the legs, but juveniles often have boldly banded hind limbs, as well as a silvery-white lip stripe or light mark below each eye.

Skin on head, body, and limbs smooth. Ventral skin granular.

Dorsal pattern highly variable; tan, brown, or yellow, with or without darker markings. There is a pronouced difference in the intensity of any marking between day and night; bold markings present at night may be faint or absent altogether during the day.

Limbs long and robust.

adult female

During the breeding season, adult males nearly uniform yellow-tan, with distinct dark brown lateral stripes.

Eardrums distinctly visible.

This species is extremely variable in dorsal coloration; generally the most consistent field mark is the presence of yellow or cream spots on a dark band between the front and hind limbs.

Adult males with bilobed vocal sac on throat; during breeding season, males have dark brown to black nuptial pads on inner fingers.

adult male

Dark-colored tubercles on the soles of the feet are a good field mark.

A medium-sized tree frog, with a fairly long, distinctly sloping snout in profile.

foot

hand

Anterior and posterior thigh surfaces uniform cream, tan, or brown; ventral surfaces yellowish-cream to white.

Hands with robust fingers that bear greatly expanded disks; most fingers with moderate webbing, but two inner fingers with only basal webbing. Feet extensively webbed; toes with enlarged rounded disks.

Similar species

Smilisca sordida (p. 440)

Has basal webbing on the hands, and has bluish or cream (not yellow) spots in the groin and flank area.

I. rivularis (p. 408)

Could be confused with juvenile *I. pseudopuma* but has a distinctively angular, truncated snout.

I. tica (p. 410)

Differs from juvenile *I. pseudopuma* in having granular skin and a bluntly rounded (not sloping) snout.

Isthmohyla rivularis
Mountain Stream Tree Frog

Critically endangered

A smallish stream-breeding tree frog. Has a pronounced, sharply angled canthal ridge and acute snout profile. Dorsum mostly uniform pale yellow, pale gray, or tan. White venter marked with dark spots. Males to 1.34 in (34 mm) in standard length, females to 1.46 in (37 mm).

In Costa Rica, *Isthmohyla rivularis* inhabits middle elevation forests of the Tilarán, Central, and Talamanca Mountain Ranges; it also occurs in adjacent western Panama; 3,935–6,890 ft (1,200–2,100 m).

Ecoregions
1 2 3 4 5

Natural history

I. rivularis is a nocturnal species. It breeds in streams and is most frequently seen in dense vegetation along cascading mountain streams. Males produce a three-noted, high-pitched *cheep-cheep-cheep* during the breeding season, which peaks at the onset of the wet season (April–June) but may continue throughout the year, if conditions are right. When they call, these frogs generally are hidden in very dense tangles of vegetation at the edge of streams or overhanging the water. Amplexus and oviposition have never been observed in this species and it is not known where eggs are deposited. It is likely that this happens under rocks in the stream or attached to vegetation overhanging the water. Tadpoles of *I. rivularis* are well-adapted to a life in rapid flowing streams and use their large oral disk to attach to rocks and avoid being washed downstream.

This species was once common throughout its range but declined dramatically since the late 1980s. In 1993, in what appeared to be the last sighting in Costa Rica, an individual was seen in Las Tablas, near the Panamanian border. In spite of repeated surveys in historic locations, no more individuals were found, and by 2007 *I. rivularis* was thought to possibly be extinct until, later that year, an individual was discovered in the

Monteverde Cloud Forest Preserve. Since then several other populations have been reported from a variety of locations in the Central and Talamanca Mountain Ranges, as well as in the western Panamanian highlands, suggesting that *I. rivularis* is making a recovery in recent years. Further research is needed to evaluate the extent of the recovery and identify the factors that enabled the resurgence of this species (other amphibians that declined simultaneously in the same habitat have yet to be rediscovered). Since many known populations of *I. rivularis* were found in protected areas with pristine forest, habitat destruction was probably not a factor in their decline, and pathogens or other environmental threats are likely culprits.

This *I. rivularis* metamorph still shows remnants of the large oral suction disk that it uses to adhere to rocks in the strong current.

Eardrums poorly defined, small, and difficult to see; a supratympanic skin fold starts behind each eye and curves downward over the top of the eardrum.

nighttime coloration

A small to moderate-sized tree frog, with a small head and a short, acutely angled snout when viewed in profile.

A few poorly defined, faint dark bands may be present on the hind limbs.

Limbs short.

Adult males have a single, external vocal sac on the throat and brown nuptial pads on the base of the inner fingers.

Capable of considerable color change. At night, the coloration of the head, body, and limbs is mostly uniform pale yellow, pale gray, or tan.

Dorsum smooth overall or with scattered low, rounded tubercles; skin on ventral surfaces granular.

Sides of head occasionally with fine, metallic-green speckling.

daytime coloration

The canthal ridge is pronounced, giving the head a very angular appearance.

The thighs, upper arms, throat, vocal sac, flanks, and venter are cream to grayish-white; ventral surfaces cream to yellowish, marked with dark pigment.

During the day the color pattern darkens to chestnut-brown, dark brown, or olive, often with green speckling on head and body (same individual as above).

foot

hand

Stocky, short fingers bear greatly enlarged rounded disks and only minimal webbing. Feet moderately webbed, with robust toes that each end in a moderately, to greatly expanded, round disk.

| *I. tica* (p. 410) | *I. angustilineata* (p. 394) | *I. debilis* (p. 398) | *I. pseudopuma* (p. 406) |

Has a strongly tuberculate dorsum and a rounded lateral snout profile.

Has a brown or green smooth dorsum with a thin white dorsolateral line on each side of the body.

Has a bright green dorsum and a distinct light suborbital spot.

Larger, has a rounded snout profile, and has dark tubercles on the soles of the feet.

Isthmohyla tica
Tico Tree Frog **Critically endangered**

A moderate-sized cryptic tree frog. Has a rounded snout in profile. Also note roughly tuberculate upper surfaces. Mottled green or brown, with distinct transverse bands on the hind limbs. Males to 1.34 in (34 mm) in standard length, females to 1.65 in (42 mm).

Ecoregions
1 2 3 **4** 5

In Costa Rica, *Isthmohyla tica* occurs at middle elevations in the Tilarán, Central, and Talamanca Mountain Ranges; it also occurs in western Panama; 3,610–5,410 ft (1,100–1,650 m).

Natural history

This frog used to be common but is now considered critically endangered. *I. tica* is a nocturnal species that inhabits the trees and shrubs lining fast moving mountain streams. Its breeding season is likely restricted to the drier months of the year (February–May), when water levels and flow rates are down, and these frogs can more safely access the streams.

The advertisement call is a cricketlike chirp that is repeated 3–5 times at 1–2 second intervals. Males call from low vegetation overhanging the water, from rocks in the stream, or from low trees several meters above the stream. The amplexus and oviposition site for this species remain unknown; eggs may be deposited under rocks in the water or on vegetation above the stream. The tadpoles develop in the stream and are well-adapted to the strong currents. Their large oral disk helps them cling to the downstream side of rocks.

In spite of repeated surveys, this species was absent from all known historic populations and was feared extinct by 2007. In 2010 a few individuals were found in historic sites in western Panama and Costa Rica, including the Monteverde Cloud Forest Preserve. These observations suggest that in a few areas *I. tica* is recovering, at least to detectable population levels. Future studies and monitoriong efforts are needed to better assess this situation, and only time will tell if populations will rebound to historic levels.

Description

Flanks marked with indistinct gray or white mottling.

A linear series of low tubercles present along the trailing edge of the arms and legs.

In most individuals, lighter markings form poorly defined, irregular dorsolateral bands that enclose a darkly mottled dorsal field.

Skin on head, body, and limbs distinctly textured with numerous low, rounded tubercles; ventral surfaces granular.

Adult males with a single, internal subgular vocal sac and rough, brown nuptial patches on inner fingers.

Eardrums distinctly visible and bordered by a supratympanic skin fold.

A fairly robust, moderate-sized tree frog with a small head.

Canthal ridge rounded and steep.

Coloration highly variable; usually with mottled green upper surfaces or brown to tan base color with scattered minute green spots.

Snout short with a rounded to obtuse profile.

Upper lip marked with some light gray, cream, or white pigment, approaching the appearance of a suborbital spot but never clearly defined.

Limbs robust and moderate in length.

Venter is grayish or white, without dark mottling.

Bold transverse dark bands usually present on the legs and lower arms; upper arms uniform brown or green.

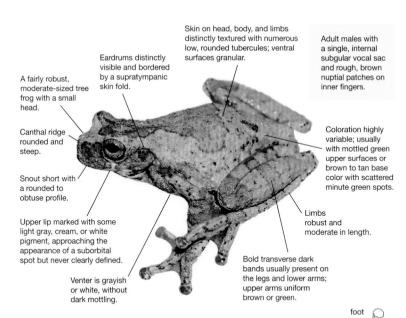

Hands with short, stocky fingers, large round disks, vestigial webbing between inner two fingers and basal webbing elsewhere. Feet extensively webbed and toes with greatly enlarged, round disks.

hand

foot

Similar species

I. rivularis (p. 408)

Similar in appearance, but has smoother skin, shorter and less robust legs, and a more angular snout.

I. angustilineata (p. 394)

Has a brown or green smooth dorsum, with a thin white dorsolateral line on each side of the body.

I. pseudopuma (p. 406)

Larger, has smoother skin, and has dark tubercles on the soles of its hands and feet.

Isthmohyla xanthosticta
Barva Tree Frog

Data deficient

A small green tree frog. Note distinct yellow spots in the groin and on the posterior surfaces of the thigh. Face with a distinct white lip stripe and a bronze stripe along the canthus rostralis; additional white stripes adorn arms and legs. The single known adult female measures 1.14 in (29 mm) in standard length.

Ecoregions
1 2 3 4 5

Isthmohyla xanthosticta is known only from the south fork of the Río las Vueltas, on the south slope of Barva Volcano, Heredia Province, Costa Rica; found at 6,890 ft (2,100 m).

Natural history

Only a single *I. xanthosticta* has ever been observed, in the late 1960s; it was found perched on low vegetation in epiphyte-laden humid upper montane forest. Even though the type locality is now protected in Braulio Carrillo National Park and the habitat is seemingly pristine, this is also one of the areas where, in the late 1980s, most amphibians declined or disappeared, possibly due to chytrid fungal infection. Repeated searches in the area have failed to turn up additional individuals, and this species may be extinct.

The taxonomic position of *I. xanthosticta* is unclear and may never be resolved if no additional individuals are recorded. *I. rivularis* and *I. pictipes* co-occurred in the same stream valley where *I. xanthosticta* was discovered.

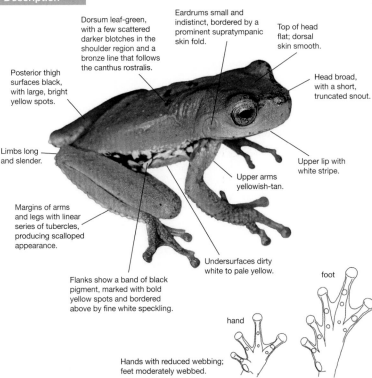

Dorsum leaf-green, with a few scattered darker blotches in the shoulder region and a bronze line that follows the canthus rostralis.

Eardrums small and indistinct, bordered by a prominent supratympanic skin fold.

Top of head flat; dorsal skin smooth.

Posterior thigh surfaces black, with large, bright yellow spots.

Head broad, with a short, truncated snout.

Limbs long and slender.

Upper lip with white stripe.

Upper arms yellowish-tan.

Margins of arms and legs with linear series of tubercles, producing scalloped appearance.

Undersurfaces dirty white to pale yellow.

foot

Flanks show a band of black pigment, marked with bold yellow spots and bordered above by fine white speckling.

hand

Hands with reduced webbing; feet moderately webbed.

I. pictipes (p. 404)

I. debilis (p. 398)

Has much smaller yellow markings and lacks a white lip stripe and white arm and leg stripes.

Has uniform yellow posterior thigh surfaces and a creamy-white lateral stripe.

Isthmohyla zeteki
Zetek's Bromeliad Treefrog   **Near threatened**

A small, relatively slender arboreal tree frog. Has limited webbing between the fingers and toes. Note large, forward facing eyes. Eardrums indistinctly visible. Front of lower jaw with a single toothlike structure (not visible when mouth is closed). Males to 0.94 in (24 mm) in standard length, females to 1.06 in (27 mm).

Ecoregions
1 2 **3 4** 5

In Costa Rica, *Isthmohyla zeteki* is known from isolated localities on the slopes of the Central and Talamanca Mountain Ranges; it also occurs in adjacent western Panama; 3,935–5,905 ft (1,200–1,800 m).

Natural history

I. zeteki is rarely seen because its entire life cycle plays out inside arboreal bromeliads. Females attach small numbers of eggs to bromeliad leaves, just above the water line at the base of the plant. Eggs may be divided over multiple leaf axils in the same rosette, or even over more than one plant. Tadpoles develop in the water-filled leaf axils and have been found with frog eggs in their stomachs, indicating that female *I. zeteki* may produce infertile, nutritive eggs to feed their developing larvae. Similar behavior has been observed in the tree frog *Anotheca spinosa* and poison dart frogs in the genus *Oophaga*. A single bromeliad may harbor adults, eggs, and tadpoles in different stages of development, suggesting a strong territoriality in this species, as well as a protracted breeding season.

Because it is hard to find this species, its population and conservation status is incompletely understood. *I. zeteki* occurs in humid montane forests; large areas of potentially suitable habitat are protected in Braulio Carrillo and La Amistad National Parks. Outside such protected areas, deforestation and habitat alteration are likely among the bigger threats to this species. In a few known localities, *I. zeteki* appears to have survived the dramatic amphibian declines seen in co-occurring species, and overall its populations seem to be stable.

I. zeteki is rarely seen because it spends much of its life high overhead, hidden in arboreal bromeliads.

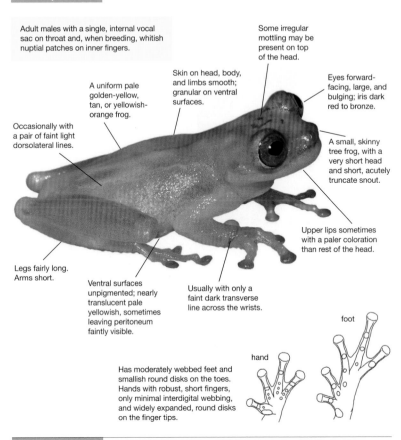

Adult males with a single, internal vocal sac on throat and, when breeding, whitish nuptial patches on inner fingers.

Some irregular mottling may be present on top of the head.

A uniform pale golden-yellow, tan, or yellowish-orange frog.

Skin on head, body, and limbs smooth; granular on ventral surfaces.

Eyes forward-facing, large, and bulging; iris dark red to bronze.

Occasionally with a pair of faint light dorsolateral lines.

A small, skinny tree frog, with a very short head and short, acutely truncate snout.

Upper lips sometimes with a paler coloration than rest of the head.

Legs fairly long. Arms short.

Ventral surfaces unpigmented; nearly translucent pale yellowish, sometimes leaving peritoneum faintly visible.

Usually with only a faint dark transverse line across the wrists.

foot

hand

Has moderately webbed feet and smallish round disks on the toes. Hands with robust, short fingers, only minimal interdigital webbing, and widely expanded, round disks on the finger tips.

I. picadoi (p. 402)

Very similar in morphology and biology. It is larger and more robust; has a pair of toothlike structures on the lower jaws; shows less webbing between the fingers; and usually has more dark markings. *I. picadoi* generally occurs at higher elevations than *I. zeteki.*

Genus *Osteopilus*

The genus *Osteopilus* currently contains eight species of Caribbean Island endemics. Several species are endangered due to habitat loss in their original small island ranges, but *O. septentrionalis* is abundant and widespread. It has spread from its native Cuba and the Bahamas to neighboring islands and into southern Florida, the Hawaiian Islands, and Puerto Limón, Costa Rica, as a stowaway on cargo. It shows a typical invasive species distribution pattern, establishing itself in port cities throughout the region and spreading from there.

Osteopilus typically are large to very large tree frogs whose skin is fused to the skull on the top of the head (co-ossified), forming a hard helmetlike casque. These frogs also typically have green bones.

Osteopilus septentrionalis
Cuban Tree Frog

Least concern

A large to very large tree frog. It has greatly expanded finger disks. Adults have co-ossified skin on top of the head. Also note green bones. Adult females reach a standard length in excess of 5.91 in (150 mm), making this potentially the largest tree frog in Costa Rica; males generally less than 2.95 in (75 mm) in standard length.

Ecoregions
1 2 3 4 5

Osteopilus septentrionalis is native to Cuba, the Bahamas, and the Grand Cayman, but has been introduced to various Caribbean islands, as well as southern Florida. It has been reported from the port city of Limón, Costa Rica, since the 1980s, and is now established there.

Natural history

Little is known about the biology of this species in Costa Rica, as it is a relatively recent invader facing new environmental challenges. In other parts of its range, *O. septentrionalis* is extremely successful at colonizing new habitats, and it is now firmly established and abundant in large parts of Florida, including in several protected areas. It also continues to show up on new islands in the Caribbean.

This species is highly adaptable. It inhabits forests, mangroves, secondary vegetation, and disturbed habitat, including areas in the immediate vicinity of humans. It is found on houses and under debris. It can survive in coastal areas and is even tolerant of brackish water.

It breeds in temporary rain pools, roadside ditches, drainage areas, swimming pools, ponds in golf courses, and any other suitably large body of water. Males produce a rasping growl during rainy nights and generally call from low vegetation near a breeding pond. Eggs are deposited in a floating film on the water, where the larvae complete their development.

O. septentrionalis is an invasive species. In other areas where it has become established, it outcompetes native amphibians and preys on anything smaller than itself. Therefore, the presence of *O. septentrionalis* in any part of the country could be harmful to the native fauna and should be of concern. Efforts to stop the spread of this species are necessary.

Care should be taken when handling these frogs; when threatened, the glandular dorsal skin can excrete noxious secretions that may irritate the mucous membranes and can cause a burning sensation on the skin of some people.

Adult females can attain enormous sizes, becoming formidable predators of just about anything that fits in their mouth.

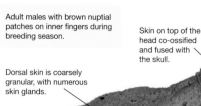

Adult males with brown nuptial patches on inner fingers during breeding season.

Skin on top of the head co-ossified and fused with the skull.

Usually with a concave depression between the eyes.

Head broad and rounded; snout with a sharply defined canthal ridge.

Dorsal skin is coarsely granular, with numerous skin glands.

Eardrums clearly visible.

Hind limbs variable, with contrasting dark bars in some individuals, but uniformly colored in others.

Coloration extremely variable; adults tend to be tan, gray, pale green, or rust-brown, either uniformly colored or with a pattern of dark elongate blotches on the dorsum.

This species has green bones.

Juveniles are generally more boldly marked than adults, with brown or green blotches on a silver-gray or tan background.

Hidden posterior surfaces of the thighs show a dark reticulum on a yellowish-cream background (this sometimes extends onto the groin).

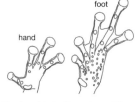

Hands and feet with very large disks on all digits. Fingers with basal webbing; moderate webbing between toes.

Trachycephalus typhonius (p. 446)

Also large and has glandular dorsal skin, but lacks the co-ossified skull.

Smilisca baudinii (p. 430)

Has a far less glandular dorsum. It also lacks the green bones and co-ossified skull.

Genus *Ptychohyla*

A genus of 13 small to medium-sized tree frogs. They have large, well-developed venterolateral glands (swollen and particularly visible in breeding males); a distinct fleshy ridge (or linear series of tubercles) along the trailing edge of the forearm; and breeding males have nuptial pads covered with distinct keratinized spines.

Most species occur in a region that spans from Mexico to central Nicaragua, but one species, *Ptychohyla legleri*, occurs disjunctly in southwestern Costa Rica and adjacent Panama.

These frogs are found in foothills and mountains. They breed in streams, and adults are normally found on rocks in or near the streams in which their tadpoles live. Little information exists on the breeding biology of *Ptychohyla legleri* in the wild; it is not known where the eggs are placed, but they are possibly deposited on or under rocks in very wet areas or placed on vegetation overhanging a stream.

Ptychohyla legleri was formerly considered a member of the *Hyla salvadorensis*-species group and is generally referred to as *Hyla legleri* in older literature.

Ptychohyla legleri
Legler's Tree Frog   **Endangered**

A moderate-sized tree frog. Olive brown overall, with red eyes and horizontally elliptical pupils. Note an irregular white lateral stripe (sometimes incomplete) and discontinuous white lip stripe. Lacks a light spot below each eye. Males to 1.46 in (37 mm) in standard length, females to 1.53 in (39 mm).

Ecoregions
1 2 3 4 5

Ptychohyla legleri occurs on the Pacific slopes of the Talamanca Mountain Range, in Costa Rica and adjacent western Panama; 2,885–5,000 ft (880–1,524 m).

Natural history

P. legleri breeds in streams; it is most often seen during the breeding season (February–July), at night. Males call from low, dense vegetation along small streams, or occasionally from vegetation-covered rocks. In a captive situation, eggs of this species were found attached to the roof of a small cavity in a rock beneath a small waterfall, just below the water line. In nature these frogs presumably place their eggs on or below rocks in a streambed, but this has not been confirmed. The tadpoles inhabit shallow streams and can be seen amid debris on the bottom.

This is still a fairly common species in humid montane forests, though it occurs in the elevational zone with the highest prevalence of chytrid fungal infections; within this zone, stream-breeding frogs are at risk of chytrid iomycosis.

Dorsal pattern somewhat variable; often uniform dark, although some individuals have numerous silver-white spots on the dorsum.

Dorsum dark brown to olive-green; either uniform, with a few silvery-white spots, or extensively mottled with light markings.

Eardrum distinctly visible, bordered by a pronounced supratympanic skin fold.

Eyes red, with horizontal pupils; lacks a light spot below the eye (as is usually the case in frogs of the genus *Duellmanohyla*).

Has a moderate-sized head and a bluntly rounded to nearly truncated snout.

Many individuals have a white lip stripe, though it is sometimes indistinct.

adult male

Skin smooth.

Thin white lines are present on the trailing edges of the forearm and legs and above the vent.

Generally has an irregular white lateral stripe that may be interrupted or indistinct in some individuals.

The dorsal surfaces of the forearm and the anterior and posterior thighs are orange to tan; these colors can extend posteriorly onto the venter, which is otherwise white.

Adult males have a single, internal vocal sac on the throat; during the breeding season, also note brown nuptial patches on the inner fingers.

foot

hand

Large hands and feet. Fingers moderately webbed, with large disks on tips. Toes with extensive interdigital webbing and enlarged disks on tips.

Duellmanohyla rufioculis
(p. 370)

Has red eyes and a horizontally elliptical pupil, but its lateral light line is expanded below the eye to form a light spot.

Duellmanohyla uranochroa
(p. 372)

Also has red eyes and a light lateral stripe, but note bright leaf-green dorsum.

Agalychnis callidryas
(p. 343)

Agalychnis species always have a vertically elliptical pupil.

Genus *Scinax*

These medium-sized frogs are gray to brown in coloration. Their body is flattened, and they have a long snout. The disks on the tips of the digits are moderately developed. The fingers are free of webbing, or have only basal webbing; feet are webbed, but the two inner toes are free of webbing. When these frogs perch on a vertical surface, typically facing downward, they have the ability to increase their grip by rotating their inner toes and fingers more than 90° so that they point upward.

Mesoamerican species of the genus *Scinax* breed during the wet season, in permanent or temporary pools. Their eggs and larvae are aquatic. The tadpoles are medium-sized, with extremely high tail fins that extend onto the body; they tend to be predatory, feeding on other tadpoles.

The genus *Scinax* is most diverse in South America. It currently contains 112 species, 3 of which occur in Costa Rica. It has undergone taxonomical revisions; species were formerly assigned to the genera *Hyla* or *Ololygon*. (There is also some debate over spelling; *Scinax elaeochroa*, for example, sometimes appears in literature as *Scinax elaeochrous*.)

Scinax boulengeri
Boulenger's Long-snouted Tree Frog **Least concern**

The only tree frog in its range with a distinctively long snout, granular dorsum, and alternating black and yellow (or green) transverse bands on its posterior thigh surfaces. Males to 1.93 in (49 mm) in standard length, females to 2.09 in (53 mm).

Ecoregions
1 2 3 4 5

On the Atlantic slope, *Scinax boulengeri* occurs from southern Nicaragua to eastern Panama; on the Pacific slope, it occurs in Costa Rica, central Panama, and northern Colombia; from near sea level to 2,295 ft (700 m).

Natural history

This is a relatively common species. It is nocturnal and arboreal. *S. boulengeri* is most frequently encountered during the wet season, when it gathers around permanent or temporary pools to breed. Males characteristically call from vertical surfaces on low vegetation surrounding the breeding pond, and position themselves head down. The inner digits of the hands and feet can rotate more than 90° and point up when the frog is in its head-down position, allowing for better grip on the substrate. The advertisement call of *S. boulengeri* is a loud, nasal *wraak*, repeated infrequently.

Breeding takes place throughout the wet season, but peaks with the first rains. On rainy nights, the calling activity starts at dusk, with the most intense calling lasting for 1–2 hours, after which it gradually tapers off. Eggs are laid in the water, and the larvae develop there. Tadpoles of this species are medium-sized, but with extremely high tail fins and muscular, heavily serrated beaks; they are predatory and feed on small crustaceans and other tadpoles, including those of their own species.

Though common throughout its range, *S. boulengeri* is rarely seen outside of the breeding season. It is usually found in small trees and bushes in open, secondary forest, but also occurs, in low density, in undisturbed, primary forest. Its mottled coloration and tuberculate skin render it nearly invisible on tree bark, and these frogs readily

escape detection except when calling. *S. boulengeri* is quite adaptable and can withstand substantial habitat alteration. No major threats to this species are currently recognized.

S. boulengeri in typical head-down calling position; note the inner finger, which is bent upward in order to increase its grip on vertical surfaces.

A well-camouflaged frog, usually with shades of gray, tan, brown, or olive-green, marked with darker blotches on the dorsum.

Dorsum often shows faint, irregular markings that are shaped like an inverted V.

Skin on head, body, and limbs rugose and tuberculate.

Snout characteristically long and flattened, with distinctly raised nostrils.

A moderate-sized tree frog with a distinctively flattened body.

Eardrums clearly visible, with the top partly covered by a supratympanic skin fold.

Ventral surfaces granular overall but ventral skin of limbs is smooth.

Limbs with alternating dark and light transverse bands.

Hidden surfaces of groin yellowish-green, with one to several black spots or with mottling.

Adult males with single, external vocal sac on throat; nuptial pads difficult to detect.

Fingers long and thin, with greatly expanded, truncated disks. Hands without interdigital webbing. Legs long and slender, with toe disks greatly expanded and truncated; extensive webbing between the long toes.

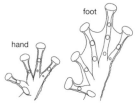

hand

foot

Other *Scinax* species lack the tuberculate dorsum and bold color pattern on the thighs.

Isthmohyla lancasteri (p. 400)

Shares alternating yellow and black bars on the posterior thigh surface but has a very short, truncated snout.

Scinax elaeochroa
Olive Long-snouted Tree Frog

Least concern

A fairly small, flattened tree frog. Dorsum nearly uniform olive-green, tan, or bright yellow (in breeding males), usually with a pattern of faint longitudinal stripes. Has visible dark green bones. Lacks webbing between the fingers and the two inner toes. Males to 1.50 in (38 mm) in standard length, females to 1.57 in (40 mm).

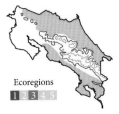

Ecoregions
1 2 3 4 5

On the Atlantic slope, *Scinax elaeochroa* occurs from eastern Nicaragua to northwestern Panama; on the Pacific slope, it occurs from central Costa Rica to southwestern Panama; from near sea level to 3,935 ft (1,200 m).

Natural history

Except for the driest months of the year (January–March), this abundant tree frog appears to be active year-round. During nights with heavy rains, males call from emergent vegetation or from low bushes near the edge of temporary ponds. In some areas, one can hear choruses of several hundred calling males. The advertisement call is series of short, harsh *wrrrink* notes, repeated frequently; it can even be heard outside of the breeding season, during which time males call from tree holes and other arboreal hiding places. Eggs are laid in a large mass that is often attached to floating water plants. Tadpoles live in the shallow margins of their breeding ponds, where they hide between the vegetation as they complete their development.

In many lowland regions in Costa Rica, *S. elaeochroa* may very well be the most common tree frog. It is quite tolerant of habitat alteration, inhabiting undisturbed rainforest but also secondary vegetation and even extremely disturbed areas, as long as there is some vegetation left. It can be found in the immediate vicinity of human settlements and frequently enters people's homes; it is regularly discovered in shower stalls or toilet tanks. Although pollution and extreme habitat alteration likely adversely affect this species, no immediate threats to *S. elaeochroa* are currently known.

During the breeding season, males develop a nearly uniform bright yellow coloration.

425

A moderately small, flat-bodied tree frog.

daytime coloration

Snout long and protruding, rounded in profile.

Dorsum smooth; venter granular.

Color variable. During the day, olive-tan to olive-green, with irregular dark linear stripes or a series of spots. At night, yellowish-green or brown overall, with bright yellow flanks.

Eardrums distinctly visible.

nighttime coloration

Adult males with single, external subgular vocal sac and indistinct, whitish nuptial pad on inner fingers.

Ventral surfaces pale green, white, or yellowish.

foot

hand

Bones dark green, visible through skin on ventral surfaces.

Posterior thigh surfaces uniform yellowish-green (with no bold markings).

Hands unwebbed; long, slender fingers end in broadly extended, bluntly rounded disks. Feet moderately webbed; no significant webbing between toes I and II; toe tips also with large bluntly rounded disks.

Similar species

S. staufferi (p. 427)

This very similar species is smaller, has a more pointed snout, and has white bones. It prefers drier habitats.

S. boulengeri (p. 423)

Has more rugose skin and alternating black and light bars on the posterior thigh surfaces.

Scinax staufferi
Dry Forest Long-snouted Tree Frog

Least concern

This small, rather smooth-skinned tree frog has irregular longitudinal dorsal stripes, a flattened body, a long snout, and white bones. It lacks webbing between its fingers. Males to 1.14 in (29 mm) in standard length, females to 1.26 in (32 mm).

Ecoregions
1 2 3 **4 5**

Scinax staufferi ranges widely throughout Central America. On the Pacific slope, it ranges from Guerrero, Mexico, to northwestern Costa Rica; on the Atlantic slope, it occurs from Tamaulipas, Mexico, to northern Nicaragua. In Costa Rica, it is found only in Guanacaste Province and on the Nicoya Peninsula; from near sea level to 1,805 ft (550 m).

Natural history

This is a very common species. It is generally found in disturbed open habitats, including pastures, fields, grassland, and savannah, as well as regenerating secondary forests; it avoids closed canopy forest. *S. staufferi* breeds in temporary rain pools during the wet season (August–December). Males emit their advertisement call, a series of short nasal w*ah* notes, repeated regularly from emergent vegetation and low bushes along the edge of the breeding pond. Males occasionally call while perched on the ground or in shallows along the edge of the pond. Eggs are deposited in small clumps in the water.

S. staufferi is a very adaptable and wide ranging species that appears to be stable throughout its range, with no immediately discernible threats to its existence.

Description

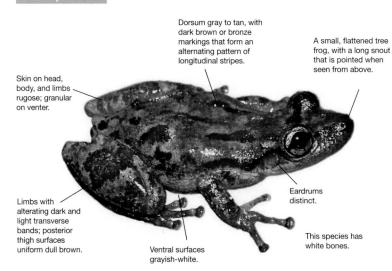

Dorsum gray to tan, with dark brown or bronze markings that form an alternating pattern of longitudinal stripes.

A small, flattened tree frog, with a long snout that is pointed when seen from above.

Skin on head, body, and limbs rugose; granular on venter.

Limbs with alternating dark and light transverse bands; posterior thigh surfaces uniform dull brown.

Ventral surfaces grayish-white.

Eardrums distinct.

This species has white bones.

adult male

When breeding, also note indistinct whitish nuptial patch on inner fingers.

Adult males with single, external vocal sac on throat. Note pale dorsal color at night.

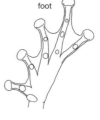

hand

foot

Hands without webbing; fingers stocky, with rounded to truncate disks. Feet with moderate webbing; toes with rounded disks.

Similar species

S. elaeochroa (p. 425)

This very similar species is slightly larger, with a more rounded snout and green bones. It prefers areas with higher humidity.

S. boulengeri (p. 423)

Larger, with rugose skin, and alternating black and light bars on its posterior thigh surfaces.

Genus *Smilisca*

This Central American genus has eight species, five of which are commonly encountered in the lowlands and foothills of Costa Rica. Of moderate to large size, they are somewhat drab, with tan, brown, or green coloration.

Frogs of the genus *Smilisca* are best distinguished from most other tree frogs in the region by the shape of the vocal sac in adult males, which is distinctly bilobed. Nearly all other Costa Rican tree frogs possess only a single vocal sac (*Isthmohyla pseudopuma* and *Trachycephalus typhonius* are the only exceptions).

All species of *Smilisca* mate through axillary amplexus. Three Costa Rican species (*S. baudinii*, *S. phaeota*, and *S. puma*) lay their eggs in temporary and permanent pools, while the remaining two (*S. sila* and *S. sordida*) breed in rocky streams. All are relatively tolerant of habitat alteration and are often found in close proximity to humans, making them among the more frequently observed tree frogs in the country.

Smilisca baudinii
Mexican Tree Frog

Least concern

A large tree frog. Upper lip has dark vertical bars. There is a light green or cream spot below each eye; a dark band runs from behind the eye, along the supratympanic fold, to the insertion of the front limbs. Also note dark transverse bands on its hind limbs. Males to 2.99 in (76 mm) in standard length, females to 3.54 in (90 mm).

On the Atlantic slope, *Smilisca baudinii* ranges from southern Texas, United Sates, to Costa Rica; and on the Pacific slope, from southern Sonora, Mexico, to Costa Rica; from near sea level to 5,280 ft (1,610 m).

Ecoregions

1 2 3 4 5

Natural history

This nocturnal and arboreal species is generally found on relatively low vegetation (e.g., shrubs, small trees). Breeding takes place during the wet season, in temporary pools, roadside ditches, flooded pastures, and even man-made basins. Calling males are usually seen on bushes or small trees near the water's edge, where they may occur in large numbers and in close proximity to one another. *Smilisca baudinii* usually calls in duets, with two males calling in an alternating fashion. It also frequently forms very large choruses during the height of the wet season; a breeding congregation of more than 45,000 *S. baudinii* was reported from a pond in Veracruz, Mexico. The mating call is a series of short, harsh *eck* notes that have a mechanical-sounding quality.

S. baudinii is an explosive breeder; it breeds in temporary rain pools, ponds, man-made basins, and even swimming pools. Pairs in amplexus are usually seen in shallow sections of their breeding pool, where they may lay more than 3,000 eggs, in a floating layer on the water surface. The tadpoles develop extremely fast and may be ready to metamorphose and leave the water in fewer than three weeks.

During the dry season, these frogs retreat into burrows underground; under debris and cover objects; as well as into tree holes or bromeliad leaf axils. Observations indicate that *S. baudinii* is able to generate a cocoonlike structure from multiple layers of shed outer skin; this provides added protection from dehydration during dry spells.

S. baudinii is predominantly a species of lowlands and foothills, in areas with a pronounced dry season. It is a common and adaptable species found in disturbed and open habitat, including places in the immediate vicinity of human settlements: on roads, near street lights, or clinging to the walls of buildings. It is often found active in irrigated areas (e.g., agricultural fields, golf courses, ornamental vegetation around buildings), even during the dry season. Its ability to survive in significantly altered habitat means that the conservation status of this species is secure.

A large tree frog. Robust, with a broad head and rounded snout.

Top of the head usually marked with a transverse dark band that connects the upper eyelids.

Coloration extremely variable; dorsum green, tan, or brown, with a pattern of dark irregular blotches that are sometimes bordered by black or dark brown lines.

Eardrums clearly visible; a distinct supratympanic skin fold lines the top of the eardrum, sometimes continuing a short distance onto the body.

Generally a dark band or line extends from behind each eye, along the supratympanic fold, to the insertion of the front limbs.

Trailing edge of lower arms marked with a linear series of raised tubercles (often white).

Dorsum smooth; skin on flanks and venter granular.

Limbs marked with alternating dark and light bars, most prominently on hind limbs (dark bands may be inconspicuous in pale individuals and juveniles).

The lips are barred; usually a light green or cream spot is present below the eye, which is bordered on either side by a dark brown vertical bar.

Ventral surfaces white or cream.

foot

hand

A brown nuptial patch present on the base of each thumb in adult males.

Hands and feet large, with long digits and moderately expanded, rounded finger and toe disks. Fingers with minimal webbing, but toes extensively webbed.

adult male

Adult males with bilobed, extendable vocal sacs on the throat, visible as grayish-brown wrinkled skin when not in use.

Uniformly pale individuals might be confused with *S. phaeota* or *S. puma*.

Dorsal coloration is quite variable in *S. baudinii*.

Similar species

Trachycephalus typhonius (p. 446)

Has green bones and distinctly glandular skin.

Osteopilus septentrionalis (p. 417)

Has a co-ossified skull and lacks a light spot below each eye.

S. phaeota (p. 433)

Has a white lip stripe. Also note dark brown eye mask. If present, green markings on the snout are not restricted to a spot below each eye.

Smilisca phaeota
Masked Tree Frog

Least Concern

A large, spectacular tree frog. Dorsum tan or green. Has a silvery white stripe on the upper lip and a dark brown eye mask that starts in front of the eyes and continues as a broad stripe backward onto the body, covering the eardrums. Often with green markings between the dark mask and the white lip stripe. Males to 2.56 in (65 mm) in standard length, females to 3.07 in (78 mm).

Ecoregions
1 2 3 4 5

On the Atlantic slope, *Smilisca phaeota* occurs from northeastern Honduras, through Nicaragua, Costa Rica, and Panama, to northern Colombia; on the Pacific slope, it is found in Costa Rica, Panama (in southwestern Panama and Coclé Province), Colombia, and western Ecuador. From near sea level to 3,605 ft (1,100 m).

Natural history

S. phaeota is a nocturnal tree frog that sleeps during the day on the upper surface of large leaves or sometimes inside rolled-up banana or heliconia leaves. On rainy nights, males descend to temporary rain pools to breed. Preferred breeding locations tend to be small bodies of water, including drainage ditches, water-filled tire ruts, pools in forested streams, water-filled depressions in rocks or mud, and even cattle foot prints. Males produce their advertisement call, a very loud and harsh *wrauk*, at dusk, while floating in a suitable breeding pool. Males tend to call while hidden amid vegation or underneath an overhanging bank; but if no cover is available, they simply float on the water surface while vocalizing.

S. phaeota lays its eggs in small rafts of up to 2,000 eggs that float on the water surface. The eggs often hatch within a day or two. Larvae develop extremely rapidly, with tadpoles generally completing their metamorphosis less than one month after hatching. The very short tadpole phase is an adaptation to life in the precarious conditions of temporary rain pools, which may quickly disappear during a dry spell. In extreme situations, the tadpoles can survive short periods of drought, and they have been observed surviving up to 24 hours out of the water, in drying mud.

S. phaeota is an adaptable species and is often one of the more common tree frogs in a region. It is found in lowland and foothill forests, where it generally prefers habitat edges; indeed, it is most commonly found in open areas bordering forest, in roadside ditches, and in altered habitats, including farms, plantations, and landscaped areas and gardens. It is sometimes found in swimming pools and may even try to breed there. Because it tolerates substantial habitat alteration, its populations are seemingly stable.

S. phaeota eggs are deposited in a floating raft.

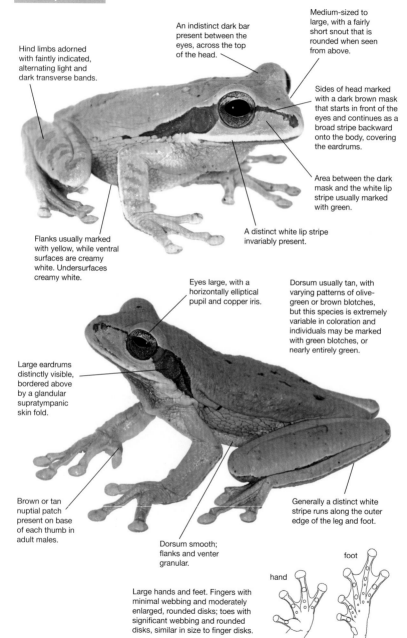

An indistinct dark bar present between the eyes, across the top of the head.

Medium-sized to large, with a fairly short snout that is rounded when seen from above.

Hind limbs adorned with faintly indicated, alternating light and dark transverse bands.

Sides of head marked with a dark brown mask that starts in front of the eyes and continues as a broad stripe backward onto the body, covering the eardrums.

Area between the dark mask and the white lip stripe usually marked with green.

A distinct white lip stripe invariably present.

Flanks usually marked with yellow, while ventral surfaces are creamy white. Undersurfaces creamy white.

Eyes large, with a horizontally elliptical pupil and copper iris.

Dorsum usually tan, with varying patterns of olive-green or brown blotches, but this species is extremely variable in coloration and individuals may be marked with green blotches, or nearly entirely green.

Large eardrums distinctly visible, bordered above by a glandular supratympanic skin fold.

Brown or tan nuptial patch present on base of each thumb in adult males.

Dorsum smooth; flanks and venter granular.

Generally a distinct white stripe runs along the outer edge of the leg and foot.

hand

foot

Large hands and feet. Fingers with minimal webbing and moderately enlarged, rounded disks; toes with significant webbing and rounded disks, similar in size to finger disks.

Adult males with bilobed, extendable vocal sacs, indicated with darker color (yellowish, cream, or gray) on throat when not in use.

Some individuals are extremely colorful.

Amplecting pair.

Similar species

S. baudinii
(p. 430)

Lacks a white lip stripe, has a distinct light blotch below each eye, and has boldly barred limbs.

S. puma
(p. 436)

Lacks a white lip stripe and never has green markings on the side of its head.

Osteopilus septentrionalis
(p. 417)

Has green bones and a co-ossified skull.

Smilisca puma
Tawny Tree Frog

Least concern

A medium-sized tree frog with a pair of olive to brown elongate bands on the dorsum, often joined in the middle of the back to form an H or ladderlike shape. Males to 1.54 in (39 mm) in standard length, females to 1.81 in (46 mm).

Smilisca puma occurs in the Caribbean lowlands of Costa Rica and adjacent Nicaragua, from near sea level to 1,705 ft (520 m).

Ecoregions
1 2 3 4 5

Natural history

S. puma is a nocturnal, arboreal frog. This secretive species is rarely seen outside the breeding season—or away from the areas in which it breeds. Reproduction takes place on rainy nights during the wet season (roughly June through November). Small numbers of males congregate around small to very small temporary pools in the forest floor and call from the edge of the water or from low vegetation surrounding the pool. Preferred breeding sites include grassy, marshy spots in the forest floor as well as weedy wet areas in secondary vegetation types, or occasionally seeps and marshy areas in open habitat such as cattle pastures.

The advertisement call of this species is a soft, low *kwowk* sometimes followed by a short series of croaking notes. Males sometimes call on humid, rainless nights during the breeding season, but females only join the male chorus during heavy rains. Calling males are usually well concealed within dense vegetation and may be difficult to locate. Eggs are deposited directly in the water, and the small tadpoles develop in the breeding pools. Like other temporary pond-breeding *Smilisca*

species (*S. baudinii* and *S. phaeota*), the larva of *S. puma* develop in just a few weeks.

S. puma is fairly adaptable, tolerating some habitat alteration, but prefers forest habitat that provides arboreal retreats. Logging and increased development for agriculture are among the biggest threats to this species.

Most individuals show a thin dark line that originates behind each eye and follows the supratympanic fold toward the insertion of the hind limbs.

Description

A moderate-sized, robust tree frog.

Eardrums are prominently visible, bordered above by a thin supratympanic skin fold that may extend some distance onto the body.

Posterior surface of the thighs is dark chocolate-brown to reddish-brown.

A distinct thin, light stripe present on the upper lip.

Bilobed, extendable vocal sacs on throat of adult males visible as brown, wrinkled skin when not in use.

A white line marks the the trailing edge of the arms and heels, and the area above the vent.

Adult males with inconspicuous tan or brown nuptial patch on base of thumbs.

Has a fairly short, rounded snout and distinctly raised nostrils.

Dorsum smooth; flanks and venter granular.

Ventral surfaces are light cream to white.

Dorsum tan, grayish-brown, or light chestnut-brown, with a characteristic pattern of two olive to brown elongate bands. Often these bands are joined in one or more places, forming an H or ladder shape.

Hind limbs marked with alternating dark and light blotches.

hand foot

Hands without webbing between the fingers; feet with only minimal webbing. Disks on fingers and toes moderately expanded and round.

Similar species

S. baudinii
(p. 430)

Much larger and typically has a light spot below each eye.

S. sordida
(p. 440)

Has powder blue, cream, or white specks on the groin and light mottling on the posterior thigh surfaces; it also has substantial webbing between its fingers.

Tlalocohyla loquax
(p. 443)

Has orange webbing between its toes and fingers and webbed hands.

Dendropsophus phlebodes
(p. 365)

Smaller. Shows uniform yellow-orange on posterior and anterior thigh surfaces.

Smilisca sila
Granular Streamside Tree Frog

Least concern

This species can be a challenge to identify. Grayish-brown, tan, or olive-green, usually with irregular, dark markings on the dorsum and sky-blue specks on the hidden surfaces of the groin and hind legs. It has tuberculate dorsal skin, and often tuberculate eardrums. Also note truncated snout in lateral profile. Males to 1.77 in (45 mm) in standard length, females can reach 2.44 in (62 mm).

Ecoregions
1 2 **3** 4 5

On the Pacific slope, *Smilisca sila* ranges from southwestern Costa Rica to western Colombia; on the Atlantic slope, it occurs from central Panama to northern Colombia; from near sea level to 1,640 ft (500 m).

Natural history

S. sila breeds in low-gradient streams and rivers, generally during the dry season, when water levels are low. Individuals can be seen at night on low vegetation overhanging the water or perched on rocks in streams. The advertisement call of males is a loud, grating *wreek* that is usually repeated two or three times in rapid succession. Males arrive at the breeding site before females and sometimes form large choruses; the females respond to this calling activity and arrive at the breeding site later in the night. The eggs are deposited in quiet rocky pools in the river, where they are generally attached to rocks or leafy debris at the bottom of the pool.

This species can be locally common. It is perhaps underreported; reports of the similar *S. sordida* may actually be of *S. sila*. Both species occur in southwestern Costa Rica, and it is unclear how these two species—with seemingly identical habitat preferences—avoid competing with one another. In areas of range overlap, careful examination is needed to distinguish between these two somewhat unremarkable and highly variable species. *S. sila* often displays varying amounts of green speckling in its dorsal coloration and has more bright blue speckling on the hidden surfaces of its hind limbs; most importantly, its dorsum is more granular.

S. sila is relatively adaptable and persists in open areas that have undergone significant habitat alteration, as long as they contain a corridor of dense vegetation along streams and rivers.

Metamorphs with yellow to tan hands and feet; often with small green spots on dorsum.

Description

male

Adult males have a bilobed, extendable vocal sac. When not in use, vocal sac appears as gray excess skin on the throat.

Dorsum often marked with varying amounts of discreet metallic-green spotting.

Eardrums distinct, bordered above by a supratympanic skin fold. Eardrums sometimes (partly) covered with enlarged tubercles.

Dorsum with scattered, enlarged tubercles; skin on venter and flanks granular.

Hidden surfaces of groin and hind limbs with sky-blue spots (visible only when hind legs are extended).

A moderate-sized tree frog, with a broad head and truncated snout in lateral profile. Females considerably larger than males.

A pronounced linear series of enlarged, often white, tubercles on the trailing edge of the forearm.

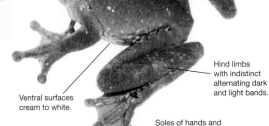

Hind limbs with indistinct alternating dark and light bands.

Ventral surfaces cream to white.

Soles of hands and feet dark brown.

female

male

Dorsal coloration highly variable; usually grayish-brown, tan, or olive-green, with irregularly shaped darker blotches on the head and body (some individuals are uniformly colored).

foot

hand

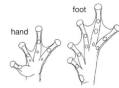

male

Adult males with a dark brown to black nuptial patch on the base of each thumb.

Hands and feet with robust digits, each tipped with a rounded, moderately expanded disk. Hands moderately webbed, but only minimal webbing between fingers I and II. Feet extensively webbed.

S. sordida
(p. 440)

Has relatively smooth dorsal skin, without scattered, enlarged tubercles. It also tends to have a more pointed snout and has more webbing between fingers I and II.

S. puma
(p. 436)

Has unwebbed hands and a diagnostic H-shaped dorsal pattern. It does not inhabit streams.

S. phaeota
(p. 433)

Larger and has a white stripe on its upper lip and a dark face mask (that often includes a green mark).

S. baudinii
(p. 430)

Much larger and typically has a light spot below each eye.

Tlalocohyla loquax
(p. 443)

Has orange webbing between its fingers and toes.

Smilisca sordida
Drab Streamside Tree Frog

Least concern

A somewhat unremarkable gray, tan, or reddish-brown tree frog, usually with irregular dark markings on its dorsum. Hidden surfaces of the groin and hind legs dark brown, with a pattern of cream and chalk-blue specks. Dorsum is relatively smooth. Also note rounded or sloping snout. Males to 2.13 in (54 mm) in standard length, females to 2.52 in (64 mm).

Ecoregions
1 2 3 4 5

Smilisca sordida is common and widespread along the Pacific and Atlantic slopes of Costa Rica. It also occurs in isolated populations in northern Honduras, central Nicaragua, central and western Panama, and possibly Colombia. From near sea level to 5,000 ft (1,525 m).

Natural history

S. sordida is a common tree frog, seen most frequently during its breeding season, which takes place during the drier months of the year, roughly January through May. It breeds in low-gradient rocky streams and rivers when the water level is at its lowest.

Advertisement call of males is a short, rattling *wrink*, usually repeated two or three times in rapid succession. Individuals usually call from a location near the water surface, while perched on rocks in the river, from gravel banks along its edge, or from low vegetation nearby. Pairs mate and lay their eggs in shallow, slow moving sections of the stream or in pools in the streambed. Eggs are attached individually to the substrate. The tadpoles can be found on the bottom of large pools in the river bed, hidden in the rocky substrate, or in collected debris; they can also be seen rasping algae off river rocks.

When not breeding, S. sordida is infrequently seen at night, as it forages on low vegetation in

forested areas near streams. Individuals sometimes sleep in exposed locations during the day and can be found on hot, sun-exposed rocks in rivers, an unlikely location for any amphibian. A protective layer of skin secretions prevents these frogs from excessive evaporative water loss and an untimely death through dehydration.

S. sordida is common throughout its range, and its populations appear to be stable. It is relatively tolerant of habitat disturbance and can be found in plantations, gardens, and urban areas, as long as a fairly intact corridor of streamside vegetation persists.

Due to its rather nondescript appearance and variable color pattern, S. sordida can be dificult to identify. Its preference for riparian habitats may help to narrow down the options. The most difficult identification challenges occur in some streams in the southern Pacific lowlands, where the range of S. sordida overlaps with that of the extremely similar S. sila.

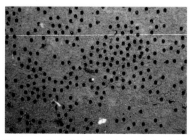

Eggs are attached to the rocky substrate at the bottom of a stream.

Pair of S. sordida in amplexus—note the size difference between male (on top) and female.

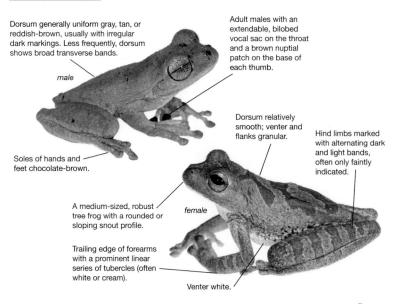

Dorsum generally uniform gray, tan, or reddish-brown, usually with irregular dark markings. Less frequently, dorsum shows broad transverse bands.

male

Adult males with an extendable, bilobed vocal sac on the throat and a brown nuptial patch on the base of each thumb.

Dorsum relatively smooth; venter and flanks granular.

Hind limbs marked with alternating dark and light bands, often only faintly indicated.

Soles of hands and feet chocolate-brown.

A medium-sized, robust tree frog with a rounded or sloping snout profile.

female

Trailing edge of forearms with a prominent linear series of tubercles (often white or cream).

Venter white.

Groin and hidden surfaces of the hind limbs dark brown, marked with cream and chalk-blue spots.

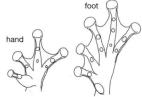

hand

foot

Robust hands and feet. Finger and toe disks moderately enlarged and rounded. Hands moderately webbed, but only basal webbing between fingers I and II; feet with extensive webbing.

S. sila (p. 438)	*S. puma* (p. 436)	*S. baudinii* (p. 430)	*Tlalocohyla loquax* (p. 443)
Has a tuberculate dorsum. Dorsum often marked with small green spots; hidden surfaces of its hind limbs have bright blue spots or mottling.	Has unwebbed hands and a diagnostic H-shaped dorsal pattern. It does not inhabit streams.	A much larger species that typically has a light spot below each eye.	Has orange webbing between its fingers and toes.

Genus *Tlalocohyla*

The newly formed genus *Tlalocohyla* was recently split off from the genus *Hyla* based on molecular characteristics. No distinctive morphological features are known that unite and characterize this genus. Four species of *Tlalocohyla* are currently recognized in northern Central America, only one of which occurs in Costa Rica.

Tlalocohyla loquax has extensive axillary membranes, folds of skin in the armpits that are distinctly visible when the arms are held away from the body. These axillary membranes are usually bright orange or red, as are the hidden surfaces of the hind legs and its finger and toe webbing.

This new genus is named after Tlaloc, the Olmec rain god. In older literature, the Costa Rican species *T. loquax* is referred to as *Hyla loquax*.

Tlalocohyla loquax
Swamp Tree Frog

Least concern

A moderate-sized, stocky tree frog. Dorsum tan to light gray (yellow during the breeding season). Has a yellow venter. Hidden surfaces of the limbs, as well as its finger and toe webbing, are bright orange or red. Males to 1.77 in (45 mm) in standard length, females to 1.89 in (48 mm).

Tlalocohyla loquax is found on the Atlantic slope, from southern Veracruz, Mexico, through Belize, Honduras, Nicaragua, and Costa Rica; from near sea level to 3,605 ft (1,100 m).

Ecoregions
1 2 3 4 5

Natural history

T. loquax is a rainforest species that breeds in ponds and marshes deep within the forest. Breeding takes place at the start of the wet season (July–August). Males call from floating aquatic vegetation while hidden on the margin of the pond, or from woody vegetation along the water's edge. The advertisement call is a loud, grating *woaack*; the sound produced by a chorus of these frogs has a very mechanical quality. Amplexus is axillary; females lay eggs in a large gelatinous mass attached to floating vegetation. The tadpoles complete their development in deep sections of the breeding pond.

This species has a discontinuous, localized distribution range; it is not often seen outside the breeding season. Wherever *T. loquax* does occur, however, it tends to be quite common, and large numbers of these animals gather around suitable breeding ponds. Although most common in lowland settings, it does range into premontane rainforest at times. It can persist in areas of secondary growth and disturbed habitats such as plantations and other agricultural areas, but requires relatively deep water with some aquatic vegetation for breeding.

Adult males with a single, internal vocal sac on throat; no nuptial patches in breeding males. During the breeding season, they turn bright yellow.

Dorsal coloration light gray to tan; at night, may darken to chestnut-brown. Dorsum usually shows an irregular pattern of small dark spots.

A fairly robust, moderate-sized tree frog.

Head short, with a rounded to bluntly rounded snout.

Has smooth skin.

There is a prominent supratympanic skin fold over the small eardrum, which is clearly visible.

Ventral surfaces are light yellow.

Finger and toe webbing, armpits, groin, and hidden surfaces of the legs all show orange or red (often concealed when frog is in resting position).

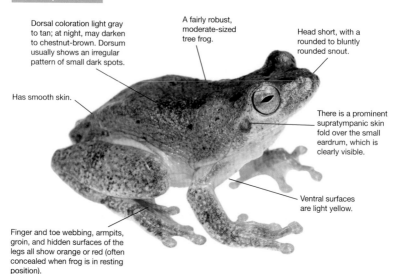

During the day, dorsal coloration may lighten to gray or tan.

An axillary membrane present in each armpit. Note orange coloration on the hidden surfaces of the armpit and groin.

hand

foot

Robust fingers and toes. All digits with moderately enlarged disks. Feet with extensive interdigital webbing; hands with limited webbing only.

Hypsiboas rufitelus (p. 391)

This is the only other Costa Rican tree frog with red webbing on hands and feet, but it is green and has a pointed tubercle on each heel.

Scinax elaeochroa (p. 425)

Breeding males can be bright yellow also, but lack red webbing on the hands and feet.

Smilisca sordida (p. 440)

Males have purplish-brown interdigital webbing (never bright red).

Dendropsophus phlebodes (p. 365)

Smaller and lacks red webbing on hands, feet, and thighs.

Genus *Trachycephalus*

This genus includes 14 species of large, casque-headed tree frogs. Species in this genus have a helmetlike skull, completely roofed and without the openings seen in the skulls of other species; they also have green bones.

Males have a pair of external vocal sacs that are large and shaped like a balloon when fully inflated; they emerge from near both corners of the mouth. Most species of *Trachycephalus* have extremely glandular skin that exudes a white, milky skin secretion that is sticky and noxious. When inhaled, it can cause sneezing and irritation of the mucous membranes; if rubbed in the eyes, it can cause sharp pain or even temporary blindness. The skin secretion is toxic to other amphibians and even to the frogs themselves in high doses, and care should be taken when handling them for field studies.

Trachycephalus typhonius, the only Central American species included in this genus, was formerly known as *Phrynohyas venulosa*. Originally described by Linnaeus in 1758 as *Rana typhonia*, this frog has since been described from numerous locations throughout its vast distribution range under a variety of names.

Trachycephalus typhonius
Milk Frog   **Least concern**

A very large tree frog, with unique thick, glandular skin on its head, back, and limbs; capable of producing copious amounts of noxious, sticky milky-white secretions. Has green bones. Calling males are recognized by the presence of a pair of lateral vocal sacs that expand, balloonlike, from both angles of the jaws when inflated. Males to 3.95 in (101 mm) in standard length, females to 4.49 in (114 mm).

Ecoregions

1 2 3 4 5

Trachycephalus typhonius has a huge distribution range. On the Atlantic slope, it extends from central Tamaulipas and southern Sinaloa, Mexico, southward to southern Nicaragua; on the Pacific slope, it extends from Mexico to eastern Panama. In Costa Rica, it is found in the Pacific lowlands and, marginally, on the Atlantic slope, in the San Juan watershed on the border with Nicaragua; from near sea level to 985 ft (300 m). Its range continues far into South America, extending across the Amazon basin, south to Brazil, Paraguay, and northern Argentina; it also occurs on Trinidad and Tobago.

Natural history

This very large tree frog is arboreal and nocturnal. It occurs mostly in areas with a prolonged dry season, where it survives the dry months hidden in relatively humid, cool places such as tree holes, leaf axils of bromeliads, and below the bark of standing trees. Limited observations indicate that secretions produced in its glandular skin may help prevent desiccation. The glands in the neck region increase in size during the drier months (roughly January through March); related species are known to cover their skin with the water insoluble secretions that their glands produce. South American relatives of *T. typhonius* reportedly even coat the inside of their arboreal retreats during the dry season to limit excessive water loss.

T. typhonius is an explosive breeder; its reproductive activity is triggered by the first rainstorms of the wet season; mating and egg-laying are restricted to a few rainy nights every year. Males aggregate in large numbers at shallow temporary or permanent ponds, where they call from partly submerged vegetation or while floating on the surface of the water to attract nearby females. The call of *T. typhonius* is a very loud and far carrying growl; the large paired vocal sacs in adult males help amplify vocalizations; each of the two sacs can be inflated to the point that they touch above the frog's head.

Amplecting pairs lay their eggs in a single floating layer that may cover an area as large as 5 ft^2 (1.5 m^2). The eggs typically hatch within one day, and the tadpoles develop very quickly (an adaptation to the temporary nature of their breeding pools). Metamorphosis takes place 37–47 days after the eggs hatch.

Under experimental conditions, adult *T. typhonius* displayed the ability to control the rate and angle of their descent after jumping from a high perch. This parachuting behavior is facilitated by using the extensive webbing between the fingers and toes as an airfoil. This may allow these frogs to quickly reach their breeding pools as soon as conditions are suitable.

The white, sticky skin secretions produced in the glandular skin of these frogs not only protects against dehydration but also provides an effective defense against predators. On several occasions, *T. typhonius* have been observed thwarting snake attacks by covering the snake's head with their skin secretions, using their legs to spread the sticky substance over the predator's eyes, mouth, and nose. As the secretions dry they adhere tightly and physically impede the snake's ability to see or swallow; in these observations, the frogs were able to escape as the snakes attempted to rub themselves clean. Handling these frogs often results in sticky mucus covering hands and glueing together fingers; once dry, these secretions are difficult to remove. In addition, they can cause local irritation, resulting in redness, itching, pain, and swelling. In severe cases the mucus may cause temporary blindness if rubbed into the eyes. These effects on humans have led to the alternative common name of pepper frog in some areas.

T. typhonius is a commonly seen species of forest, forest edges, savanna, and secondary growth. It is often more common in open habitat types than in forested areas. It persists in disturbed habitats and plantations, and is regularly found in or near human dwellings.

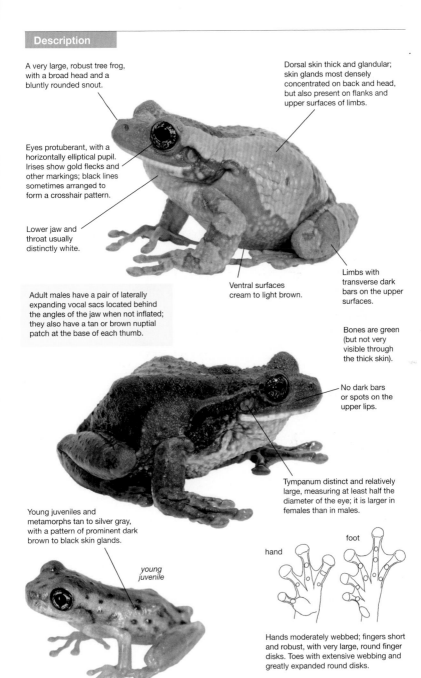

A very large, robust tree frog, with a broad head and a bluntly rounded snout.

Dorsal skin thick and glandular; skin glands most densely concentrated on back and head, but also present on flanks and upper surfaces of limbs.

Eyes protuberant, with a horizontally elliptical pupil. Irises show gold flecks and other markings; black lines sometimes arranged to form a crosshair pattern.

Lower jaw and throat usually distinctly white.

Adult males have a pair of laterally expanding vocal sacs located behind the angles of the jaw when not inflated; they also have a tan or brown nuptial patch at the base of each thumb.

Ventral surfaces cream to light brown.

Limbs with transverse dark bars on the upper surfaces.

Bones are green (but not very visible through the thick skin).

No dark bars or spots on the upper lips.

Tympanum distinct and relatively large, measuring at least half the diameter of the eye; it is larger in females than in males.

Young juveniles and metamorphs tan to silver gray, with a pattern of prominent dark brown to black skin glands.

young juvenile

hand

foot

Hands moderately webbed; fingers short and robust, with very large, round finger disks. Toes with extensive webbing and greatly expanded round disks.

Dorsum generally gray, tan, reddish, or brown, with a large dark dorsal field that covers much of the top of the head and the center of the back. Additonal dark spots sometimes outlined in black. Occasionally (as in this individual) the dorsal markings are indistinct and the frog appears relatively uniform in coloration.

The thick glandular skin, expanded finger and toe disks, and crosshair pattern in the iris are diagnostic for *T. typhonius.*

Similar species

Smilisca baudinii (p. 430)

Has a pattern of vertical bars on its lips, usually a light spot below each eye, and lacks extremely glandular skin.

Osteopilus septentrionalis (p. 417)

Lacks distinctly glandular skin.

Family Leptodactylidae
(Foam-nest Frogs)

Taxonomic revision of the family Leptodactylidae in recent years has led to the removal of several large groups of species that are now each placed in their own families (e.g., Craugastoridae, Eleutherodactylidae, and Strabomantidae). The newly constituted, smaller family Leptodactylidae still contains more than 200 species in 15 genera. Six species, in 2 genera (*Engystomops* and *Leptodactylus*), occur in Costa Rica. The one Costa Rican species of *Engystomops*, *E. pustulosus*, is vaguely toadlike in its overall appearance, while all *Leptodactylus* species superficially resemble true frogs of the family Ranidae.

Costa Rican *Leptodactylus* and *Engystomops* have a free-swimming tadpole stage. Their eggs are deposited in a foamy nest composed of mucus, air, and water that the male whips up with his hind legs during mating. Generally the nest floats near the edge of a river or pond, often hidden under vegetation, or it may be located in a water-filled burrow. On hatching, tadpoles leave the nest and complete their development in water.

Genus *Engystomops*

This predominantly South American genus comprises nine species of small, toadlike frogs. *Engystomops pustulosus*, the túngara frog, is the only species that resides in Costa Rica.

During the past decade, this genus has undergone dramatic taxonomic rearrangement; it was removed from the family Leptodactylidae, together with six other South American genera, and placed in the newly resurrected family Leiuperidae as part of a large-scale review of the amphibian tree of life in 2006. Subsequent revisions in 2011 re-classified the family Leiuperidae as a sub-family of Leptodactylidae, and moved *Engystomops* back into the latter taxon. Also, some authors considered frogs in the genus *Engystomops* to be part of a larger group, *Physalaemus*, with insufficient data to warrant separating the two. In 2011 molecular data confirmed that *Physalaemus* and *Engystomops* are sister taxa and the latter should now be considered the proper genus for the only Costa Rican species, *Engystomops pustulosus*.

Due to these taxonomic exercises, *Engystomops pustulosus* is often reported in literature as *Physalaemus pustulosus*, and sometimes placed in the family Leiuperidae.

Engystomops pustulosus
Túngara Frog   **Least concern**

Small to medium-sized frog. Superficially toadlike in appearance, with prominent parotoid glands, but lacks bony crests on the head and webbing between its toes and fingers. Has a dark throat and chest; a white venter marked with bold black spots; and a light stripe that runs down the middle of the throat and chest. Males to 1.34 in (34 mm) in standard length, females to 1.38 in (35 mm).

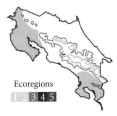

Ecoregions
1 2 **3 4** 5

On the Atlantic slope, *Engystomops pustulosus* occurs in Veracruz and Campeche, Mexico. On the Pacific slope, it occurs from Oaxaca, Mexico, to northwestern Costa Rica; it also occurs disjunctly, in southwestern Costa Rica, throughout Panama, and in Colombia and Venezuela. It is found from near sea level to 5,050 ft (1,540 m).

Natural history

E. pustulosus is a very common species, often found in close proximity to human settlements. This nocturnal frog calls on rainy nights; its advertisement call, an explosive *mew*, can be heard over considerably distance. Males call from rain puddles, potholes in roads, roadside ditches, cattle hoof prints in flooded pastures, gardens, or small permanent ponds and catchment basins—basically any natural or artificial shallow basin that contains water and has some emergent vegetation. Calling males float on the water and expand their enormous inflatable vocal sac during calls. These frogs are preyed on by bats such as the common fishing bat (*Trachops cirrhosus*), which locates calling males based on their call. *E. pustulosus* produces a wide range of call variations, frequency modulations, and call patterns to confuse these predators.

Amplexus in *E. pustulosus* is axillary (male clasps the female around her waist) and eggs are laid in the water. The male will whip air into the egg mass by kicking his hind legs to create a foam nest roughly the size of a tennis ball. The nest floats on the water surface but is often lodged between emerging vegetation. Larvae will develop inside the foam nest and drop into the water below when hatching. If the breeding pond dries up before the tadpoles hatch, they can survive inside the moist environment of the foam nest for several days as they wait for rain. Reproductive activity peaks in the wet season; when not breeding, *E. pustulosus* retreats into leaf litter during the day. In arid parts of its range, these animals will retreat into underground burrows for much of the dry season.

Due to its small size and abundance, these frogs fall prey to a wide variety of predators. Although it has glandular skin, little information exists on how skin secretions possibly avert predation.

This very adaptable frog survives in severely altered habitats and does not appear to face any major threats.

Vocal sac on adult males can extend to very large proportions when calling.

Amplecting pair in the process of laying eggs and creating a foam nest.

Adult males have brown nuptial patches on the base of each thumb.

Eyes large, with copper or bronze irises and horizontally elliptical pupils.

Dorsal coloration variable; generally shades of brown or gray, with or without additional markings. Common pattern elements include a thin, light middorsal line, dorsolateral light stripes, or irregular blotches.

Throat and chest with dark mottling; bisected by a light stripe that runs down the middle of the throat and chest.

Skin on head, body, and limbs extremely warty and glandular; skin glands scattered over entire body, enlarged parotoid glands present in the neck region.

male

Some dorsal tubercles orange, reddish, or pink.

Limbs with poorly defined, irregular bars; front limbs are often marked with a broad light band.

Eardrums, hidden beneath glandular skin, are nearly invisible.

A stocky, toadlike frog with a short snout that appears pointed when viewed from above (rounded in profile).

female

foot

hand

Flanks dirty white, with dark mottling.

Venter and undersides of hind limbs white with bold black spots.

Fingers and toes long and slender; they lack interdigital webbing and enlarged disks.

Craugastor rugosus (p. 280)

Young toads of the genus *Incilius* (family Bufonidae) sometimes appear similar, but generally have bony crests on the head and basal webbing between their toes. In addition, no *Incilius* have the bold black-and-white ventral pattern seen in *E. pustulosus*.

Juveniles sometimes appear warty, but have a very broad head (with crests) and lack parotoid glands.

452

Genus *Leptodactylus*

This is a large group of mostly South American species (77 species currently recognized). The five species found in Costa Rica are medium-sized to very large frogs that are superficially similar to true frogs in the family Ranidae. They have a streamlined body, pointed snout, large eyes, and long, powerful legs. Most have distinct longitudinal glandular ridges on their dorsal surfaces that are either intact or broken up.

Species in the genus *Leptodactylus* are distinguished from similar looking species in the region by their unwebbed hands and feet and the absence of finger and toe disks. Their dorsal coloration is variable and not always a good field mark; the color pattern on the hidden posterior surface of the thighs can be helpful in identification.

Species in Costa Rica breed during the wet season, along the grassy edges of permanent ponds or in a variety of temporary wetlands, usually in relatively open habitat. Many *Leptodactylus* produce foam nests that contain their eggs (and tadpoles) during the early stages of larval development. Eventually the tadpoles drop into the water below and become free-swimming. If their breeding site dries up, *Leptodactylus* eggs and larvae can survive for considerable time while protected inside the humid climate of the nest, an adaptation to life in precarious temporary wetlands.

All five Costa Rican *Leptodactylus* occur in the Pacific lowlands; only the variable foam frog (*L. melanonotus*) and the Central American bullfrog (*L. savagei*) are also widespread on the Caribbean slope. The very recent discovery of two *L. fragilis* in the Sarapiquí region of northeastern Costa Rica indicates that this species also may be expanding its range into open, disturbed habitats of the Atlantic slope.

Leptodactylus fragilis
White-lipped Foam-nest Frog   **Least concern**

A medium-sized, streamlined frog that lacks enlarged finger and toe disks and has no interdigital webbing. Typically it has two pairs of glandular dorsolateral skin folds and dense concentrations of tubercles on the soles of its feet and lower legs. Color pattern is variable but generally includes a white lip stripe and a longitudinal light stripe on the posterior surface of the thighs. Males to 1.42 in (36 mm) in standard length, females to 1.57 (40 mm).

Ecoregions
1 2 3 4 5

Leptodactylus fragilis occurs from extreme southern Texas, United States, through Mexico and Central America, to northern Colombia and Venezuela. In Costa Rica it is found in the Pacific lowlands and, locally, in the lowlands of the north Atlantic slope; from near sea level up to about 1,800 ft (550 m).

Natural history

L. fragilis is an abundant and widespread species, found in a wide variety of habitats. Originally, it probably occurred exclusively in swamps and wet situations in forest clearings; it is now usually found in man-made habitats such as plantations, grasslands, cultivated fields, roadside ditches, and other wet areas with a herbaceous or grassy vegetation cover. During the day, this frog can be found under logs or rocks, or in piles of debris. *L. fragilis* actively hunts for insect prey at night.

Breeding occurs primarily during the wet season (May–December). Reproductive males excavate a subterranean burrow and call from its entrance. During amplexus, as the female lays her eggs, the male kicks up a frothy mass of water, mucus, and eggs to create a foam nest; the eggs hatch inside the humid environment of the nest, which slowly disintegrates during subsequent heavy

rains. Hatching tadpoles leave the nest and swim out of—or get washed out of—the burrow into an adjacent temporary pool, often a flooded pasture or other type of rain pool in an open setting. Like other temporary-pool-breeding amphibians in Costa Rica, larval development is exceedingly rapid, and tadpoles can undergo metamorphosis and leave the water in less than three weeks.

Due to its ability to adapt to extensive habitat alteration, *L. fragilis* does not face major threats. Recently, it was found in the Sarapiquí region of northeastern Costa Rica, which suggests that it may be expanding its range into the Atlantic lowlands.

There is a lot of confusion over the correct name of this species. In much of the literature it is described erroneously as *L. labialis*, which is in fact a synonym for a different species, *L. mystacinus,* which inhabits southern South America.

A white lip stripe, bordered above and below by dark pigment, usually present; in some individuals it extends past the tympanum, toward the insertion of the front limbs.

A pair of parallel glandular dorsolateral skin folds present on each side of the body; usually broken up to form linear series of tubercles. Additional scattered tubercles adorn the body and limbs; head generally with smooth skin.

Dorsal coloration gray, tan, or bronze, usually with a pattern of dark brown spots.

Dorsolateral skin folds often cream, tan, or orange, contrasting with dorsal coloration.

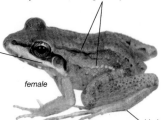

female

Medium-sized and streamlined, with a long pointed snout.

Limbs long and powerful.

Eardrums large and prominently visible.

female

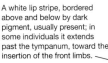

male

Adult males lack swollen arms and black thumb or chest spines (present in some *Leptodactylus*), but do have a pair of expandable, dark vocal sacs, located at the corners of the mouth.

Venter uniform white to cream or pale yellow.

foot

hand

Posterior surfaces of the thighs marked with a longitudinal light (white, cream, or yellow) stripe.

Hands and feet with long, slender digits. Lacks interdigital webbing, but toes are lined with narrow skin fringes (not visible without magnification). There are no enlarged finger or toe disks. The soles of the feet and the lower tarsus are adorned with several series of small, white tubercles.

Rana forreri (p. 476)

Species of *Rana* (family Ranidae) can be similar in morphology and color pattern but invariably have extensive toe webbing.

L. poecilochilus (p. 460)

Sometimes has multiple pairs of dorsolateral skin folds and a light longitudinal stripe on the back of the thigh, but is generally larger and lacks the white tubercles on the sole of the foot and lower tarsus.

L. insularum (p. 456)

A much larger species; has mottled posterior thigh surfaces (not a longitudinal light stripe).

Leptodactylus insularum
Spotted Foam-nest Frog

Least concern

A large frog with prominent dorsolateral skin folds, long legs, and a streamlined body. Has fleshy skin fringes lining the toes, but lacks toe webbing. This is a large, dark-spotted frog with a prominent white lip stripe. Its posterior thigh surfaces are dark brown, with cream to light yellow spots. Males to 3.74 in (95 mm) in standard length, females to 4.72 in (120 mm).

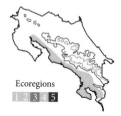

Ecoregions
1 2 3 4 5

On the Pacific slope, *Leptodactylus insularum* is found at low elevations from northern Costa Rica, through Panama, into Colombia and Venezuela; it is also found on Trinidad and Tobago; from near sea level to at least 1,500 ft (455 m).

Natural history

L. insularum is a nocturnal species, most frequently seen hunting for invertebrate prey at night in relatively open areas. An adaptable species, it can be encountered in a variety of habitats, ranging from dry forest to swamps, but also in drainage areas and on lawns. Adults can be seen perched at the entrance of a burrow in late afternoon or early evening.

The advertisement call of males is a short, melodic *whroop*, often made from shallow sections of permanent or temporary ponds or swamps. Breeding takes place mostly in the wet season (May–December); like many other species in the genus, the female lays her eggs in a floating foam nest. An amplectant pair floats in the water, and as the female lays eggs, the male fertilizes them and aerates the egg mass with kicking motions of his hind legs,

forming a foamy mass. When the outside of the nest is exposed to air, it dries to form a tough coat that maintains a humid interior environment. The nest is often hidden in emergent vegetation, under a log, or at the entrance to a burrow. Tadpoles will develop inside the nest; during heavy rains it disintegrates, releasing the tadpoles into the water, where they complete their development.

L. insularum is a common and adaptable species, seemingly unaffected by habitat alteration. No immediate threats have been identified.

This wide-ranging species has long been referred to as *L. bolivianus*, and is included in most literature under that name. However, a recent study concluded that *L. insularum* is the appropriate name for Central American populations.

When threatened, these frogs inflate their body and raise themselves off the ground on extended legs to appear bigger to potential predators.

Skin on head, body, and limbs smooth; but note a prominent, glandular dorsolateral skin fold on each side of the body. A parallel, linear series of tubercles usually present below each skin fold.

Medium-sized to large, with a robust body and long snout.

Dorsum gray to brown, with rounded dark spots that have light borders. Top of snout, in front of the eyes, generally unpatterned.

female

Dorsolateral skin folds usually bordered below by a dark stripe.

A stripe on each side of the head forms a dark eye mask that offsets a prominent white to cream lip stripe.

Eardrums clearly visible.

A few scattered tubercles may be present on the flanks and hind limbs.

Limbs long and powerful.

Limbs marked with dark spots or bands.

Venter generally uniform white to cream, but sometimes suffused with dark pigment.

Adult males have distinctly swollen arms and a pair of black, keratinized spines on the base of each thumb. They have a small internal vocal sac on their throat.

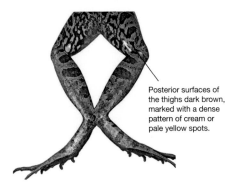

Posterior surfaces of the thighs dark brown, marked with a dense pattern of cream or pale yellow spots.

hand

foot

Hands and feet with long, slender digits. No webbing between fingers and toes, but toes are lined with distinct skin fringes. There are no enlarged finger or toe disks. Soles of the feet are mostly smooth, but lower tarsus marked with a patch of small dark-tipped tubercles.

Rana forreri (p. 476)

Has extensively webbed feet.

L. fragilis (p. 454)

Much smaller, with a longitudinal light stripe on the posterior surfaces of its thighs.

L. poecilochilus (p. 460)

Smaller and lacks prominent toe fringes; it has a longitudinal light stripe on its posterior thigh surfaces.

L. savagei (p. 462)

Lacks white lip stripe and spotted dorsal pattern.

Leptodactylus melanonotus
Variable Foam-nest Frog

Least concern

A stocky, smallish frog. Lacks dorsolateral skin folds (but often has several indistinct rows of glandular tubercles on the dorsum). Also lacks expanded digital disks and webbing between fingers and toes, but note fleshy fringes lining the toes. Adult males with a pair of sharp, black keratinized spines on the base of each thumb. Limbs, flanks, and dorsum have numerous glands with minute black tips. Males to 1.77 in (45 mm) in standard length, females to 2.17 in (55 mm).

Ecoregions
1 2 3 4 5

On the Atlantic and Pacific slopes, *Leptodactylus melanonotus* ranges from parts of Mexico to Ecuador; from near sea level to 4,265 ft (1,300 m).

Natural history

L. melanonotus is abundant throughout its range. Although it tolerates a great range of climatic conditions, it prefers drier habitats, where it inhabits small to very small pools and puddles on the margins of a larger body of water. It is regularly found in water-filled cattle hoof prints in wet pastures and also inhabits roadside ditches and man-made basins, just about any small body of water with sufficient herbaceous or grassy cover available. These frogs are encountered singly or sometimes in very large congregations; the size of the population seems to be determined primarily by the presence of sufficient hiding places.

L. melanonotus is nocturnal; calling males can be heard vocalizing on humid nights. But, when multiple males congregate, they sometimes emit territorial calls, even during the day. Their call is a distinctive, slow-paced metallic *tuc-tuc-tuc*, similar to the sound of a metal object softly tapping a dinner plate. Calling males can be difficult to locate since they are invariably hidden in dense vegetation or in a burrow underneath a rock or log. The sound produced by a large chorus of these frogs can travel a great distance and somewhat resembles the noise made by an idling engine.

Like many other species in its genus, *L. melanonotus* breeds in shallow water or on the periphery of larger bodies of water. The yellow-and-black eggs are deposited in a foam nest that is created by the male through the kicking action of his hind limbs during amplexus. While the female lays her eggs, he fertilizes them

and kicks up a frothy mass of water, mucus, and eggs. The foam nest protects the eggs and tadpoles from drying out; the exposed exterior of the nest becomes viscous or may even harden somewhat, but the interior remains damp. Sometimes the water around the nest may recede temporarily, and the nest will remain out of water for a few days. The humid climate inside the nest ensures continued development of the eggs. Even after the eggs have hatched, the tadpoles may stay inside the nest until heavy rains destroy it and refill the breeding pond. Free-swimming tadpoles of this species often form large schools and move in synchrony, thus appearing like a much larger animal.

L. melanonotus is a very adaptable species with no major known threats.

Typically has many enlarged tubercles on its limbs and flanks; some are prominently black-tipped.

Dorsum tan, gray, or brown, either uniform or with a pattern of obscure dark blotches, bands, or stripes.

Medium-sized and robust, with stocky limbs. Has a relatively short, rounded snout.

Eardrums distinct.

female

Limbs relatively short and stocky.

Posterior surfaces of the thighs mottled dark brown.

male

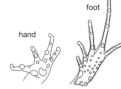

Lips marked with faint alternating dark and light bars; often several white or cream spots present on upper and lower lips.

Upper surfaces of limbs marked with dark spots or bands.

Often a light stripe or a light-bordered triangular marking present on the top of the head, between the eyes.

Adult males lack distinctly swollen arms, but have a pair of sharp, black keratinized spines on the base of each thumb. Small internal vocal sac on throat.

Venter smooth. Cream or dirty-white, mottled with dark pigment.

hand

foot

Hands and feet with moderately long, slender digits. No webbing between fingers and toes, but toes are lined with distinct skin fringes. There are no enlarged finger or toe disks. The soles of the feet bear numerous tubercles, often black-tipped.

L. fragilis (p. 454)

Has paired dorsolateral skin folds. Usually has a white lip stripe. Posterior surfaces of its thighs marked with a longitudinal light stripe.

L. poecilochilus (p. 460)

Has dorsolateral skin folds, lacks tubercles on the soles of its feet, and has a longitudinal light stripe on the posterior surfaces of its thighs.

L. insularum (p. 456)

Males also have a pair of black, keratinized spines on the base of their thumb. This is a much larger species, however, with paired dorsolateral skin folds.

459

Leptodactylus poecilochilus
Brown Foam-nest Frog

Least concern

A medium-sized, streamlined frog. Lacks enlarged finger and toe disks and interdigital webbing. Typically it has at least one glandular dorsolateral skin fold on each side of the body, and usually one or two longitudinal glandular ridges on its lateral surfaces. The soles of its feet are smooth and lack tubercles. There is a cream or yellow longitudinal stripe on the posterior surface of its thighs. Males to 1.93 in (49 mm) in standard length, females to 2.17 in (55 mm).

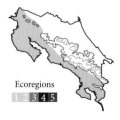

Ecoregions
1 2 3 4 5

On the Pacific slope, *Leptodactylus poecilochilus* ranges from northwestern Costa Rica through Panama into northern Colombia and northern Venezuela; it crosses (marginally) onto the Atlantic slope in the area surrounding Lake Arenal, Costa Rica; from near sea level to 3,775 ft (1,150 m).

Natural history

This is a common species of relatively open areas in lowlands and foothills. *L. poecilochilus* can be found in forest clearings, marshes, grasslands, agricultural fields, ornamental gardens, and roadside ditches. It is terrestrial and nocturnal.

Reproduction takes place during the wet season, commencing immediately after the first heavy rains of the season (May–June). This is an explosive breeder that uses permanent ponds and temporary rain pools to deposit its eggs. Amplecting pairs gather during rainy nights in the shallow, densely vegetated borders of a wetland or flooded field. The male fertilizes the eggs as they emerge from the female's body, kicking vigorously with his hind limbs to create a foamy mass of eggs, water, and skin secretions. The next morning, the outside of the nest dries upon exposure to air and sunlight, forming a barrier against dehydration. The eggs hatch inside the humid environment of the foam nest, which gradually breaks down during subsequent rains. Hatching tadpoles leave the nest either by swimming out or being washed out by rains, reaching an adjacent larger body of water, where they become free-swimming creatures. *L. poecilochilus* tadpoles can survive a substantial amount of time out of the water, a beneficial trait for amphibians that inhabit a precarious aquatic environment that can dry up quickly.

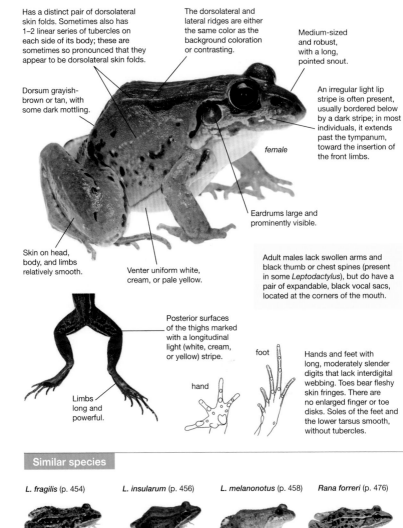

Has a distinct pair of dorsolateral skin folds. Sometimes also has 1–2 linear series of tubercles on each side of its body; these are sometimes so pronounced that they appear to be dorsolateral skin folds.

The dorsolateral and lateral ridges are either the same color as the background coloration or contrasting.

Medium-sized and robust, with a long, pointed snout.

Dorsum grayish-brown or tan, with some dark mottling.

An irregular light lip stripe is often present, usually bordered below by a dark stripe; in most individuals, it extends past the tympanum, toward the insertion of the front limbs.

female

Eardrums large and prominently visible.

Skin on head, body, and limbs relatively smooth.

Venter uniform white, cream, or pale yellow.

Adult males lack swollen arms and black thumb or chest spines (present in some *Leptodactylus*), but do have a pair of expandable, black vocal sacs, located at the corners of the mouth.

Posterior surfaces of the thighs marked with a longitudinal light (white, cream, or yellow) stripe.

foot

hand

Limbs long and powerful.

Hands and feet with long, moderately slender digits that lack interdigital webbing. Toes bear fleshy skin fringes. There are no enlarged finger or toe disks. Soles of the feet and the lower tarsus smooth, without tubercles.

L. fragilis (p. 454)

Has white tubercles on the soles of its feet and on its lower legs.

L. insularum (p. 456)

Larger and has cream or yellow spots (rather than a longitudinal stripe) on the back of its thighs.

L. melanonotus (p. 458)

Lacks dorsolateral skin folds and has mottled, brown posterior thigh surfaces.

Rana forreri (p. 476)

Has extensively webbed feet.

Leptodactylus savagei
Central American Bullfrog, Smoky Jungle Frog   **Least concern**

A huge frog with a distinct dorsolateral skin fold on each side of the body; also note a second, lateral glandular skin fold that wraps over the top of the eardrum and continues part way toward the insertion of each hind limb. Its hands and feet are not webbed, but young individuals have clearly visible toe fringes that disappear with age. Lips and hind limbs boldly barred or spotted. Males to 6.97 in (177 mm) in standard length, females to 7.28 in (185 mm).

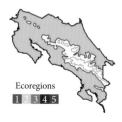

Ecoregions
1 2 3 4 5

On the Atlantic slope, *Leptodactylus savagei* occurs from northern Honduras to Colombia; on the Pacific slope it ranges from northern Nicaragua to Colombia. In Costa Rica, it occurs on both slopes; from near sea level to 3,935 ft (1,200 m).

Natural history

Male *L. savagei* produce a loud, somewhat ominous *wrrrooop* at dusk and into the early evening. Males have been observed calling from the edge of ponds and swamps but also from the entrance of burrows located beneath a log or rock, often in areas with dense vegetation. Reproduction takes place during the wet season. Amplecting pairs generally use permanent ponds or temporary wetlands and swamps to breed in. Eggs are laid in a foam nest that is often hidden in, and anchored to, dense vegetation. Occasionally, the foam nests are found in or at the entrance of the burrow, a short distance away from a body of water. The tadpoles drop into the water below or wash into a nearby body of water during heavy rains. They can survive up to 156 hours out of the water, and are able to survive brief dry spells even if the area around their foam nests temporarily dries up. Larval *L. savagei* are opportunistic and carnivorous, often preying on tadpoles of other and their own species, including the toxic eggs and tadpoles of the cane toad (*Rhinella marina*). However, they are able to complete their development on an entirely vegetarian diet.

Adult *L. savagei* are predators of anything that is small enough to eat; apart from the usual insect prey common in other frogs, this species is also known to consume scorpions and a variety of small vertebrates such as rodents, bird chicks, snakes, and other frogs. *L. savagei* possesses several anti-predator mechanisms. It produces a slippery, noxious skin secretion that is toxic to some animals, including the infamous fer-de-lance (*Bothrops asper*), a large pit-viper that regularly consumes amphibians. In humans, the skin secretions can cause sneezing, skin rashes, a burning sensation, and pain, especially when they enter the bloodstream via cuts or when accidentally rubbed into eyes. When seized, individuals may emit a loud, spine-chilling scream that rarely fails to startle an unsuspecting frog collector, or a potential predator.

L. savagei is a very common species with no known major threats. In some parts of its range, these frogs are collected for human consumption, but this practice is not very widespread and does not appear to have a pronounced effect on its populations in Costa Rica.

This form was recently split from *L. pentadactylus*, which is now thought to be restricted to South America; Costa Rican populations are often included under that name in older literature.

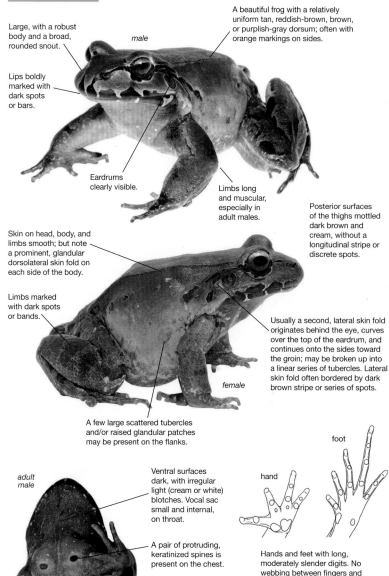

A beautiful frog with a relatively uniform tan, reddish-brown, brown, or purplish-gray dorsum; often with orange markings on sides.

Large, with a robust body and a broad, rounded snout.

male

Lips boldly marked with dark spots or bars.

Eardrums clearly visible.

Limbs long and muscular, especially in adult males.

Posterior surfaces of the thighs mottled dark brown and cream, without a longitudinal stripe or discrete spots.

Skin on head, body, and limbs smooth; but note a prominent, glandular dorsolateral skin fold on each side of the body.

Limbs marked with dark spots or bands.

Usually a second, lateral skin fold originates behind the eye, curves over the top of the eardrum, and continues onto the sides toward the groin; may be broken up into a linear series of tubercles. Lateral skin fold often bordered by dark brown stripe or series of spots.

female

A few large scattered tubercles and/or raised glandular patches may be present on the flanks.

adult male

Ventral surfaces dark, with irregular light (cream or white) blotches. Vocal sac small and internal, on throat.

A pair of protruding, keratinized spines is present on the chest.

Adult males have distinctly swollen arms and a single, large black spine on the base of each thumb.

foot

hand

Hands and feet with long, moderately slender digits. No webbing between fingers and toes, but toes are lined with distinct skin fringes (most prominent in juveniles; difficult to discern in large adults). There are no enlarged finger or toe disks. Soles of feet smooth.

463

metamorph

The bold lip markings are prominent in all but the smallest juveniles.

juvenile

Juveniles more boldly patterned, with dark-outlined squarish marks and bars on the dorsum; often with reddish-brown or orange dorsal coloration and darker sides.

juvenile

Similar species

L. insularum (p. 456)

Smaller and has a spotted dorsal pattern and discrete light spots on the posterior surfaces of its thighs.

Rana vailanti (p. 480)

Like other *Rana* species, has extensively webbed feet.

Family Microhylidae
(Narrow-mouthed Frogs)

This very diverse group of frogs currently contains 572 described species that occur primarily in tropical and subtropical regions. Costa Rica is home to three species of microhylid. Recent re-interpretation of the taxonomic placement of these species has led to name changes that are not reflected in older literature.

All Costa Rican narrow-mouthed frogs are small to moderate-sized frogs with a small head and a plump body. A key diagnostic feature is a transverse skin fold across the head, just behind the eyes. In addition, they have hardened, sharp shovel-like tubercles on their feet and short, powerful legs that allow them to burrow backwards into the soil.

Microhylids are secretive, nocturnal frogs with a fossorial lifestyle. During the day, they hide in leaf litter, under cover objects, or in burrows; at night, they surface and actively hunt for small insects. Their diet mainly consists of ants and mites.

The reproductive season coincides with periods of heavy rains. Females are attracted to males that call from the water surface of temporary ponds. Amplexus is axillary, and the two frogs are held together firmly, partly because of the male's resolute grip, but also due to sticky secretions produced by skin glands in some species. Some members of this family, found outside of Costa Rica, are known to produce tadpoles that develop directly within terrestrial eggs. The three Costa Rican species, however, produce aquatic eggs that hatch into tadpoles that complete their development in the breeding pool. Costa Rican microhylid tadpoles lack keratinized, hard mouthparts but instead have flaps of skin surrounding the mouth opening. Rather than scrape algae off the substrate or otherwise actively forage, these tadpoles filter suspended food particles out of the water column as they swim.

Genus *Ctenophryne*

This primarily South American group of six microhylid frogs is represented in Costa Rica by a single species, *Ctenophryne aterrima*. It was formerly placed in the genus *Nelsonophryne*, and, before that, in *Glossostoma*. A recent study consolidated several species from different genera into the genus *Ctenophryne*.

These forest-dwelling frogs lead a secretive life in burrows underground, or hidden underneath logs or rocks. They are mostly observed when they emerge to breed, during torrential nighttime rains. Eggs are deposited in water-filled depressions in the forest floor. Their tadpoles are morphologically unique in that their spiracle, the tube-like projection that releases water after it has passed over the tadpole's gills, is located on the underside of the body (it is located on the side of the body in most other species). In addition, they lack the typical hard, keratinized mouth parts (beak and toothlike denticles) normally associated with the feeding apparatus of other larval frogs.

Ctenophryne aterrima
Black Narrow-mouthed Frog

Least concern

A glossy, uniform dark (brown, gray, black) frog with a stocky body, small head, and a distinct transverse skin fold behind its eyes. Males to 2.40 in (61 mm) in standard length, females to 2.83 in (72 mm).

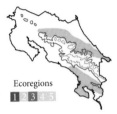

Ecoregions

1 2 3 4 5

Ctenophryne aterrima ranges from northern Costa Rica to the mountains of Colombia and northwestern Ecuador. In Costa Rica, it is only known from scattered localities in the lowlands and foothills of the Atlantic Coast and the Pacific southwest; from near sea level to 5,250 ft (1,600 m).

Natural history

Little is known about *C. aterrima* because it is rarely seen, in part because it spends much of its time underground or hidden beneath leaf litter or cover objects (logs or rocks). Most sightings of this species have occurred in clearings or road cuts in dense forest. Individuals have been observed emerging from flooded burrows during heavy rains; others were found during excavations for construction work.

Reproduction takes place during the wet season (June–November), in swamps or temporary rain pools on the forest floor. A release call has been observed in a Panamanian individual after it was picked up by the observer, but no advertisement call has been documented, and indeed perhaps there is none.

The nuptial glands seen in breeding males occupy the entire chest and abdomen, and extend onto the ventral surfaces of the upper arms. These structures help the male to remain in amplexus with the slippery female, and skin secretions produced by the nuptial glands make his venter sticky, helping him adhere to the female's back. Amplecting pairs are not easily dislodged.

The eggs are black-and-yellow; a single clutch may contain as many as 70 eggs. The relatively large tadpoles lack hardened, keratinized mouthparts (beak, denticles) and have a midventral spiracle.

Although rarely observed, *C. aterrima* is likely not rare; suitable habitat exists in several protected areas.

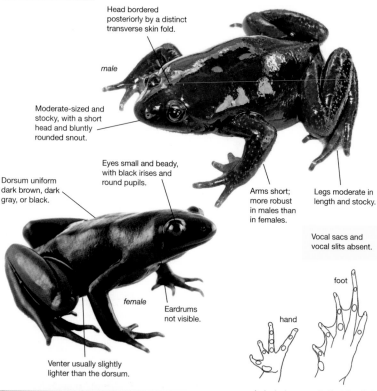

Head bordered posteriorly by a distinct transverse skin fold.

male

Moderate-sized and stocky, with a short head and bluntly rounded snout.

Eyes small and beady, with black irises and round pupils.

Dorsum uniform dark brown, dark gray, or black.

Arms short; more robust in males than in females.

Legs moderate in length and stocky.

Vocal sacs and vocal slits absent.

female

Eardrums not visible.

foot

hand

Venter usually slightly lighter than the dorsum.

A single, large spade-like tubercle is present at the base of the innermost toe on each foot. Toes have basal webbing and rounded tips. Fingers without webbing.

adult male

Breeding females with white pustules around the cloacal area only; adult males with white pustules on the limbs, chin, finger edges, toes, and toe webs. Breeding males also have whitish nuptial patches on the inner fingers and a large nuptial gland that covers most of the ventral surfaces.

Hypopachus pictiventris (p. 470)

Has a similar morphology, but it is never solid black.

Genus *Hypopachus*

This genus was recently revised to include two Central American species previously placed in the genus *Gastrophryne* (*Gastrophryne pictiventris* and *G. usta*). As currently defined, *Hypopachus* now contains four species, two of which occur in Costa Rica. Like the other microhylid in the country, *Ctenophryne aterrima*, these frogs have an egg-shaped, globular body, a small head, and a distinct transverse skin fold that crosses the top of the head posterior to the eyes. However, unlike *Ctenophryne aterrima*, Costa Rican *Hypopachus* are never solid black.

These are small to medium-sized frogs. Nocturnal and fossorial, they are infrequently observed outside the breeding season.

Frogs in the genus *Hypopachus* are very adept at using the spade-like protruberances on their hind feet for digging. With their hind legs making shuffling motions, they "sink" backwards into the substrate as they burrow.

Hypopachus species are explosive breeders that reproduce in shallow flooded areas after the initial heavy rains of the wet season. Their eggs are laid in a floating mass on the water surface. *Hypopachus* tadpoles typically lack hard, keratinized beaks and denticles. Instead, protruding from their upper lip is a pair of scalloped flaps that cover the mouth.

Hypopachus pictiventris
Southern Narrow-mouthed Frog

Least concern

A microhylid with just a single, spade-like tubercle on each foot. Also note bold white spots on dark (brown or gray) venter. Males to 1.22 in (31 mm) in standard length, females to 1.50 in (38 mm).

Ecoregions

1 2 3 4 5

Hypopachus pictiventris is known from the Atlantic lowlands of northeastern Costa Rica and adjacent Nicaragua; from near sea level to 1,650 ft (500 m).

Natural history

This is a relatively common but rarely seen species that hides in the leaf litter during the day and forages for insect prey at night. Its cryptic color pattern closely resembles a dead leaf and individuals sitting motionlessly on the forest floor blend in so well that they easily escape detection. *H. pictiventris* is an explosive breeder; individuals gather in large numbers at rain-filled depressions in the forest floor at the onset of the rainy season. As they float on the water surface, males produce a nasal *baaaaa* call, not unlike a bleating sheep. Eggs are deposited in a floating surface layer and the larvae hatch and complete their development in the water.

H. pictiventris inhabits undisturbed wet forests and does not tolerate habitat alteration, especially logging activities that open up the forest canopy. Its breeding sites are occasionally in edge situations such as pastures or plantations, but generally border natural forest habitat.

This species is included in older literature as *Gastrophryne pictiventris*.

Amplecting pair in process of laying eggs; note the black-and-white eggs floating on the water surface.

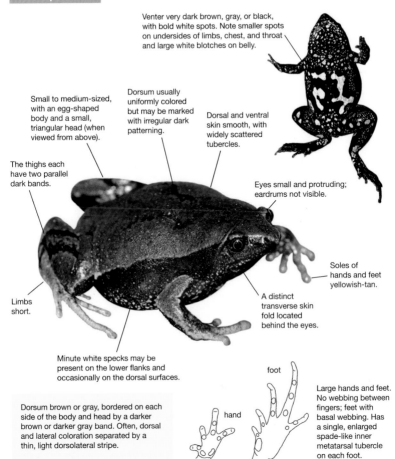

Venter very dark brown, gray, or black, with bold white spots. Note smaller spots on undersides of limbs, chest, and throat and large white blotches on belly.

Small to medium-sized, with an egg-shaped body and a small, triangular head (when viewed from above).

Dorsum usually uniformly colored but may be marked with irregular dark patterning.

Dorsal and ventral skin smooth, with widely scattered tubercles.

The thighs each have two parallel dark bands.

Eyes small and protruding; eardrums not visible.

Limbs short.

Soles of hands and feet yellowish-tan.

A distinct transverse skin fold located behind the eyes.

Minute white specks may be present on the lower flanks and occasionally on the dorsal surfaces.

Dorsum brown or gray, bordered on each side of the body and head by a darker brown or darker gray band. Often, dorsal and lateral coloration separated by a thin, light dorsolateral stripe.

hand

foot

Large hands and feet. No webbing between fingers; feet with basal webbing. Has a single, enlarged spade-like inner metatarsal tubercle on each foot.

Ctenophryne aterrima
(p. 467)

Similar in body shape but has solid black coloration.

H. variolosus
(p. 472)

Lacks the boldly spotted venter; no range overlap.

Rhinophrynus dorsalis
(p. 487)

Much larger and has only four toes on each hind limb; no range overlap.

471

Hypopachus variolosus
Sheep Frog

Least concern

Hypopachus variolosus is a small to moderate-sized microhylid with two prominent spade-like tubercles on each foot. Males to 1.53 in (39 mm) in standard length, females to 2.09 in (53 mm).

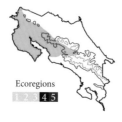

Ecoregions
1 2 3 **4 5**

On the Atlantic slope, *Hypopachus variolosus* ranges from southern Texas, US, to northwestern Costa Rica. Within Costa Rica, it occurs in the Pacific northwest, in the foothills of the Tilarán Mountain Range, and in the Central Valley; from near sea level to 5,250 ft (1,600 m).

Natural history

This is a common but secretive species. It is strictly nocturnal and hides in burrows or underneath rocks or logs during much of the year. It surfaces after heavy rains, and is often seen on the warm pavement of roads on rainy nights. *H. variolosus* reproduces during the rainy months (June–November) in temporary, rain-filled depressions, including flooded pastures, drainage ditches, and in cement wash basins and other man-made structures.

Adult males produce a nasal, bleating advertisement call that lasts about three seconds; call sites are often hidden from view in burrows or vegetation-covered depressions along the edge of the pool, making it difficult to find these calling males. When additional males and females arrive at the breeding pool, activity becomes more frenetic and males regularly relocate to more exposed calling areas, or even call while swimming. Skin secretions produced by the nuptial glands of an amplecting male help adhere his ventral surface to the female's back, and pairs are not easily dislodged. Several hundred

eggs are laid in a floating surface film and hatching occurs within 24 hours. *H. variolosus* tadpoles complete their development rapidly, metamorphosing into froglets within about one month; this is an adaptation to life in ephemeral rain pools.

H. variolosus mainly eats ants and termites. It possesses glandular skin that can secrete irritating substances to deter ant and termite soldiers from stinging the frog when it enters their nests. These skin secretions can also irritate human mucous membranes and may be harmful to other amphibians in high doses. Handling these frogs can cause eye irritations, burning sensations on the skin, and sneezing in some people.

H. variolosus favors open, dry areas with grassy vegetation, and it is commonly found in cultivated areas such as pastures and plantations, even occasionally in urban settings. This species is adaptable and appears to not suffer from any major threats, although exact population dynamics are difficult to assess because of its limited surface activity.

Typically with a gray, olive, brown, or reddish-brown dorsal color, marked with scattered dark spots and a prominent, thin middorsal yellow or orange line.

Small to moderate-sized, with a stocky, egg-shaped body.

Head small and triangular when seen from above; a distinct transverse skin fold present across the top of the head, behind the eyes.

Skin generally smooth; dorsum on breeding males is sometimes tuberculate.

Limbs short and robust.

Dark dorsal markings often outlined in red, in particular on the hidden rear surface of the hind limbs.

Sides of the head marked with a white diagonal stripe that runs from the eye toward the insertion of the forelimb.

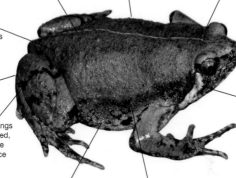

Venter with dark mottling on a white or gray background, and with a light midventral stripe.

A faint to very prominent broad dark brown or gray lateral band starts behind each eye and runs along the flanks to the groin and the hind limbs.

foot

hand

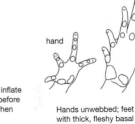

Individuals sometimes inflate their body before calling or when threatened, making these frogs appear even more egg-shaped.

Hands unwebbed; feet with thick, fleshy basal webbing. A pair of enlarged metatarsal tubercles on the sole of each foot.

Ctenophryne aterrima (p. 467)

Larger and solid black; no known range overlap.

Rhinophrynus dorsalis (p. 487)

Much larger and only has four toes on each hind limb.

H. pictiventris (p. 470)

Has a boldly marked black and white venter. It only ocurs on the Atlantic slope; there is no range overlap.

473

Family Ranidae
(True Frogs)

The family Ranidae currently contains 378 species that occur in most of the world except for Australia, the Sahara Desert region, and in extreme southern South America. It is represented in the New World by a single genus: *Rana*. A recent analysis of the amphibian tree of life led several taxonomists to assign all American species of *Rana* to the genus *Lithobates*. Old World and New World ranids are geographically distant sister groups that do not necessarily differ morphologically. The author prefers to follow other biologists and assign New World ranids to the subgenus *Lithobates*, which is embedded in the genus *Rana,* in order to minimize taxonomic confusion. Five species of ranid frog are currently recognized in Costa Rica but poorly defined taxa such as *Rana (Lithobates) taylori* will likely be split into multiple species at some point.

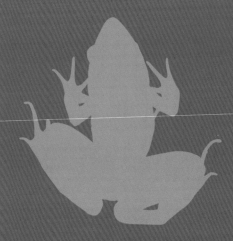

Genus *Rana*
(subgenus *Lithobates*)

All Costa Rican ranid frogs are characterized by a pair of glandular dorsolateral skin folds. In addition, these frogs typically have long, muscular legs, extensively webbed feet, and hands completely free of webbing.

With their streamlined body, long legs, smooth skin, and aquatic habits, members of this genus are what most people consider "typical" frogs, similar to those North Americans or Europeans might see at home. Sometimes referred to as true frogs or pond frogs, these are the archetypical denizens of ponds, lakes, and slow-moving sections of rivers, where they float in the water or hide in vegetation lining the banks. Their powerful legs allow for long jumps on land and rapid propulsion when swimming.

Most male ranids have internal, paired vocal sacs and produce croaking advertisement calls. A few species lack vocal sacs but are still able to produce soft advertisement calls; female ranids never vocalize. Amplexus is typically axillary, with the male clasping a female near her armpits. Females lay eggs in the water, in a globular mass that is usually attached to vegetation. In Costa Rica, tadpoles of the genus *Rana* can be seen in many ponds, slow-moving rivers, and streams, especially in the lowlands. These are not only among the most commonly observed tadpoles in the country but also easily among the largest, some exceeding 4 in (100 mm) in total length.

The five Costa Rican species of *Rana* can be divided into two morphologically distinct groups: the *Rana palmipes* species-group, with a uniformly colored dorsum, and the *Rana pipiens* species-group, whose species have distinct rows of rounded dorsal spots.

Rana (Lithobates) forreri
Dry Forest Leopard Frog
Least concern

Males to 3.54 in (90 mm) in standard length; females to 4.49 in (114 mm). A streamlined, long-legged frog with powerful limbs, extensively webbed feet, bold pattern of dark spots on a light background, and distinct uninterrupted dorsolateral skin folds.

Ecoregions
1 2 3 4 5

Rana forreri inhabits the seasonally dry northwest of Costa Rica (Guanacaste and Nicoya Peninsula) and ranges south along the Pacific coast at least as far as Dominical, Puntarenas Province, from near sea level to 2,800 ft (850 m). Its range extends northward to southwestern Sonora, Mexico.

Natural history

This species is commonly found both during the day and at night, in vegetation along the water's edge of canals, marshes, and other stagnant or slow-moving wetlands. When disturbed, these frogs escape into the water with a powerful leap, followed by a rapid swim to the bottom, where they hide in the sediment or below vegetation or debris. *R. forreri* is an adaptable species that readily colonizes man-made ponds and basins; it is regularly found in the immediate vicinity of humans.

R. forreri reproduces in temporary or permanent ponds during the wet season. Males emit a raspy croak while floating on the water surface at night. The vocal sacs in this species are paired and emerge from the edge of the lower jaw below the tympanum; they are characteristically darkly pigmented. Eggs are laid in floating masses, often attached to emergent or submerged vegetation. Each clutch may contain several hundred black-and-white eggs. Tadpoles complete their development in the breeding pond; tadpoles are very large (to 3.23 in [82 mm] in total length), with a long, muscular tail and low fins that are heavily marked with dark blotches.

Description

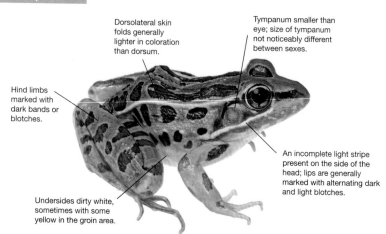

Dorsolateral skin folds generally lighter in coloration than dorsum.

Tympanum smaller than eye; size of tympanum not noticeably different between sexes.

Hind limbs marked with dark bands or blotches.

An incomplete light stripe present on the side of the head; lips are generally marked with alternating dark and light blotches.

Undersides dirty white, sometimes with some yellow in the groin area.

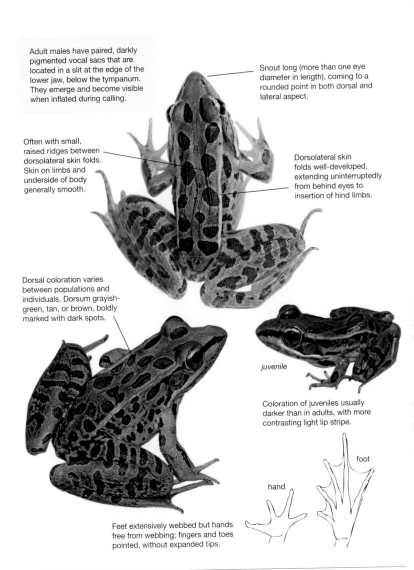

Adult males have paired, darkly pigmented vocal sacs that are located in a slit at the edge of the lower jaw, below the tympanum. They emerge and become visible when inflated during calling.

Snout long (more than one eye diameter in length), coming to a rounded point in both dorsal and lateral aspect.

Often with small, raised ridges between dorsolateral skin folds. Skin on limbs and underside of body generally smooth.

Dorsolateral skin folds well-developed, extending uninterruptedly from behind eyes to insertion of hind limbs.

Dorsal coloration varies between populations and individuals. Dorsum grayish-green, tan, or brown, boldly marked with dark spots.

juvenile

Coloration of juveniles usually darker than in adults, with more contrasting light lip stripe.

hand

foot

Feet extensively webbed but hands free from webbing; fingers and toes pointed, without expanded tips.

Similar species

R. taylori (p. 478)

Dorsolateral skin folds discontinuous; inset posteriorly.

R. vaillanti (p. 480)

Lacks the boldly spotted appearance.

Leptodactylus **sp.** (p. 453)

Leptodactylus species lack webbing between their toes.

477

Rana (Lithobates) taylori
Montane Leopard Frog

Least concern

Males to 3.07 in (78 mm) in standard length; females to 3.46 in (88 mm). A streamlined frog with extensively webbed feet, powerful but relatively short legs, bold pattern of large, elongated dark spots on a lighter background, and distinct dorsolateral skin folds that are interrupted near the hind limbs and offset so that they do not form a single line.

Ecoregions

1 2 3 4 5

Rana taylori is primarily found in the foothills and mountains of Costa Rica (Central Valley and surrounding Central Montane Range); it has also been found in a few areas in the Atlantic lowlands, from extreme eastern Nicaragua to southeastern Costa Rica; in Costa Rica, there is an isolated population in the extreme highlands of the southern Talamanca Mountain Range that may cross into adjacent Panama; from near sea level to 10,500 ft (3,200 m).

Natural history

This species is made up of a complex of cryptic forms that are still poorly understood and defined. The Atlantic lowland (below 3,600 ft [1,100 m]) populations of this species are possibly different from the Central Mountain Range (4,900–6,000 ft [1,500–1,850 m]) populations. Individuals from the isolated population in the highlands of the southern Talamanca Mountain Range (5,575–10,500 ft [1,700–3,200 m]) appear smaller and sufficiently different in overall appearance that they may represent a distinct species. Lowland and high elevation frogs of this complex are rarely seen and most of what is known about the species is derived from the more frequently encountered populations in and around the Central Valley.

These frogs are most commonly seen at night, perched on floating vegetation or on the banks of ponds and streams.

More research and more observations are needed to clarify this situation.

Rana taylori is semi-aquatic and is active (at night and during the day) along the vegetation-choked edges of ponds, swamps, and slow-moving streams. Relatively tolerant of habitat alteration, some populations are found in manmade ponds in close proximity to humans. When approached, these frogs escape into the water with a powerful jump and hide below debris at the bottom. They are strong swimmers and spend considerable time in the water, especially at night.

Reproduction takes place during the wetter months of the year (May–December). At night, males emit a short, raspy call while floating at the water surface; amplexus is axillary. Large egg clumps are attached to submerged or emergent vegetation and show black-and-white embryos during the earliest stages of their development. Tadpoles are very large (to 3.23 in [82 mm] in total length) and dark brown, with a relatively patternless tail and fins; they complete their development in the water.

Because of the confusing taxonomy of *R. taylori*, it is difficult to assess threats to populations of this species. Most Costa Rican populations occur at middle and high elevations and are therefore more susceptible to chytridiomycosis. Some populations found in relatively intact forest environments were difficult to detect during the 1990s and early 2000s but seem to be increasing in numbers, while other populations have disappeared since the 1990s and have not recovered.

Populations differ significantly in morphology, coloration, and pattern and may represent different species. A and B are from Central Valley populations; C is from an Atlantic lowland population.

Large dorsal spots are dark brown and usually elongate or rounded; spots usually outlined with a lighter color.

Head pointed, with a long snout (longer than one eye diameter) that is rounded in dorsal and lateral aspect.

Short parallel ridges may be present between the dorsolateral skin folds.

Dorsolateral skin folds distinct but discontinuous, and generally interrupted near the groin area. The posterior portion may be offset medially.

There is no dark eye mask in this species but a light lip stripe may be present; some individuals show dark mottling or green markings on each side of the head.

A boldly patterned frog with dark discrete spots on a grayish-tan, bronze, or grayish-green background.

Tympanum distinct and roughly the same size as the eye or slightly smaller; no apparent difference in tympanum size between sexes.

Legs and flanks variously marked with dark bands, spots, or mottling.

Undersurfaces are white.

foot

Note that adult males have paired, unpigmented lateral vocal sacs that can be extended during calling. The sacks emerge from a slit located under the tympanum, at the edge of the lower jaw.

Feet extensively webbed, but no webbing between fingers; digits are relatively pointed, without expanded tips.

hand

R. forreri (p. 476)

Has continuous dorsolateral skin folds.

R. vaillanti (p. 480)

Lacks the boldly spotted appearance.

Leptodactylus sp. (p. 453)

Leptodactylus species never have webbing between their toes.

479

Rana (Lithobates) vaillanti
Vaillant's Frog   **Least concern**

Males to 3.70 in (94 mm) in standard length; females to 4.92 in (125 mm). A large green-headed frog with fully webbed feet and dark-outlined dorsolateral skin folds.

In Costa Rica, *Rana vaillanti* is found in the Atlantic and north Pacific lowlands and foothills; 0–2,900 ft (0–880 m). This is the most widespread ranid in the neotropics. It ranges along the Atlantic coastal plain, from southern Veracruz, Mexico, south to northern Colombia. It also occurs on the Pacific slope, in southeastern Oaxaca and northeastern Chiapas, Mexico, in southern Nicaragua and adjacent northwestern Costa Rica, and in Colombia and northwestern Ecuador.

Ecoregions
1 2 3 4 5

Natural history

In spite of the significant variation in its color pattern throughout its range, its green head makes it relatively easy to identify. The amount and intensity of the green coloration varies with age and between sexes; juveniles tend to have a bright green head only, while in adult males the green coloration may spread onto other parts of the body, more so than in adult females, which tend to be mottled brown with a green head.

One of the more aquatic frogs in Costa Rica, it is a common sight along many lowland ponds, streams, and rivers. During the day and at night it can be observed floating in the water or perched in dense vegetation on the banks of quiet waters. When startled, it escapes into the water, swims a short distance, and hides on the bottom or beneath vegetation. As they escape, these frogs may emit a high-pitched, shrill one-note alarm call.

Although variable in coloration, this common lowland frog is easily recognized by its green head and brownish body.

Like other species of the genus *Rana*, this species reproduces in water. The male attracts a female with his advertisement call, a series of short croaks or grunts, during an extended breeding season that intensifies in the rainy months (May–December), though breeding can happen during rainy spells at other times of the year as well. Amplexus is axillary and females deposit a very large clump of eggs in the water, which may contain many hundreds of eggs. The tadpoles of this species are very large, to 3.15 in (80 mm) in total length, with a dark body and boldly spotted tail.

These ambush hunters sit motionless for prolonged periods and quickly nab invertebrate prey that comes within range. Recorded prey items also include a variety of small vertebrates: birds, fish, and other amphibians. In turn, it falls prey to other large frogs, fish, several species of frog-eating snakes, as well as co-occurring Central American snapping turtles (*Chelydra acutirostris*), spectacled caiman (*Cayman crocodilus*), and American crocodiles (*Crocodylus acutus*).

R. vaillanti is adaptable and very common throughout its range. It is able to withstand relatively high levels of habitat disturbance and pollution, and readily persists in the immediate vicinity of humans.

This wide-ranging species likely represents a complex of several cryptic species. In older literature, Costa Rican populations were referred to as *Rana palmipes* but that species is now thought to be restricted to South America.

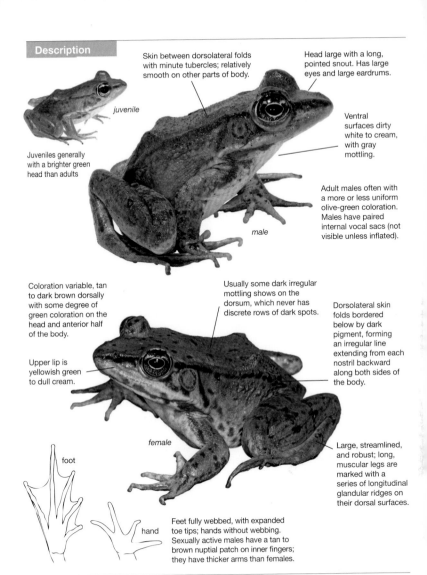

juvenile

Skin between dorsolateral folds with minute tubercles; relatively smooth on other parts of body.

Head large with a long, pointed snout. Has large eyes and large eardrums.

Ventral surfaces dirty white to cream, with gray mottling.

Juveniles generally with a brighter green head than adults

Adult males often with a more or less uniform olive-green coloration. Males have paired internal vocal sacs (not visible unless inflated).

male

Coloration variable, tan to dark brown dorsally with some degree of green coloration on the head and anterior half of the body.

Usually some dark irregular mottling shows on the dorsum, which never has discrete rows of dark spots.

Dorsolateral skin folds bordered below by dark pigment, forming an irregular line extending from each nostril backward along both sides of the body.

Upper lip is yellowish green to dull cream.

female

Large, streamlined, and robust; long, muscular legs are marked with a series of longitudinal glandular ridges on their dorsal surfaces.

foot

hand

Feet fully webbed, with expanded toe tips; hands without webbing. Sexually active males have a tan to brown nuptial patch on inner fingers; they have thicker arms than females.

| *R. warszewitschii* (p. 484) | *R. vibicaria* (p. 482) | *R. forreri* (p. 476) | *R. taylori* (p. 478) | *Leptodactylus* sp. (p. 453) |

Has red and yellow markings on the hind limbs.

Occurs at higher elevations and has green eyes.

Both may co-occur locally but are distinguished by their boldly spotted pattern.

Leptodactylus species never have webbing between their toes.

Rana (Lithobates) vibicaria
Green-eyed Frog

Vulnerable

Males to 2.87 in (73 mm) in standard length; females to 3.62 in (92 mm). A fairly large highland frog that has a pair of dorsolateral skin folds, webbed feet, green eyes, and extensive red coloration on the hands, feet, and hind limbs (most prominent in juveniles). Its powerful limbs are long and slender.

Ecoregions
1 2 3 4 5

Rana vibicaria was previously widespread in montane areas of the Tilarán, Central, and Talamanca Mountain Ranges of Costa Rica, ranging southeast to Chiriquí Volcano in western Panama; 4,900–8,850 ft (1,500–2,700 m). It now persists in only a handful of isolated populations.

Natural history

This beautiful semi-aquatic species occurs in montane humid forests. It inhabits swamps and streams in densely forested areas, but has also been observed near ponds in clearings, pastures, and areas of secondary growth that have intact forest nearby. Its extended breeding season peaks during the wetter months of the year. Even though males of this species lack a vocal sac or vocal slits, they still produce an advertisement call, a low-pitched chuckle. The quality of this call is very soft and does not carry; indeed, a chorus of these frogs is barely detectable from even nearby. Eggs are deposited directly in the water, attached to vegetation in a globular mass measuring about 4 in (100 mm) in diameter. Egg masses have been observed from November to May, during which time tadpoles of varying sizes were recorded, suggesting that R. vibicaria undergoes a prolonged larval development period. Tadpoles attain a very large size, measuring up to 2.76 in (70 mm) in total length; they are brown, with densely spotted tail fins. Tadpoles approaching metamorphosis display the bright green dorsal coloration seen in juveniles.

Until the early 1990s, R. vibicaria was locally common throughout its Costa Rican range, but it then declined precipitously and was not detected for more than a decade. In 2002, a single individual was discovered in Monteverde, followed by the discovery of a breeding population in the area. In subsequent years other remnant populations were found in isolated localities within the species' historic range. Some populations such as those in the Monteverde Conservation League Children's Eternal Rainforest or in Juan Castro Blanco National Park are now seemingly healthy and recovering. Other populations have been discovered in recent years (e.g., on Cerro Chompipe and in the Pérez Zeledón area), but those are still small and tenuous. In spite of recent searches, no individuals have been been reported from the southern Talamanca Mountain Range, either in Costa Rica or Panama.

Since R. vibicaria disappeared even from protected, seemingly undisturbed, areas, chytrid fungal infections are thought to be a main driver of its decline, though other factors may be in play. The recovering population at Juan Castro Blanco National Park reportedly has recently shown signs of deformities that may have been caused by agrochemicals, indicating that the future of this species is precarious. Nevertheless, the encouraging population trends in recent years led the International Union for Conservation of Nature (IUCN) in 2013 to downgrade the conservation status for R. vibicaria from Critically Endangered to Vulnerable.

Juvenile R. vibicaria are among the more strikingly colored frogs in Costa Rica.

Description

Head moderate in size; long, bluntly rounded snout, large eyes, and clearly visible eardrums.

Skin relatively smooth with raised, paired dorsolateral skin folds. Dorsal pattern marked with varying amounts of black spotting.

Sexually mature males with tan to light brown nuptial patch on inner fingers and with thicker arms than females. Males lack vocal slits or vocal sacs.

Dorsolateral skin folds highlighted with light line; bordered below by a poorly defined dark brown to black band.

Eyes with diagnostic green iris.

A prominent white lip stripe invariably present.

Dorsal coloration in adults bronze to dark golden-brown, occasionally green.

juveniles

Juveniles with bright green dorsum and head. Flanks, posterior surfaces of thighs, and soles of hands and feet bright orange-red. These bright colors often fade with age but may be retained in some adults.

In juveniles, dark band completely covers flanks and sides of head, creating a dark mask. Dark flank coloration gradually lightens with age.

foot

hand

Feet with extensively webbed toes and slightly expanded toe tips. Hands without webbing.

Similar species

R. taylori (p. 478)

Has a boldly spotted dorsal pattern.

R. vaillanti (p. 480)

Lacks green eyes and generally occurs at lower elevations

R. warszewitschii (p. 484)

Has bright yellow spots on the posterior thigh surface.

483

Rana (Lithobates) warszewitschii
Brilliant Forest Frog

Least concern

Males to 2.05 in (52 mm) in standard length; females to 2.48 in (63 mm). A small to moderate-sized, streamlined frog with dorsolateral skin folds, extensively webbed feet, and a long, pointed snout. Shows a light lip stripe, red on the undersides of its hind limbs, and a series of bright yellow spots on the hidden surfaces of the legs.

Ecoregions
1 2 3 4 5

Rana warszewitschii occurs on the Atlantic slope, from northeastern Honduras to eastern Panama, and on the Pacific slope of central and southern Costa Rica and western Panama; 0–5,700 ft (0–1,740 m).

Natural history

R. warszewitschii is one of the least aquatic species among Costa Rican *Rana*. Adults are usually found in leaf litter in dense rainforest areas but normally do not venture too far from the forest ponds or slow-moving streams in which they reproduce year-round. In spite of the absence of vocal slits and a vocal sac, males of this species do produce a soft rasping trill as an advertisement call. Their eggs, which are attached to the undersides of rocks, hatch into some of the largest tadpoles found in the region, occasionally reaching the impressive size of 4.53 in (115 mm) in total length. Adults are active during daylight hours and at night.

Adult female *R. warszewitschii* typically have yellow mottling on their ventral surfaces; the undersides of males are off-white.

Metamorphosing juveniles, however, are generally observed crawling on land at night, and young individuals are more frequently encountered hopping along the forest floor at night.

This is a common species throughout much of its range. Usually these frogs are seen when they jump away from underfoot and dive into the leaf litter on the forest floor. When motionless, *R. warszewitschii* relies on its excellent camouflage; the bright red and yellow flash colors on the normally hidden surfaces of the legs are suddenly revealed when it jumps, which presumably confuses and/or deters potential predators.

R. warszewitschii is somewhat tolerant of habitat alteration; it can persist in fragmented forest areas as long as relatively high humidity levels are maintained at the forest floor level. Reportedly, it even occurs locally in some urban areas. Nevertheless, habitat alteration and deforestation are the biggest threats to this species. Even though it is still common in many lowland areas, significant declines have been reported at higher altitudes, and *R. warszewitschii* has disappeared altogether from protected areas such as Tapantí National Park and Monteverde. The declines in these areas may be attributed to chytridiomycosis, but recent studies have shown that this species is capable of surviving significant levels of chytrid infection in some areas. An interaction between disease incidence and prevailing temperatures or climate possibly plays a role. Nevertheless, population levels seem to be rising again in recent years, even in areas where the species was not detected for a period of time.

A slender, moderate-sized frog; has a long, pointed snout, distinctly visible eardrums, and large eyes.

Dorsum bronze to brown, speckled with irregular blue or green spots.

A thin light (cream, white, or gold) dorsolateral stripe, running from the tip of the snout to the groin, highlights the dorsolateral skin folds.

This frog lacks vocal slits and vocal sacs. Sexually active males have swollen arms and tan nuptial patches on inner fingers.

Limbs moderate in length and slender.

Diffuse dark brown blotches often present on the dorsum and flanks; hind limbs marked with transverse dark bands.

Skin relatively smooth.

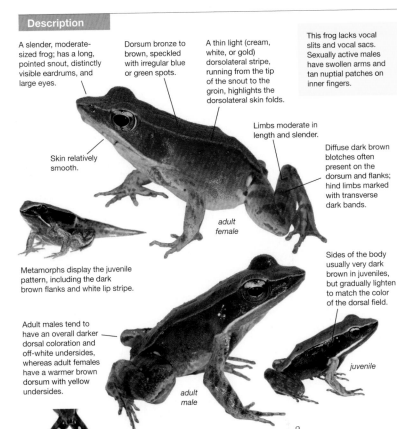

adult female

Metamorphs display the juvenile pattern, including the dark brown flanks and white lip stripe.

Sides of the body usually very dark brown in juveniles, but gradually lighten to match the color of the dorsal field.

Adult males tend to have an overall darker dorsal coloration and off-white undersides, whereas adult females have a warmer brown dorsum with yellow undersides.

juvenile

adult male

Underside of limbs and toe webbing bright red; hidden surfaces on back of thighs red, with a diagnostic pattern of 1–4 bold, bright yellow spots. Sometimes with additional yellow spots in groin area.

foot

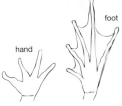

hand

Feet with fairly extensive webbing between toes; hands without webbing; tips of fingers and toes somewhat expanded.

R. vaillanti (p. 480)

Usually has a green head; never has bright yellow spots on the back of the thighs.

R. taylori (p. 478)

Has a boldly spotted dorsal pattern.

Craugastor fitzingeri (p. 246)

Has yellowish or cream mottling on the back of the thighs but lacks dorsolateral skin folds.

485

Family Rhinophrynidae

Although several fossil forms have been assigned to this primitive family, it contains just a single living species, *Rhinophrynus dorsalis*, and for that reason the family and genus descriptions have been combined here.

Several primitive traits define this genus. *Rhinophrynus dorsalis* lacks a breastbone, ribs, and teeth; it has four toes on each hind foot (rather than the five digits found in all other Costa Rican anura); and it is the only anuran whose tongue is attached to the back of its mouth (other frogs are either tongue-less or have their tongue attached to the front of their mouth).

Rhinophrynus dorsalis
Mexican Burrowing Toad

Least concern

A bizarre frog, with a globular, purplish-brown body, small pointed head, and calloused snout. Each hind foot has a pair of enlarged spade-like tubercles and only four toes. Males to 2.95 in (75 mm) in standard length, females to 3.51 in (89 mm).

Ecoregions
1 2 3 4 **5**

On the Atlantic slope, *Rhinophrynus dorsalis* ranges from extreme southern Texas, US, to northwestern Honduras; on the Pacific slope, it ranges from Guerrero, Mexico, to northwestern Costa Rica. This species also occurs in an isolated population on the Mosquito Coast of northeastern Nicaragua. In Costa Rica it is known only from the seasonally dry forests in Guanacaste Province; sea level to 985 ft (300 m).

Natural history

R. dorsalis is generally associated with seasonally flooded lowland areas. It has been found in tropical dry and moist forests, as well as in thorn scrub, savannah, and cultivated areas. These frogs remain underground in burrows and chambers that they construct using the spade-like tubercles on the hind feet; they only emerge during periods of heavy rain, when they breed in temporary pools. *R. dorsalis* can survive long periods in a dormant state in its underground retreat; a captive individual reportedly survived nearly two years without any water or food in a burrow in its terrarium.

Its diet consists entirely of ants and termites, which are foraged in their underground nests. Using the keratinized, spade-like tubercles on the hind feet these frogs burrow backwards into soil and, with their powerful front legs, dig their way into termite and ant nests. The calloused snout provides protection against stings and bites when it is inserted into an ant or termite tunnel. Prey are caught with the tip of the rodlike tongue, which is extended through a groove in the front of the mouth.

This species is an explosive breeder; during the first heavy rains of the wet season (late May–early June), many individuals will emerge simultaneously from their burrows. Pairs engage in inguinal amplexus and lay 2,000–8,000 eggs. Hatching and larval development occurs very rapidly, a common strategy for amphibians breeding in temporary pools. Tadpoles tend to form large aggregations that move in synchronized motion as if they were a single large animal. They are unique among Costa Rican tadpoles in that they lack hard, keratinized beaks and denticles, and sport thin barbels around

the oral opening, which gives them a catfishlike appearance. During the day, tadpoles appear black but may be translucent at night.

Male *R. dorsalis* call at night from the water surface in large unorganized choruses. The advertisement call is a very loud, moaning *uwooooa* that has been likened to the sound of a person vomiting, hence its common name in Spanish, *sapo borracho* (drunk toad). While calling, the male inflates his paired internal vocal sacs to such an extent that the animal resembles a floating balloon. When threatened, these frogs also balloon up to discourage predators or to tightly wedge themselves inside their burrow.

R. dorsalis plays an important role in Mayan mythology, in which the legendary frog *uo* summons the rain god Chac to release water onto parched lands at the end of the dry season.

Amplexus in *R. dorsalis* is inguinal, meaning that the male clasps the female around the waist.

487

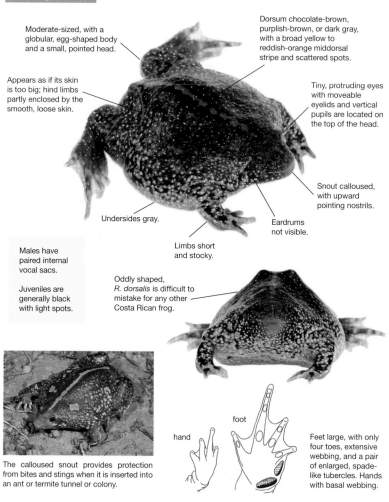

Description

Moderate-sized, with a globular, egg-shaped body and a small, pointed head.

Dorsum chocolate-brown, purplish-brown, or dark gray, with a broad yellow to reddish-orange middorsal stripe and scattered spots.

Appears as if its skin is too big; hind limbs partly enclosed by the smooth, loose skin.

Tiny, protruding eyes with moveable eyelids and vertical pupils are located on the top of the head.

Snout calloused, with upward pointing nostrils.

Undersides gray.

Eardrums not visible.

Males have paired internal vocal sacs.

Juveniles are generally black with light spots.

Limbs short and stocky.

Oddly shaped, *R. dorsalis* is difficult to mistake for any other Costa Rican frog.

hand

foot

Feet large, with only four toes, extensive webbing, and a pair of enlarged, spade-like tubercles. Hands with basal webbing.

The calloused snout provides protection from bites and stings when it is inserted into an ant or termite tunnel or colony.

Similar species

Hypopachus variolosus (p. 472)

It is difficult to confuse *R. dorsalis* with any other Costa Rican frog.

Shares a somewhat similar morphology but has five toes on its hind limbs.

Family Strabomantidae

Once considered the largest genus of vertebrate animals, *Eleutherodactylus* was recently found to consist of three evolutionary groups—one a primarily Caribbean group (now family Eleutherodactylidae), one a Central American group (now family Craugastoridae), and one mostly South American group (now family Strabomantidae). As currently understood, the family Strabomantidae contains 626 species. In spite of it being primarily a South American family, 2 genera (*Pristimantis* and *Strabomantis*) range into Central America; Costa Rica is home to 9 species of *Pristimantis* and 1 species of *Strabomantis*.

Frogs in the family Strabomantidae are grouped together based on molecular characteristics, and members of this very large and diverse group are not easily recognized by shared diagnostic field characteristics. The 9 Costa Rican *Pristimantis* are small to moderate-sized cryptic frogs that lack webbing on their hands and feet; the disks on the fingers and toes are of various sizes and shapes. The single *Strabomantis* species in the country is a toadlike, robust species that lacks the paired parotoid glands and other features of true toads (family Bufonidae). All members of the family Strabomantidae are thought to reproduce through direct development.

Genus *Pristimantis*

A large group of 471 mostly South American frogs that were formerly assigned to the genus *Eleutherodactylus*. The genus *Pristimantis* is defined by molecular characteristics and its members can be difficult to distinguish based on morphological features. The nine Costa Rican species all lack webbing between their fingers and toes but are otherwise quite diverse morphologically; this group includes a small, green canopy specialist, several cryptically brown leaf litter species, and a striking case of possible poison-dart-frog mimicry. The diversity among the various subgroups contained in this large genus indicates likely future taxonomic splits.

Pristimantis altae
Coral-spotted Rain Frog

Near threatened

A tiny dark brown to almost black frog with a distinctly truncated snout; has bold yellowish-orange to bright red spots in the groin and on the hidden surfaces of the thighs. Males to 0.94 in (24 mm) in standard length, females to 1.06 in (27 mm).

Ecoregions
1 2 3 4 5

Pristimantis altae is known from scattered locations on the Atlantic slope, in Costa Rica and adjacent northwestern Panama; from near sea level to 1,250 ft (1,250 m).

Natural history

This rarely seen but seemingly widespread frog inhabits pristine low elevation rainforests. Individuals have been observed perched on low vegetation at night. During the day, *P. altae* retreats into the leaf litter layer on the forest floor or into arboreal bromeliads. It likely also deposits its eggs in these places. Reproduction takes place through direct development; juveniles hatch directly from terrestrial eggs.

Too few records exist to have a clear idea about the population trends and conservation needs of *P. altae*. Since it inhabits undisturbed forest habitats, this species is likely adversely affected by habitat alteration and deforestation. Given its lowland distribution, *P. altae* may be less impacted by the chytrid fungal pathogens that threaten higher elevation species.

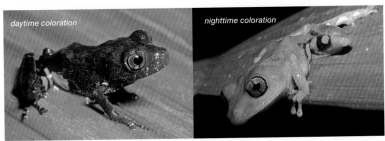

daytime coloration

nighttime coloration

Adults mostly uniform dark brown. Capable of color change, often appearing considerably darker during the day than at night.

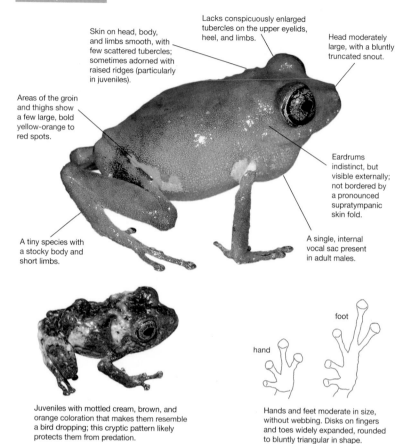

Skin on head, body, and limbs smooth, with few scattered tubercles; sometimes adorned with raised ridges (particularly in juveniles).

Lacks conspicuously enlarged tubercles on the upper eyelids, heel, and limbs.

Head moderately large, with a bluntly truncated snout.

Areas of the groin and thighs show a few large, bold yellow-orange to red spots.

Eardrums indistinct, but visible externally; not bordered by a pronounced supratympanic skin fold.

A tiny species with a stocky body and short limbs.

A single, internal vocal sac present in adult males.

foot

hand

Juveniles with mottled cream, brown, and orange coloration that makes them resemble a bird dropping; this cryptic pattern likely protects them from predation.

Hands and feet moderate in size, without webbing. Disks on fingers and toes widely expanded, rounded to bluntly triangular in shape.

P. cruentus (p. 497)

Dark individuals can be nearly uniform dark brown, with bold, bright yellow spots in the groin area; but *P. cruentus* is larger than *P. altae*, has a conspicuously enlarged, pointed tubercle on top of each upper eyelid, and a row of pointed tubercles along the trailing edge of the lower legs.

P. pardalis (p. 504)

Similar in appearance but note silvery white markings on the thigh and groin. No range overlap.

Pristimantis caryophyllaceus
Leaf-breeding Rain Frog

Near threatened

A small, smooth-skinned frog with a distinctively long, pointed snout, a prominent tubercle on the edge of each upper eyelid, and a triangular tubercle on each heel. Males to 0.94 in (24 mm) in standard length, females to 1.02 in (26 mm).

Ecoregions
1 2 3 4 5

On the Atlantic slope, *Pristimantis caryophyllaceus* occurs from northern Costa Rica to eastern Panama; on the Pacific slope it is found from extreme southwestern Costa Rica to western Panama; from near sea level to 6,235 ft (1,900 m). Individuals from eastern Panama and adjacent Colombia represent different, undescribed species.

Natural history

This nocturnal frog of forest interiors is most frequently encountered at night on low vegetation. Like other *Pristimantis*, it reproduces through direct development. Eggs are deposited in a small clump on the upper surfaces of leaves, and turn white shortly after they are laid; the clutch is tended by the female.

P. caryophyllaceus occurs in undisturbed rainforests. Although it occurs over a wide range of elevations, it is much more common at intermediate altitudes and rarely found in lowland settings. During the mid-1990s, when many amphibian populations in the country declined, *P. caryophyllaceus* disappeared from most lowland areas in Costa Rica but persisted at higher elevations. This pattern is puzzling, since amphibian declines were generally more pronounced at middle and high elevation sites; in Panama, populations of this species did indeed decline dramatically in some highland sites. Currently, populations in Costa Rica seems to be recovering in areas above 2,265 ft (800 m), but *P. caryophyllaceus* remains absent from many sites where it occurred historically.

Molecular studies show that this species contains at least three lineages, one of which, *P. educatorius*, was recently described as a separate species that ranges from extreme southeastern Costa Rica to Panama and Colombia. However, the current consensus is that *P. educatorius* has not been adequately defined to warrant species status at this point, and for now it remains tentatively contained within the highly diverse species *P. caryophyllaceus*.

Color and pattern quite variable. Usually tan, brown, pink, or (more rarely) green. Can be relatively uniform, mottled, or marked with bold dorsolateral stripes, chevrons, or transverse bars; sometimes has a thin middorsal stripe.

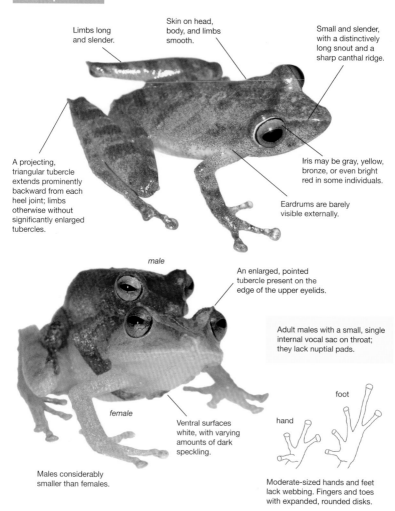

Limbs long and slender.

Skin on head, body, and limbs smooth.

Small and slender, with a distinctively long snout and a sharp canthal ridge.

A projecting, triangular tubercle extends prominently backward from each heel joint; limbs otherwise without significantly enlarged tubercles.

Iris may be gray, yellow, bronze, or even bright red in some individuals.

Eardrums are barely visible externally.

male

An enlarged, pointed tubercle present on the edge of the upper eyelids.

Adult males with a small, single internal vocal sac on throat; they lack nuptial pads.

female

foot

hand

Ventral surfaces white, with varying amounts of dark speckling.

Males considerably smaller than females.

Moderate-sized hands and feet lack webbing. Fingers and toes with expanded, rounded disks.

P. ridens (p. 506)

Shows red coloration in the groin and on the posterior surfaces of the thighs. Lacks both a pointed heel tubercle and the flaplike tubercle on the edge of the upper eyelid (although it does have several enlarged tubercles on the middle of the upper eyelid).

Pristimantis cerasinus
Clay-colored Rain Frog

Least concern

A moderate-sized, long-limbed frog. Has a prominent, pointed tubercle on each heel joint and a row of tubercles along the trailing edge of the tarsus. Also note beautiful "sunset" eyes and a distinct W-shaped ridge that extends from the back of the head onto the shoulder region. Males to 0.98 in (25 mm) in standard length, females to 1.38 in (35 mm).

Ecoregions
1 2 3 4 5

On the Atlantic slope, *Pristimantis cerasinus* ranges from southern Honduras to eastern Panama; it also occurs on the Pacific slope in northwestern Costa Rica; from near sea level to 3,940 ft (1,200 m).

Natural history

This is a species of relatively undisturbed humid forests. During the day, it retreats into the leaf litter, but at night it becomes active and can be seen perched on low vegetation, to about 5 ft (1.5 m) above the ground, often along the banks of a stream, along a forest trail, or in some other edge situation. It is now commonly found at certain middle elevation sites, but, interestingly, it was only detected at those sites in the mid-1990s. Even though populations of this species have locally shown a remarkable increase in the past 20 years, populations at high elevations have declined dramatically, or disappeared altogether. Since *P. cerasinus* has even declined in protected areas where habitat alteration is not an issue, disease or other environmental factors may be responsible for these declines.

P. cerasinus presumably reproduces through direct development, but no information on the specifics of its breeding biology is available. A short, single *tick* call has been reported for males of this species, but it is not known if this is an advertisement or territorial call. Eggs are probably deposited in the leaf litter.

Although *P. cerasinus* is variable in coloration and pattern, its "sunset" irises are unique.

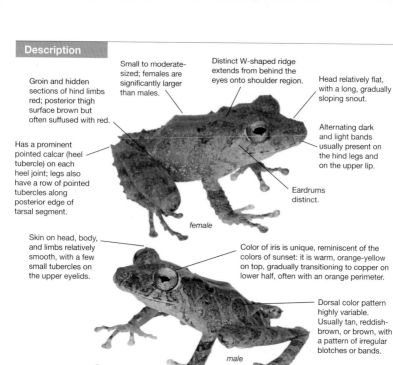

Small to moderate-sized; females are significantly larger than males.

Distinct W-shaped ridge extends from behind the eyes onto shoulder region.

Head relatively flat, with a long, gradually sloping snout.

Groin and hidden sections of hind limbs red; posterior thigh surface brown but often suffused with red.

Alternating dark and light bands usually present on the hind legs and on the upper lip.

Has a prominent pointed calcar (heel tubercle) on each heel joint; legs also have a row of pointed tubercles along posterior edge of tarsal segment.

Eardrums distinct.

female

Skin on head, body, and limbs relatively smooth, with a few small tubercles on the upper eyelids.

Color of iris is unique, reminiscent of the colors of sunset: it is warm, orange-yellow on top, gradually transitioning to copper on lower half, often with an orange perimeter.

Dorsal color pattern highly variable. Usually tan, reddish-brown, or brown, with a pattern of irregular blotches or bands.

male

Skin on venter coarsely granular.

Ventral surfaces dirty white to cream.

hand

foot

Hands unwebbed; small, rounded disks on two inner fingers and greatly enlarged truncated disks on two outer fingers. Feet large and unwebbed. Toe disks slightly enlarged and round on two inner toes; moderately to greatly enlarged, rounded to truncate on three outer toes.

P. ridens (p. 506)

Shares red coloration in the groin and on the posterior thigh surfaces, but lacks the W-shaped ridges in the shoulder region and has distinctly enlarged tubercles on top of the upper eyelids.

P. cruentus (p. 497)

Has a flaplike, pointed tubercle protruding from the edge of the upper eyelid. It generally has bright cream or yellow spots in the groin and lacks the red coloration on the back of the thighs. Its irises are marked with a striking dark reticulum.

Pristimantis cruentus
Golden-spotted Rain Frog

Least concern

Extremely variable in coloration and pattern. Groin area usually marked with cream, yellow, greenish-yellow, or orange spots. Has enlarged disks on fingers and toes but lacks webbing on hands and feet; a distinctly enlarged tubercle is present on the heel joint and on top of the upper eyelids. Iris variably colored but always with a characteristic, intricate black reticulum. Males to 1.26 in (32 mm) in standard length, females to 1.73 in (44 mm).

Ecoregions
1 2 3 4 5

Pristimantis cruentus occurs on both slopes, from northern Costa Rica to eastern Panama; from near sea level to 6,400 ft (1,950 m).

Natural history

P. cruentus occurs in a broad range of habitats, from lowland to high elevation forests; it occasionally occurs in areas of secondary vegetation but is most often found in relatively undisturbed, wet forest habitats. It is most commonly observed at elevations from 1,640–3,600 ft (500–1,100 m), and is relatively uncommon at low or higher elevations.

This appears to be a primarily nocturnal, arboreal frog; individuals are generally found active at night, on leaves 2–7 ft (0.6–2 m) above the ground. It reproduces through direct development, independent of water, although individuals are most often found in association with riparian vegetation along stream corridors. Its relatively large, white eggs are deposited in humid places, in crevices of tree trunks or in thick moss banks; tadpoles develop completely inside the egg and hatch as small frogs. Nothing is known about the breeding biology of this frog; its call has never been recorded and it may be mute.

Some populations have declined in past decades, but recovery has been documented at other sites. This is now a common frog in Rara Avis

Rainforest Reserve, for example, although it was not detected there during a monitoring study from 1993 to 2000. The exact reasons for these localized population fluctuations are not known; across its range, *P. cruentus* populations currently seem to be stable or increasing.

P. cruentus is highly variable and identifying it can be a challenge. The striking black reticulum on the eyes is often a good diagnostic field mark for this species.

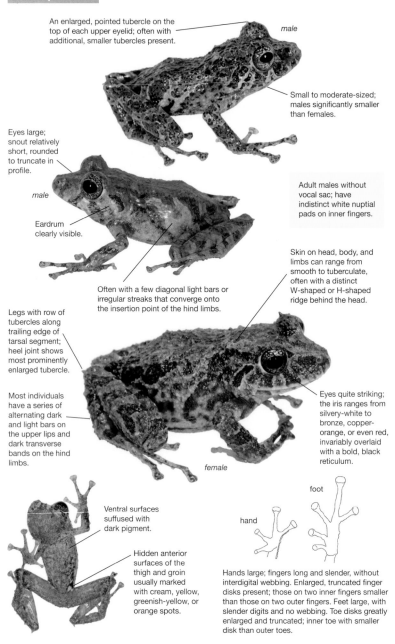

An enlarged, pointed tubercle on the top of each upper eyelid; often with additional, smaller tubercles present.

male

Small to moderate-sized; males significantly smaller than females.

Eyes large; snout relatively short, rounded to truncate in profile.

male

Eardrum clearly visible.

Adult males without vocal sac; have indistinct white nuptial pads on inner fingers.

Skin on head, body, and limbs can range from smooth to tuberculate, often with a distinct W-shaped or H-shaped ridge behind the head.

Often with a few diagonal light bars or irregular streaks that converge onto the insertion point of the hind limbs.

Legs with row of tubercles along trailing edge of tarsal segment; heel joint shows most prominently enlarged tubercle.

Most individuals have a series of alternating dark and light bars on the upper lips and dark transverse bands on the hind limbs.

Eyes quite striking; the iris ranges from silvery-white to bronze, copper-orange, or even red, invariably overlaid with a bold, black reticulum.

female

foot

Ventral surfaces suffused with dark pigment.

hand

Hidden anterior surfaces of the thigh and groin usually marked with cream, yellow, greenish-yellow, or orange spots.

Hands large; fingers long and slender, without interdigital webbing. Enlarged, truncated finger disks present; those on two inner fingers smaller than those on two outer fingers. Feet large, with slender digits and no webbing. Toe disks greatly enlarged and truncated; inner toe with smaller disk than outer toes.

Defies easy description due to extreme variation between populations in skin texture, color pattern, and size. Multiple cryptic forms are likely contained in what is currently considered one species. Generally gray, tan, or chocolate-brown; either relatively uniform or heavily mottled with dark spots; sometimes shows a broad middorsal stripe.

Atlantic slope foothills.

Atlantic slope foothills.

Atlantic slope lowlands.

Atlantic slope lowlands.

Central Mountain Range highlands.

Talamanca Mountain Range highlands.

Similar species

P. cerasinus (p. 495)

Lacks enlarged pointed tubercles on top of each eyelid; also lacks reticulated iris.

Craugastor melanostictus (p. 257)

Larger. Shows a pattern of red, orange, yellow, or pink bands alternating with black vertical bars on its posterior thigh surfaces.

499

Pristimantis gaigei
Gaige's Rain Frog

Least concern

A possible poison dart frog mimic, with a black dorsum, a pattern of paired yellow, orange, or red dorsolateral stripes, and varying amounts of sky blue speckling on the flanks and limbs. It has a copper or bronze iris. Lacks paired scutes on the finger and toe tips (unlike the two Costa Rican *Phyllobates*). Males to 1.22 in (31 mm) in standard length, females to 1.73 in (44 mm).

Ecoregions
1 2 3 4 5

On the Atlantic slope, *Pristimantis gaigei* occurs from extreme southeastern Costa Rica to eastern Panama; from near sea level to 985 ft (300 m). It also occurs on the Pacific slope, in central Colombia.

Natural history

P. gaigei is uncommon in Costa Rica; it is most frequently encountered in the southern parts of its range, on the banks of forest rivers and streams and in floodplains. It inhabits leaf litter and debris on the floor of lowland rainforests; it can also be found in secondary forest and is seemingly fairly tolerant of habitat alteration.

The natural habits of *P. gaigei* are poorly known. Absence of a vocal sac and vocal slits in this species suggests that it may be silent (although this is not always the case with other species that lack these). Regardless, so far no calls have been positively attributed to this species.

Axillary amplexus takes places prior to oviposition; the female lays a clutch of eggs hidden under leaf litter or a cover object or tucked within the root system of a plant. Reportedly, the male and female parent both tend to the eggs. In captivity, egg clutches contained 22–37 eggs, each measuring about 0.2 in (5 mm) in diameter; clutches were produced at intervals of 6–8 weeks. The larvae complete their development entirely within the egg, and hatch as small frogs with a standard length of approximately 0.2 in (5 mm). Peak breeding activity occurred during periods of drought.

This unusual species has a color pattern reminiscent of the highly toxic poison dart frogs in the genus *Phyllobates*, and it may co-occur locally with *Phyllobates lugubris.* The presence of both species in the same place—one highly toxic, the other not—suggests that mimicry may be at work, with *P. gaigeae* gaining protection from potential predators by resembling a dangerously toxic species.

Description

Head moderate in size, with a fairly long snout that comes to a rounded point; canthal ridge distinct.

Eardrums prominent.

Dorsal skin finely granular, without conspicuously enlarged tubercles.

Dorsum black, with a prominent pair of yellow, orange, or reddish dorsolateral stripes.

Iris copper to bronze.

Venter usually gray, light brown, or orange.

Belly, flanks, and limbs with varying amounts of sky blue to turquoise speckling. Some individuals with blue mottling on hind limbs and arms.

No vocal sac or nuptial pads recorded for this species.

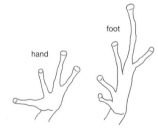

hand

foot

Sometimes has an indistinct lateral blue stripe that extends partially onto the upper lip. Coloration can be quite variable in this species.

Hands and feet without webbing. Fingers with slightly truncated disks on two outer digits and rounded tips on two inner fingers. Toes with slightly expanded disks, similar in size or smaller than largest finger disks.

Similar species

Phyllobates lugubris (p. 305)

Has black iris (*P. gaigei* has a bronze to copper iris). May co-occur locally.

Phyllobates vittatus (p. 307)

Similar in coloration but has black iris. Does not co-occur.

Pristimantis moro
Green Rain Frog

Least concern

A tiny green frog, often with a reddish-brown head and/or a cream or white band covering the upper eye-lids and crossing over the top of the head. Has rounded disks on its fingers and toes, and lacks interdigital webbing. Males to 0.79 in (20 mm) in standard length, females to 0.98 in (25 mm).

Ecoregions
1 2 3 4 5

Pristimantis moro is known from very few specimens found in isolated locations throughout its presumably extensive range. On the Atlantic slope, it occurs in central Costa Rica; on the Pacific slope, it occurs in Panama and Colombia. In Costa Rica it is only known from the Bajo la Hondura area, San José Province, at 4,085 ft (1,245 m).

Natural history

Little is known about the biology of this species; its biology, tiny size, and mostly inaccessible habitat, make *P. moro* difficult to detect. The only known individuals were found hiding in arboreal bromeliads during the day. It is likely that this is a nocturnal canopy frog.

This species may be more common and widespread than is currently understood, but no information on its population and conservation status exists. No *P. moro* have been reported from Costa Rica in almost 50 years, and it may have suffered from habitat alteration and chytrid fungal pathogens, although the only known site lies within Braulio Carrillo National Park, which has suitable habitat. *P. moro* was found more recently in Panama, including in areas that have been impacted by chytridiomycosis. This may mean that its population status is relatively secure, and that its perceived rarity more likely results from the difficulty of observing this species.

Adult males have a single internal vocal sac on their throat.

Color pattern unique. The few known individuals either have uniform green head, body, and limbs without transverse markings, or a green dorsum with a reddish-brown head sometimes marked with a light transverse band across the top of the head and upper eyelids.

Tiny and stocky, with a small head and a short, rounded snout.

Dorsal skin smooth; lacks obviously enlarged tubercles on dorsum, on top of the eyelids, or on limbs.

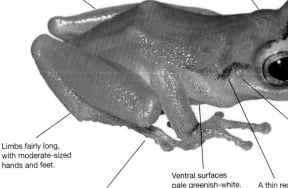

Eardrums poorly defined and very small.

Limbs fairly long, with moderate-sized hands and feet.

Hands and feet lack webbing. Finger and toe disks small and rounded; only the two outer fingers and four outer toes have disks that are expanded beyond the width of the digit.

Ventral surfaces pale greenish-white.

A thin reddish-brown line on each side of the head connects the nostril with the eye; this line continues from the back of eye and curves over the top of the eardrum.

P. ridens (p. 506)

P. ridens and other small, co-occurring members of this genus are never uniformly bright green.

Isthmohyla debilis (p. 398)

Has webbing between its toes and distinctly enlarged, rounded finger and toe disks.

Cochranella euknemos (p. 194)

Has webbing between its toes and a strikingly sloping snout.

503

Pristimantis pardalis
Silver-spotted Rain Frog

Near threatened

A tiny black or dark brown frog with a pattern of bold silvery white spots in the groin area and on the anterior and posterior surfaces of its thighs. Males to 0.75 in (19 mm) in standard length, females to 1.14 in (29 mm).

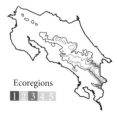

Ecoregions
1 2 3 4 5

In Costa Rica, *Pristimantis pardalis* is known from isolated populations in the foothills of the Fila Costeña in the central Pacific region and in the foothills and middle elevations of the southern Talamanca Mountain Range; 1,150–4,750 ft (350–1,450 m). It also occurs in a few lowland areas in central and eastern Panama.

Natural history

A poorly known species that is infrequently seen; it may be more common and more widespread than the limited observations suggest, however. It is possible that this is a complex of cryptic species; populations in the isolated localities that are currently known may in fact represent genetically isolated taxa.

P. pardalis is nocturnal. It is most often observed in dense humid premontane and montane forest. During the day, individuals retreat into the leaf litter layer on the forest floor, where raking through the dead leaves may uncover significant numbers of these frogs. Even within suitable forest habitat, this species is unevenly distributed; it is usually encountered in discrete populations, often in areas with a few large trees, an understory of palms and broad-leaved plants, and a thick layer of leaf litter. *P. pardalis* reproduces though direct development and can be found far from streams or ponds.

A small juvenile was found during the dry season, in late February. Its dark brown dorsum was marked with obscure white and cream mottling. This pattern, combined with the relatively rugose dorsal skin, makes these small frogs resemble bird droppings, adding to their camouflage.

Little information exists on the exact population or conservation status of this species. The biggest threat is likely destruction of its preferred forest habitat, although *P. pardalis* is still regularly encountered throughout its range and occurs in several protected areas. An individual of this species recently encountered in Panama was infected with large carnivorous fly larvae that were consuming the frog's intestines while it was still alive. It is not known what impact parasites and pathogens have on wild populations of this species.

daytime coloration

nighttime coloration

This species is capable of color change, sometimes appearing darker during the day than at night.

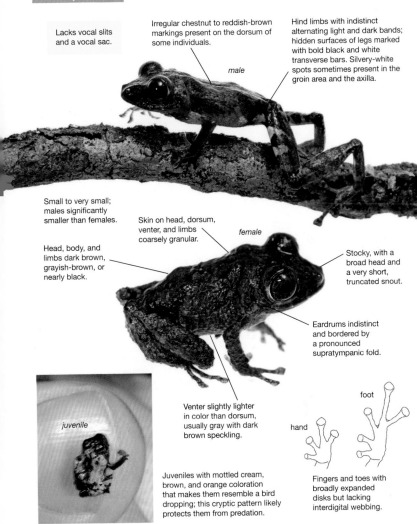

Lacks vocal slits and a vocal sac.

Irregular chestnut to reddish-brown markings present on the dorsum of some individuals.

male

Hind limbs with indistinct alternating light and dark bands; hidden surfaces of legs marked with bold black and white transverse bars. Silvery-white spots sometimes present in the groin area and the axilla.

Small to very small; males significantly smaller than females.

Skin on head, dorsum, venter, and limbs coarsely granular.

female

Head, body, and limbs dark brown, grayish-brown, or nearly black.

Stocky, with a broad head and a very short, truncated snout.

Eardrums indistinct and bordered by a pronounced supratympanic fold.

juvenile

Venter slightly lighter in color than dorsum, usually gray with dark brown speckling.

Juveniles with mottled cream, brown, and orange coloration that makes them resemble a bird dropping; this cryptic pattern likely protects them from predation.

foot

hand

Fingers and toes with broadly expanded disks but lacking interdigital webbing.

P. altae (p. 491)

Has coral red to orange spots in the groin and on the anterior thigh surfaces (*P. pardalis* has silvery-white markings). No known range overlap.

Pristimantis ridens
Pygmy Rain Frog

Least concern

This tiny frog shows great variation in color and pattern. Thighs, calves, and feet often marked with red; a brown spot covers the top of each eardrum. It lacks finger and toe webbing and has rounded disks on all digits. Nostrils conspicuously raised and snout ends in a fleshy bump; when observed from above, the snout looks as though it has three lobes. Males to 0.79 in (20 mm) in standard length, females to 1.10 in (28 mm).

Ecoregions
1 2 3 4 5

Pristimantis ridens ranges from northern Honduras to western Colombia. In Costa Rica it is widespread on the Atlantic slope and on the central and southern Pacific slope; from near sea level to 5,250 ft (1,600 m).

Natural history

This abundant species occurs in a variety of habitats, including humid low elevation forests, montane forests, secondary growth, and plantations and gardens.

During the day, individuals are found in the leaf litter or hidden in moss mats and arboreal bromeliads up to 100 ft (30 m) above the ground. However, this species is most frequently seen at dusk or at night, when it actively forages on low vegetation. Males call nearly year-round, though activity is most intense during dry spells. The increased thickness of the leaf litter layer during drier months may boost the availability of suitable habitat and create optimal conditions for oviposition. Males produce a chuckling trill sound, reminiscent of a rattling high-pitched laugh (*ridens* means laughing). Mated pairs engage in axillary amplexus and deposit eggs in a moist place in the leaf litter or other suitable site. Larvae undergo direct development, and the young hatch as fully developed tiny frogs.

P. ridens feeds on tiny invertebrates. In turn, these small frogs fall prey to a variety of vertebrate and invertebrate predators, including katydids, spiders, and the large bullet ant (*Paraponera clavata*). The small size, relative abundance, and high nutritional value for their size make these froglets an important component of the rainforest's food web. In spite of the high threat of predation to these frogs, their Costa Rican populations seem stable.

Description

male

male

male

Dorsum with a variety of markings that can include a series of thin longitudinal stripes, a pair of broad irregular dorsolateral stripes, or a single, light middorsal pinstripe. Often a W-shaped series of isolated tubercles present in the shoulder region, behind head, with a tubercle marking each point of the letter.

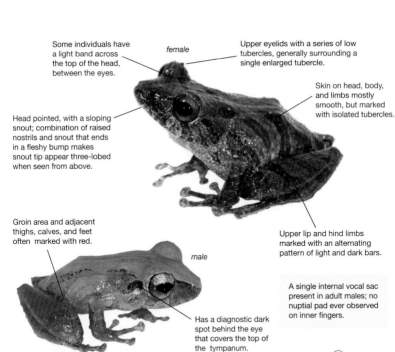

Some individuals have a light band across the top of the head, between the eyes.

female

Upper eyelids with a series of low tubercles, generally surrounding a single enlarged tubercle.

Skin on head, body, and limbs mostly smooth, but marked with isolated tubercles.

Head pointed, with a sloping snout; combination of raised nostrils and snout that ends in a fleshy bump makes snout tip appear three-lobed when seen from above.

Groin area and adjacent thighs, calves, and feet often marked with red.

male

Upper lip and hind limbs marked with an alternating pattern of light and dark bars.

A single internal vocal sac present in adult males; no nuptial pad ever observed on inner fingers.

Has a diagnostic dark spot behind the eye that covers the top of the tympanum.

female

male

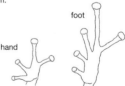

hand

foot

Hands and feet without webbing. Fingers with greatly enlarged, rounded disks on three outer fingers; inner finger with a small disk. Toes with enlarged, rounded disks; innermost toe has a significantly smaller disk.

Ventral surfaces granular. Venter cream to pinkish-white, marked with fine dark brown specks. Adult males often with yellow on the venter.

Similar species

P. taeniatus
(p. 508)

Lacks enlarged tubercle on top of each eyelid; also lacks red coloration in the groin.

P. caryophyllaceus
(p. 493)

Has a long pointed snout and a well-developed heel tubercle. Lacks red coloration on the legs and enlarged tubercles on the upper eyelid.

P. cerasinus
(p. 495)

Lacks enlarged tubercles on the upper eyelid and has a prominent row of pointed tubercles along the trailing edge of its lower leg.

Eleutherodactylus coqui
(p. 329)

Has very distinctly demarcated eardrums, bordered by an obvious tympanic ring (absent in *P. ridens*), a greatly enlarged vocal sac, and a very different call.

507

Pristimantis taeniatus
Banded Rain Frog

Least concern

A fairly small, cryptic frog with broadly expanded finger and toe disks, a thin W-shaped skin fold on the back of the head, and a distinct dark brown blotch covering the supratympanic skin fold and the top half of each eardrum. Its upper lip and hind limbs are marked with a pattern of alternating dark and light bars. Males to 1.06 in (27 mm) in standard length, females to 1.46 in (37 mm).

Ecoregions
1 2 3 4 5

The distribution of *Pristimantis taeniatus* is not well understood. It occurs in disjunct populations in central Panama and is widespread in Colombia west of the Andes, from near sea level to 4,600 ft (1,400 m). It likely occurs in other areas between these currently known populations and was recently reported from southern Costa Rica.

Natural history

The biology of *P. taeniatus* has not been thoroughly studied. It inhabits the forest interior, where it is found in areas of dense leaf litter. When motionless, these cryptic brown frogs blend in perfectly, only becoming visible when they launch themselves from underfoot.

This seemingly adaptable species is commonly encountered in older-growth secondary forest, shaded plantations, and also in relatively undisturbed forest habitat. It is reportedly common in Colombia and is regularly encountered in eastern Panama; a recent addition to the herpetofauna of Costa Rica, no information exists on its population status or distribution range in the country; observations of this species should be reported.

Description

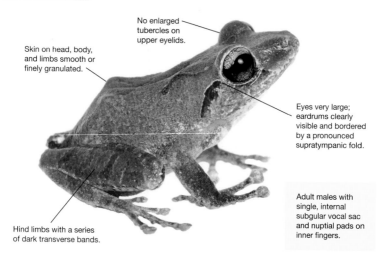

No enlarged tubercles on upper eyelids.

Skin on head, body, and limbs smooth or finely granulated.

Eyes very large; eardrums clearly visible and bordered by a pronounced supratympanic fold.

Adult males with single, internal subgular vocal sac and nuptial pads on inner fingers.

Hind limbs with a series of dark transverse bands.

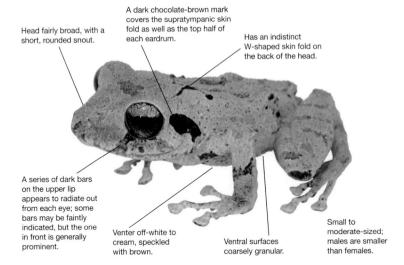

Head fairly broad, with a short, rounded snout.

A dark chocolate-brown mark covers the supratympanic skin fold as well as the top half of each eardrum.

Has an indistinct W-shaped skin fold on the back of the head.

A series of dark bars on the upper lip appears to radiate out from each eye; some bars may be faintly indicated, but the one in front is generally prominent.

Venter off-white to cream, speckled with brown.

Ventral surfaces coarsely granular.

Small to moderate-sized; males are smaller than females.

Coloration variable; generally dorsal surfaces with shades of tan, brown, or reddish-brown. Dorsum usually marked with an irregular pattern of small dark blotches. Often with W-shaped or V-shaped dark markings.

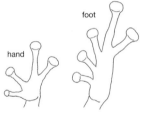

Hands lack webbing, fingers with broadly expanded disks (inner fingers with smaller disks than fingers II-IV). Disks broader than long. Finger I shorter than finger II. Feet without interdigital webbing; toe disks equal in size and shape to largest finger disks.

Similar species

P. ridens (p. 506)

Has an enlarged tubercle and several smaller ones on each upper eyelid; has red coloration in the groin.

P. cerasinus (p. 495)

Has a conspicuous row of pointed tubercles along the trailing edge of the lower leg and usually red coloration in the groin area.

Genus *Strabomantis*

These robust, broad-headed riparian frogs were previously included in the genus *Eleutherodactylus* (family Leptodactylidae), in the *Eleutherodactylus bufoniformis*-species group. A revision of this exceedingly diverse genus led to the placement of the Central American species in the new family Craugastoridae. Subsequently, the 16 toadlike frogs of the former *Eleutherodactylus bufoniformis*-species group were removed from the family Craugastoridae, placed in the newly formed family Strabomantidae, and assigned to their own genus, *Strabomantis*.

    Strabomantis bufoniformis is the sole species in this genus that occurs in Costa Rica. Due to its convoluted taxonomic history, it has been included in recent literature as *Eleutherodactylus bufoniformis*, *Craugastor bufoniformis*, and *Strabomantis bufoniformis*.

Strabomantis bufoniformis
Rusty Robber Frog

Least concern

This is a large, toadlike species with a very broad head; disti nctly tuberculate skin; toe and finger disks that are barely expanded or not expanded; and toes with only basal webbing and fingers with no webbing. Males to 2.32 in (59 mm) in standard length, females to 3.70 in (94 mm).

Ecoregions
1 2 3 4 5

On the Atlantic slope, *Strabomantis bufoniformis* ranges from extreme southeastern Costa Rica to western Colombia. From sea level to 165 ft (50 m).

Natural history

This is a nocturnal species that is mostly found along streams in humid lowland forest types. It reproduces through direct development and has no free-swimming tadpoles. The call of *S. bufoniformis* has been described as a series of low, short barks followed by even shorter knocklike notes.

Individuals are generally found in undisturbed forests, which suggests that this species is not tolerant of habitat alteration. Known Costa Rican localites are within national parks, with relatively secure habitat.

S. bufoniformis is an uncommon species in the northern parts of its range and is only known from a few localities in extreme southern Limón Province, Costa Rica. In spite of repeated searches, it has not been found in Costa Rica since 1978. However, this species is still regularly encountered in Panama and Colombia and may eventually turn up in appropriate habitat in Costa Rica.

The unique color pattern of this species' irises provides a good field mark.

Adult males with single, internal vocal sac in throat.

A robust frog, with a very broad head. Width of head equals or exceeds 50% of the standard length.

Head, body, and limbs with pointed tubercles that are each tipped with cream or white glandular tissue.

Iris distinctly dark bronze to brown, with several light rays radiating outward from center.

Lips boldly barred.

Fingers lack expanded disks and webbing; toes with slightly expanded disks and basal webbing.

Head bears bony, cranial crests that may be obscured by the extremely tuberculate condition of the skin.

Cryptically colored, mottled tan to brown.

Eardrums small and distinct, but somewhat obscured by rugose skin.

Limbs short and robust, with moderate-sized hands and feet.

Undersurfaces cream to white; individuals sometimes with brown "leopard" spots (with pale centers) on flanks.

Limbs with pattern of alternating light and dark brown transverse bands.

Rhinella marina
(p. 188)

Because of its toadlike appearance, *S. bufoniformis* could be confused with *Rhinella marina*, but it lacks paired parotoid glands.

Craugastor megacephalus
(p. 255)

Has pronounced dorsal and cranial crests and lacks toe webbing; range may overlap with *S. bufoniformis*.

Craugastor gulosus
(p. 253)

Has cranial crests and relatively smooth skin; does not co-occur.

Craugastor rugosus
(p. 280)

Has cranial and dorsal crests and lacks toe webbing; does not co-occur.

Glossary

advertisement call. The species-specific vocalization of male frogs used to attract females and establish territorial boundaries.

alkaloid. An active chemical compound found in many animal toxins.

amplexus. The sexual embrace of frogs and toads, in which the male holds on to the female with his arms. In **axillary amplexus**, the male grabs the female under the armpits. In **inguinal amplexus**, the male grabs the female around the waist.

annular folds. Skin folds that encircle the body of all caecilians.

anterior. *adj.* Pertaining to the front (head) end of an animal's body.

autotonomy. The ability to voluntarily break off part or all of the tail, an anti-predator strategy found in many lizards. In **pseudoautotonomy**, the tail breaks off only when it is restrained by a predator. This ability is seen in salamanders and some snakes.

axillary amplexus. See **amplexus**.

Batrachochytrium dendrobatidis. A species of chytrid fungal pathogen that has been implicated in the global decline of amphibian populations.

Bd. *abbreviation.* Refers to the potentially deadly amphibian pathogen *Batrachochytrium dendrobatidis*.

bufotoxin. Any toxic secretion produced by toads.

caecilian. An order of wormlike, limbless amphibians.

canopy. The ecosystem formed by the overhanging branches of the trees in a forest.

canthal ridge. The area between the snout tip and the eye that separates the top of the head from the sides of the snout.

canthus rostralis. See **canthal ridge**.

chemoreception. A faculty of sense perception widespread in amphibians and reptiles.

chytrid fungus. See ***Batrachochytrium dendrobatidis***.

chytridiomycosis. The potentially lethal disease caused by the chytrid fungus *Batrachochytrium dendrobatidis*.

cloaca. A body cavity where both the urogenital and digestive tracts terminate.

cloud forest. A high elevation forest characterized by cool temperatures and high annual rainfall.

co-ossified. Describes a morphological feature in which skin is fused to underlying bone.

costal fold. Skin fold between the ribs of salamanders.

cranial. Referring to the cranium, or skull.

cryptic. *adj.* Serving to conceal.

direct development. An amphibian reproductive strategy; larval development takes place entirely within the confines of the egg and the young hatch after metamorphosis. There is no free-swimming larval stage.

dorsal. *adj.* Pertaining to the back of an animal.

dorsal stripe. A stripe running along the back of an animal. A **dorsolateral stripe** runs between the back and the sides of the animal. A **lateral** stripe runs along the sides of the animal.

dorsolateral stripe. See **dorsal stripe**.

ectothermous. *adj.* Referring to animals that rely on external sources of heat to maintain their body temperature. All amphibians and reptiles are ectothermous. Endothermous animals generate body heat through the internal process of metabolism.

edge situation. The boundary between two habitats.

endemic. A species that only inhabits a restricted geographical area.

epiphyte. Any plant that obtains nutrients and water from the air instead of the soil. Epiphytic plants usually grow on other plants.

fossorial. *adj.* Having burrowing, secretive habits.

glandular skin. Skin containing glands.

granular skin. Skin covered with grainy bumps.

herpetofauna. The group of fauna that consists of amphibians and reptiles.

herpetology. The study of amphibians and reptiles.

humeral hook. a bony projection on the upper arm (humerus).

inguinal amplexus. See **amplexus**.

keratin. A hard, hornlike material that is the main component of nails and the top layer of turtle shells. Keratin is also found in human nails and hair.

keratinized. Hardened, covered with **keratin**.

lateral. *adj.* Pertaining to either side of an animal.

lateral stripe. See dorsal stripe.

larva (pl. larvae). The immature stage of an animal prior to metamorphosis. Frog and toad larvae are commonly referred to as tadpoles.

leaf axil. The point on the main stem of a plant where a leaf or leaf stem joins it.

leaf litter. The accumulated dead leaves that cover the forest floor.

lichenous. *adj.* Displaying a lichenlike pattern.

longitudinal. *adj.* Running parallel to the axis of the body. Compare with **transverse**.

metamorphosis. The transformation from larva to adult undergone by salamanders, frogs, and toads.

neotropics. The New World tropics (found in Central and northern South America).

oviparous. adj. Egg-laying. **viviparous.** *adj.* Live-bearing. **ovoviviparous.** *adj.* Producing eggs that are incubated inside the mother's body. The young hatch and leave the female's body as fully formed juveniles.

oviposition. The act of laying eggs.

ovoviviparous. See **oviparous**.

paramó. *Spanish.* A high elevation habitat with shrublike, alpine vegetation.

phallodeum. The copulatory organ of male caecilians.

pheromone trail. The trail of volatile chemical compounds left by some animals.

posterior. *adj.* Pertaining to the rear (tail) end of an animal's body.

premontane. *adj.* Pertaining to foothills.

prepollex. A bony projection extending from the base of the inner finger (thumb) in some amphibians; may end in a **prepollical spine**.

prepollical spine. See **prepollex**.

primary forest. Undisturbed, old-growth forest. Compare **secondary forest**.

pseudoautonomy. See **autotonomy**.

pustule. A large wartlike structure.

rainforest. Any forest that receives more than 98 in (2,500 mm) of annual rainfall.

reticulum. A weblike pattern. A term used to describe the eye or skin pattern of some amphibians and reptiles.

secondary forest. Partially logged or otherwise disturbed forest, in some stage of regeneration. Compare **primary forest**.

species group. An informal grouping of similar species.

speciose. *adj.* Containing a large number of species.

tarsus. One of four segments of the hind limb of frogs and toads. The other segments are the thigh, lower leg, and foot.

taxon (pl. taxa). Generic name for a taxonomic grouping such as species or genus.

tetrapod. Any animal with four limbs.

transverse. *adj.* Running perpendicular to the axis of the body. Compare with **longitudinal**.

tubercle. Any wartlike protuberance.

understory. The vegetation growing below a forest's canopy.

vent. The opening that allows products from the urogenital and reproductive systems to leave the cloaca (and the body).

ventral. adj. (n. venter). Pertaining to the underside of an animal, the belly.

versant. The slope or drainage system of a mountain range.

vertebral. *adj*. Located on the spine.

viviparous. See **oviparous**.

vocal sac. A balloonlike, inflatable structure, located on the floor of the mouth in most male frogs. During calling, the vocal sac functions as a resonating chamber that expands from the throat or, in a few species, from the corners of the mouth.

Bibliography

Those with a keen interest in neotropical amphibians can find information beyond the scope of this field guide in a variety of academic resources. A great place to start is Savage (2002), an enormous tome on the herpetofauna of Costa Rica, filled cover to cover with fascinating information. Another good starting point is AmphibiaWeb (www.amphibiaweb.org), which provides regular updates on changes in taxonomy and biology. Linked with AmphibiaWeb is BerkeleyMapper, which is a good resource for distribution maps. The gold standard for conservation status designations is the International Union for the Conservation of Nature (IUCN) Red List, which is accessible online at www.iucnredlist.org and is regularly updated.

Acosta-Chaves, V., & G. Granados (2015). Nature notes: *Smilisca sordida*. Predation by a Tropical Screech Owl (*Megascops choliba*). Mesoamerican Herpetology 2(1):105-106.

Acosta,V., & O. Morún. 2014. Predation of *Craugastor podiciferus* (Anura: Craugastoridae) by *Catharus frantzii* (Passeriformes: Turdidae) in a Neotropical cloud forest. Bol. Assoc. Herpetol. Española 25: 1–2.

Acosta-Chaves, V., G. Chaves, J.G. Abarca, A. García-Rodríguez & F. Bolaños (2015). A checklist of the amphibians and reptiles of Río Macho Biological Station, Provincia de Cartago, Costa Rica. Check list 11(6):1784.

Arias, E., G. Chaves, A.J. Crawford & G. Parra-Olea (2016). A new species of the *Craugastor podiciferus* species group (Anura: Craugastoridae) from the premontane forest of southwestern Costa Rica. Zootaxa 4132(3):347-363.

Batista, A., A. Hertz, K. Mebert, G. Köhler, S. Lotzkat, M. Ponce & M. Vesely (2014). Two new fringe-limbed frogs of the genus *Ecnomiohyla* (Anura: Hylidae) from Panama. Zootaxa 3826(3):449-474.

Blaustein, A.R., P.D. Hoffman, D.G. Hokit, J.M. Kiesecker, S.C. Walls & J.B. Hays (1994). UV repair and resistance to solar UV-B in amphibian eggs: A link to population declines? Proc. Nat. Acad. Sci. 91:1791-1795.

Bolaños, F., & D.B. Wake (2009). Two new species of montane web-footed salamanders (Plethodontidae: *Bolitoglossa*) from the Costa Rica-Panamá border region. Zootaxa 1981: 57-68.

Boza-Oviedo, E., S.M. Rovito, G. Chaves, A. García-Rodríguez, L.G. Artavia, F. Bolaños & D. Wake (2012). Salamanders from the eastern Cordillera de Talamanca, Costa Rica, with descriptions of five new species (Plethodontidae: *Bolitoglossa*, *Nototriton*, and *Oedipina*) and natural history notes from recent expeditions. Zootaxa 3309:36-61.

Brame, A.H. (1968). systematics and evolution of the Mesoamerican salamander genus *Oedipina*. J. Herpetol. 2:1-64.

Brame, A.H., & W.E. Duellman (1970). A new salamander (Genus *Oedipina*) of the *uniformis* group from western Panama. Los Angeles County Mus. Contr. Sci. 201:1-8.

Brodie, E.D., Jr. & P.K. Ducey (1991). Antipredator skin secretions of some tropical salamanders (*Bolitoglossa*) are toxic to snake predators. Biotropica 23(1):58-62.

Brown, C.W. (1968). Additional observations on the function of the nasolabial grooves of plethodontid salamanders. Copeia 1968(4): 728-731.

Brust, D.G. (1993). Maternal brood care by *Dendrobates pumilio*: A frog that feeds its young. J. Herpetol. 27(1):96-98.

Bunnell, P. (1973). Vocalizations in the territorial behavior of the frog *Dendrobates pumilio*. Copeia 1973:277-284.

Caldwell, J.P. (1996). The evolution of myrmecophagy and its correlates in poison frogs (Family Dendrobatidae). J. Zool. 240:75-101.

Campbell, J.A., & J.M. Savage (2000). Taxonomic reconsideration of Middle American frogs of the *Eleutherodactylus rugulosus* group (Anura: Lepodactylidae):

a reconnaissance of subtle nuances among frogs. Herpetological Monographs 14:186-292.

Chaves, G., F. Bolaños, J.E. Rodriguez & Y. Matamoros (eds.) 2014. Actualización de las Listas Rojas nacionales de Costa Rica. Anfibios y reptiles. Taller Escuela de Biología, San Pedro, Costa Rica. 50 pp.

Chaves, G., A. García-Rodríguez, A. Mora & A. Leal (2009). A new species of dink frog (Anura: Eleutherodactylidae: *Diasporus*) from Cordillera de Talamanca, Costa Rica. Zootaxa 2088:1-14.

Chaves, G., H. Zumbado-Ulate, A. García-Rodríguez, E. Goméz, V.T. Vredenburg & M. Ryan (2014). Rediscovery of the critically endangered streamside frog, *Craugastor taurus* (Craugastoridae), in Costa Rica. Trop. Cons. Sci. 7(4): 628-638.

Cocroft, R.B., R.W. McDiarmid, A.P. Jaslow & P.M. Ruiz-Carranza (1990). Vocalizations of eight species of *Atelopus* (Anura: Bufonidae) with comments on communications in the genus. Copeia 1990(3):631-643.

Crump, M.L. (1986). Homing and site fidelity in a Neotropical frog, *Atelopus varius* (Bufonidae). Copeia 1986(2):438-444.

Crump, M.L. (1988). Aggression in harlequin frogs: male-male competition and a possible conflict of interest between the sexes. Anim. Behav. 36:1064-1077.

Crump, M.L., & R.H. Kaplan (1979). Clutch energy partitioning of tropical treefrogs. Copeia 1979:626-635.

Crump, M.L., & J.A. Pounds (1989). Temporal variation in the dispersal of a tropical anuran. Copeia 1989(1):209-211.

Donnelly, M.A. (1994). Amphibian diversity and natural history. In: McDade, L.A., K.S. Bawa, H.A. Hespenheide, & G.S. Hartshorn (eds.) La Selva. Ecology and natural history of a Neotropical rain forest. Pp. 199-209. Univ. Chicago Press, Chicago.

Duellman, W.E. (1967a). Courtship isolating mechanisms in Costa Rican hylid frogs. Herpetologica, 23: 169-183

Duellman, W.E. (1967b). Social organization in the mating calls of some Neotropical anurans. Am. Midl. Nat. 77:156-163.

Duellman, W.E. (1970). The Hylid Frogs of Middle America. Monogr. Mus. Nat. Hist., Univ. Kansas, No. 1, 753 pp., 2 vols.

Duellman, W.E. (2001). The Hylid Frogs of Middle America. New. Ed., Ithaca, NY. Society for the Study of Amphibians and Reptiles.1159 pp., 2 vols.

Duellman, W.E., & L. Trueb (1986). Biology of Amphibians. McGraw Hill, xvii + 670 pp.

Dunn, E.R. (1941). Notes on *Dendrobates auratus*. Copeia 1941(2):88-93.

Eaton, T.H., Jr. (1941). Notes on the life history of *Dendrobates auratus*. Copeia 1941(2):93-95.

Eisenberg, T., & T. Leenders (2007). *Eleutherodactylus ridens* (Pygmy Rainfrog). Predation. Herpetol. Rev. 38(3):323.

Fabing, H.D., & J.R. Hawkins (1956). Intravenous injection of bufotenine in the human being. Science 123:886-887.

Fouquette, M.J., Jr. (1960). Isolating mechanisms in three sympatric treefrogs in the Canal Zone. Evolution 14:484-497.

Franzen, M. (1987). Froschlurche aus dem Monteverde-Nebelwaldreservat, Costa Rica. Herpetofauna 9(48):18-23.

Frost, D., T. Grant, J. Faivovich, R.H. Bain, A. Haas, C.F.B. Haddad, R.O. de Sá. A. Channing, M. Wilkinson, S. Donnellan, C.J. Raxworthy, J.A. Campbell, B.L. Blotto, P. Moler, R.C. Drewes, R.A. Nussbaum, J.D. Lynch, D.M. Green & W.C. Wheeler (2006). The amphibian tree of life. Bull. Am. Mus. Nat. Hist. 297:1-370.

Fund, W. (2012). Ecoregions of Costa Rica. Retrieved from http://www.eoearth.org/view/article/180305

García-Rodríguez, A., E. Arias & G. Chaves (2016). Multiple lines of evidence support the species status of the poorly known *Diasporus tigrillo* and the recently described *Diasporus citrinobapheus* (Anura: Eleutherodactylidae). Neotropical Biodiversity 2(1):59-68.

García-Rodríguez, A., G. Chaves, C. Benavides-Varela & R. Puschendorf (2012). Where are the survivors? Tracking relictual populations of endangered frogs in Costa Rica. Diversity Distrib. 18:204-212.

Grant, T., D.F. Frost, J.P. Caldwell, R. Gagliardo, C.F.B. Haddad, P.J.R. Kok, C.B. Means, B.P. Noonan, W.E. Schargel & W.C. Wheeler (2006). Phylogenetic systematics of dart-poison frogs and their relatives (Amphibia: Athesphatanura: Dendrobatidae). Bull. Am. Mus. Nat. Hist. 299:1-79.

Gray, A.R. & A.W. Bland (2016). Notes on the reproduction of the endemic Costa Rican toad, *Incilius chompipe* (Anura: Bufonidae). Mesoamerican Herpetology 3(2):464-467.

Gregory, P.T. (1983). Habitat structure affects diel activity pattern in the neotropical frog *Leptodactylus melanonotus*. J. Herp. 17(2):179-181.

Guayasamin, J.M., S. Castroviejo-Fisher, L. Trueb, J. Ayarzagüena, M. Rada & C. Vilà (2009). Phylogenetic systematics of glassfrogs (Amphibia: Centrolenidae) and their sister taxon *Allophryne ruthveni*. Zootaxa 2100:1-97.

Guyer, C. (1994). The reptile fauna: diversity and ecology. In: McDade, L.A., K.S. Bawa, H.A. Hespenheide, & G.S. Hartshorn (eds.) La Selva. Ecology and natural history of a Neotropical rain forest. Pp. 210-216. Univ. Chicago Press, Chicago.

Guyer, C., & M.A. Donnelly (2005). Amphibians and reptiles of La Selva, Costa Rica, and the Caribbean slope: a comprehensive guide. University of California Press, Berkeley, CA. 299 pp.

Hayes, M.P. (1991). A study of clutch attendance in the Neotropical frog *Centrolenella fleischmanni* (Anura: Centrolenidae). Ph.D. dissertation, Univ. of Miami, Miami. 239 pp.

Hayes, M.P., J.A. Pounds & W.W. Timmerman (1989). An annotated list and guide to the amphibians and reptiles of Monteverde, Costa Rica. Herpetological Circular # 17. Society for the Study of Amphibians and Reptiles. University of Texas, Tyler. 67 pp.

Hedges, S.B., W.E. Duellman & M.P. Heinicke (2008). New World direct-developing frogs (Anura: Terrarana): Molecular phylogeny, classification, biogeography, and conservation. Zootaxa 1737:1-182.

Hedman, H. D., & M. C. Hughey. 2015. Body size, humeral spine size, and aggressive interactions in the Emerald Glass Frog, *Espadarana prosoblepon* (Anura: Centrolenidae) in Costa Rica.

Mesoamerican Herpetology 1(1): 500–508.

Hernández-Cuadrado, E.E., & M.H. Bernal (2009). *Engystomops pustulosus* (Tungara Frog) and *Hypsiboas crepitans* (Colombian Tree Frog). Predation on Anuran embryos. Herpetol. Rev. 40(4):431-2.

Hertz, A. S., Lotzkat, A. Carrizo, M. Ponce, G. Köhler & B. Streit (2012). Field notes on findings of threatened amphibian species in the central mountain range of western Panama. Amphib. Reptile Conserv. 6(2):10-30.

Heyer, W.R. (1967). A herpetofaunal study of an ecological transect through the Cordillera de Tilarán, Costa Rica. Copeia 1967(2):259-271.

Heyer, W.R. (1970). Studies on the genus *Leptodactylus* (Amphibia: Leptodactylidae). II. Diagnosis and distribution of the Leptodactylus of Costa Rica. Rev. Biol. Trop. 16(2):171-205.

Hillis, D.M., & R. de Sa (1988). Phylogeny and taxonomy of the *Rana palmipes* group (Salientia: Ranidae). Herpetol. Monogr. 2:1-26.

Huertas, J.A., & A. Solórzano (2014). Nature notes: *Micrurus alleni*. Predation. Mesoamerican Herpetology 1(1):160-161.

Ibañez, D.R., & E.R. Smith (1995). Systematic status of *Colostethus flotator* and *C. nubicola* (Anura: Dendrobatidae) in Panama. Copeia 1995(2):446-456.

Jacobson, S.K., & J.J. Vandenberg (1991). Reproductive ecology of the endangered Golden Toad (*Bufo periglenes*). J. Herpetol. 25(3):321-327.

Janzen, D.H. (1962). Injury caused by toxic secretions of *Phrynohyas spilomma*. Copeia 1962:651.

Jaramillo, F.E., C.A. Jaramillo & D.R. Ibanez (1997). The tadpole of *Hyalinobatrachium colymbiphyllum* (Anura: Centrolenidae). Rev. Biol. Trop. 45(2): 867-870.

Jaslow, A.P., & E.E. Lombard (1996). Hearing in the neotropical frog, *Atelopus chiriquiensis*. Copeia 1996:428-432.

Jiménez, C.E. (1994). Utilization of *Puya dasylirioides* (Bromeliaceae: Pitcairnoidea) as foraging site by *Bolitoglossa subpalmata* (Plethodontidae: Bolitoglossinii). Rev. Biol. Trop. 42(3):703-710.

Jörgens, D. (1994). Zur kenntnis der gelb-oliven Farbvarietät von *Dendrobates granuliferus* Taylor, 1958. Sauria 16(4):15-17.

Jungfer, K.-H. (1987). Beobachtungen an *Ololygon boulengeri* (Cope, 1887) und anderen "Knickzehenlaubfröschen." Herpetofauna 9(46):6-12.

Jungfer, K.-H. (1988). Froschlurche von Fortuna, Panama. 1. Microhylidae, Ranidae, Bufonidae, Hylidae (1). Herpetofauna 10(54):25-34.

Jungfer, K.-H. (1996). Reproduction and parental care of the coronated treefrog, *Anotheca spinosa* (Steindachner, 1864) (Anura: Hylidae). Herpetologica 52(1):25-32.

Kim, Y.H., G.B. Brown, H.S. Mosher & F.A. Fuhrman (1975). Tetrodotoxin: occurrence in atelopid frogs of Costa Rica. Science 189:151-152.

Kluge, A.G. (1981). The life history, social organization and parental behavior of *Hyla rosenbergi* Boulenger, a nest-building gladiator frog. Misc. Publ. Mus. Zool. Univ. Michigan, 160:1-170.

Köhler, G. (2011). Amphibians of Central America. Herpeton Verlag.Offenbach.379 pp.

Köhler, G., & J. Sunyer (2004). A new species of rain frog (Genus *Craugastor*) of the *fitzingeri* group from Río San Juan, southeastern Nicaragua (Amphibia, Anura, Leptodactylidae). Senckenbergiana Biol. 86(2):261-266.

Kolby, J.E., S.D. Ramirez, L. Berger, D.W. Griffin, M. Jocque & L. Skerratt. (2015). Presence of amphibian chytrid fungus (*Batrachochytrium dendrobatidis*) in rainwater suggests aerial dispersal is possible. Aerobiologia

Kubicki, B. (2007). Ranas de vidrio Costa Rica glass frogs. INBio. Costa Rica.312 pp.

Kubicki, B., & S. Salazar (2015). Discovery of the golden-eyed fringe-limbed treefrog, *Ecnomiohyla bailarina* (Anura: Hylidae), in the Caribbean foothills of southeastern Costa Rica. Mesoamerican Herpetology 2(1):76-86.

Kubicki, B., S. Salazar & R. Puschendorf (2015) A new glassfrog, genus *Hyalinobatrachium* (Anura: Centrolenidae), from the Caribbean foothills of Costa Rica. Zootaxa 3920(1):069-084.

Lahanas, P.J., & J.M. Savage (1992). A new species of Caecilian from the Península de Osa of Costa Rica. Copeia 1992(3):703-708.

Leenders, T. (2001). A guide to amphibians and reptiles of Costa Rica. Distribuidores Zona Tropical S.A., San José, Costa Rica. 305 pp.

Leenders, T.A.A.M., & G.J. Watkins-Colwell (2003). Morphological and behavioral adaptations in *Bolitoglossa colonnea* (Caudata, Plethodontidae) in relation to habitat use and daily activity cycle. Phyllomedusa 2(2): 101-104.

Leenders, T.A.A.M., & G.J. Watkins-Colwell (2004). *Gymnopis multiplicata*. Geographic Distribution. Herpetol. Rev. 35(2):185.

Licht, L.E. (1968). Death following possible ingestion of toad eggs. Toxicon 5:141-142.

Licht, L.E., & B. Low (1968). Cardiac response of snakes after ingestion of toad paratoid venom. Copeia 1968(3):547-551.

Lips, K.R. (1993a). Geographical distribution: *Bolitoglossa compacta*. Herp. Rev. 24(3):107.

Lips, K.R. (1993b). Geographical distribution: *Bolitoglossa minutula*. Herp. Rev. 24(3):107.

Lips, K.R. (1993c). Geographical distribution: *Bolitoglossa nigrescens*. Herp. Rev. 24(3):107.

Lips, K.R. (1993d). Geographical distribution: *Oedipina grandis*. Herp. Rev. 24(3):107.

Lips, K.R. (1996). New treefrog from the Cordillera de Talamanca of Central America with a discussion of systematic relationships in the *Hyla lancasteri* group. Copeia 1996(3):615-626.

Lips, K.R., & J.M. Savage (1996a). A new species of rainfrog, *Eleutherodactylus phasma* (Anura:Leptodactylidae), from montane Costa Rica. Proc. Biol. Soc. Washington 109(4):744-748.

Livezey, R.L. (1986). The eggs and tadpoles of *Bufo coniferus* Cope in Costa Rica. Rev. Biol. Trop. 34(2):221-224.

Lourenco, W.R. (1995). Neotropical frog *Leptodactylus pentadactylus* eats scorpions. Alytes 12:191-192.

Lynch, J.D. (1975). A review of the broad-headed Eleutherodactyline frogs of South

America (Leptodactylidae). Occ. Pap. Mus. Nat. Hist. Univ. Kansas, 38:1-46.

Lynch, J.D. (1980). Systematic status and distribution of some poorly known frogs of the genus *Eleutherodactylus* from the Chocoan lowlands of South America. Herpetologica 36(2):175-189.

Lynch, J.D., & C.W. Myers (1983). Frogs of the *fitzingeri* group of *Eleutherodactylus* in eastern Panamá and Chocoan South America. Bull. Am. Mus. Nat. Hist. 175:481-572.

McCaffery, R., C.L. Richards-Zawacki & K.R. Lips (2015). The demography of *Atelopus* decline: Harlequin frog survival and abundance in central Panama prior to and during a disease outbreak. Global Ecology and Conservation 4(2015)232-242.

McDiarmid, R.W., & K. Adler (1974). Notes on territorial and vocal behavior of Neotropical frogs of the genus *Centrolenella*. Herpetologica 30:75-78.

McDiarmid, R.W., & M.S. Foster (1975). Unusual sites for two Neotropical tadpoles. J. Herpetol. 9:264-265.

McVey, M.E., R.G. Zahary, D. Perry, & J. MacDougal (1981). Territoriality and Homing Behavior in the Poison Dart Frog (*Dendrobates pumilio*). Copeia 1981(1):1-8.

Miyamoto, M.M. (1982). Vertical habitat use by *Eleutherodactylus* frogs (Leptodactylidae) at two Costa Rican localities. Biotropica 14(2):141-144.

Miyamoto, M.M. (1983). Biochemical variation in the frog *Eleutherodactylus bransfordii*: geographical patterns and cryptic species. Syst. Zool. 32:43-51.

Myers, C.W., J.W. Daly, H.M. Garraffo, A. Wisnieski & J.F. Cover, Jr. (1995). Discovery of the Costa Rican poison frog *Dendrobates granuliferus* in sympatry with *Dendrobates pumilio*, and comments on taxonomic use of skin alkaloids. Amer. Mus. Novit. 3144:1-21.

Myers, C.W., & W.E. Duellman (1982). A new species of *Hyla* from Cerro Colorado, and other tree frog records and geographical notes from western panama. Amer. Mus. Novit. 2752:1-32.

Morris, M.R. (1991). Female choice of large males in the treefrog *Hyla ebraccata*. J. Zool. 223:371-378.

Nelson, C.E. (1962). Size and secondary sexual characters in the frog *Glossostoma aterrimum* Gunther. Transactions of the Kansas Academy of Science 65(1):87-88.

Nishida, K. (2006) Encounter with *Hyla angustilineata* Taylor, 1952. (Anura: Hylidae) in a cloud forest of Costa Rica. Brenesia 66:79-81.

Nori, J., & R. Loyola (2015). On the worrying fate of data deficient amphibians. PLoS ONE 10(5):e0125055.

Novak R.M., & D.C. Robinson (1975). Observations on the reproduction and ecology of the tropical montane toad, *Bufo holdridgei* in Costa Rica. Rev. Biol. Trop. 23:213-237.

Núñez Escalante, R., & C. Barrio-Amorós (2014). Nature notes. *Oscaecilia osae*. Predation and habitat. Mesoamerican Herpetol. 1(2):283-284.

Parra-Olea, G., M. García-París & D. Wake (2004). Molecular diversification of salamanders of the tropical American genus *Bolitoglossa* (Caudata: Plethodontidae) and its evolutionary and biogeographical implications. Biol. J. Linnean Soc. 81:325-346.

Pounds, J.A., & M.L. Crump (1987). Harlequin frogs along a tropical montane stream. Biotropica 19:306-309.

Pounds, J.A., & M.L. Crump (1994). Amphibian declines and climate disturbance: the case of the Golden Toad and the Harlequin Frog. Cons. Biol. 8(1):72-85.

Pounds, J.A., M.P.L. Fogden, J.M. Savage & G.C. Gorman (1997). Tests of null models for amphibian declines on a tropical mountain. Cons. Biol. 11(6):1307-1322.

Pounds, J.A., M.P.L. Fogden & J.H. Campbell (1999). Biological response to climate change on a tropical mountain. Nature 398:611-615.

Pröhl, H. (1997). Los anfibios de Hitoy Cerere, Costa Rica. Proyecto Namasöl, Cooperación Técnica Bilateral Holanda-Costa Rica. 66 pp.

Pyburn, W.F. (1970). Breeding behavior of the leaf-frogs *Phyllomedusa callidryas* and *Phyllomedusa dacnicolor* in Mexico. Copeia 1970(2):209-218.

Pyron, R.A., & J.J. Wiens (2011). A large-scale phylogeny of amphibia including over 2,800 species, and a revised classification of extant frogs, salamanders, and caecilians. Molec. Phylogen. Evol. 61:543-583.

Reid, F.A., T. Leenders, J. Zook & R. Dean (2010). The Wildlife of Costa Rica, A Field Guide. Zona Tropical/Comstock/Cornell University Press. 267 pp.

Roberts, W.E. (1994). Explosive breeding aggregations and parachuting in a Neotropical frog, *Agalychnis saltator* (Hylidae). J. Herpetol. 28(2):193-199.

Robinson, D.C. (1983a). *Rana palmipes*. In: Janzen, D.H. (ed.) Costa Rican Natural History. Pp. 415-416. Univ. Chicago Press, Chicago.

Ruiz-Carranza, P.M., & J. D. Lynch (1991). Ranas Centrolenidae de Colombia. I. Propuesta de una nueva clasificación genérica. Lozania 57:1-30.

Ryan, M.J., N.J. Blea, I.M. Latella & M.A. Kull (2010). *Leptodactylus savagei* (Smoky Jungle Frog) Antipredator defense. Herpetol. Rev. 41(3):337-8.

Ryan, M.J., J.M. Savage, K.R. Lips & T. Giermakowski (2010). A new species of the *Craugastor rugulosus* series (Anura: Craugastoridae) from west-central Panama. Copeia 2010(3):405-409.

Ryan, M.J., N.J. Scott, J.A. Cook, B. Willink, G. Chaves, F. Bolaños, A. García-Rodríguez, I.M. Latella & S.E. Koerner (2015). Too wet for frogs: changes in a tropical leaf litter community coincide with La Niña. Ecosphere 6(1):1-10.

Salazar, S. (2015). Redescubrimiento de la rana marsupial *Gastrotheca cornuta* (Anura: Hemiphractidae) en Costa Rica. Brenesia 83-83:81-82.

Sánchez Paniagua, K., & J. G. Abarca (2016). Nature Notes. Thanatosis in four poorly known toads of the genus *Incilius* (Amphibia: Anura) from the highlands of Costa Rica. Mesoamerican Herpetology 3(1): 135–140.

Sasa, M., & A. Solórzano (1995). The reptiles and amphibians of Santa Rosa National Park, Costa Rica, with comments about the herpetofauna of xerophytic areas. Herp. Nat. Hist. 3(2):113-126.

Savage, J.M. (1966). An extraordinary new toad (*Bufo*) from Costa Rica. Rev. Biol. Trop. 14(2):153-167.

Savage, J.M. (1968). The Dendrobatid frogs of Central America. Copeia 1968(4):745-776.

Savage, J.M. (1972). The harlequin frogs, genus *Atelopus*, of Costa Rica and Western Panama. Herpetologica 28(1):77-94.

Savage, J.M. (1974). On the leptodactylid frog called *Eleutherodactylus palmatus* (Boulenger) and the status of *Hylodes fitzingeri* O. Schmidt. Herpetologica 30:289-299.

Savage, J.M. (1986). Systematics and distribution of the Mexican and Central American rainfrogs of the *Eleutherodactylus gollmeri* group (Amphibia: Leptodactylidae). Fieldiana Zool. 33(1375):1-57.

Savage, J.M. (1997). A new species of rainfrog of the *Eleutherodactylus diastema* group from the Alta Talamanca region of Costa Rica. Amphibia-Reptilia 18(3):241-247.

Savage, J.M. (2002). The amphibians and reptiles of Costa Rica. A herpetofauna between two continents, between two seas. The University of Chicago Press. Chicago & London. 934 pp.

Savage, J.M., & F. Bolaños (2009). An enigmatic frog of the genus *Atelopus* (Family Bufonidae) from Parque Nacional Chirripó, Cordillera de Talamanca, Costa Rica. Rev. Bio. Trop. 57(1-2):381-386.

Savage, J.M., & S.B. Emerson (1970). Central American frogs allied to *Eleutherodactylus bransfordii* (Cope): A problem of polymorphism. Copeia 1970:623-644.

Savage, J.M., & W.R. Heyer (1969). The treefrogs of Costa Rica: diagnosis and distribution. Rev. Biol. Trop. 16:1-127.

Savage J.M., & A. Kluge (1961). Rediscovery of the strange Costa Rican toad *Crepidius epioticus*. Rev. Biol. Trop. 9:39-51.

Savage, J.M., & B. Kubicki (2010). A new species of fringe-limb tree frog, genus *Ecnomiohyla* (Anura: Hylidae), from the Atlantic slope of Costa Rica, Central America. Zootaxa 2719:21-34.

Savage, J.M., & P.H. Starrett (1967). A new Fringe-limbed Tree-Frog (Family

Centrolenidae) From Lower Central America. Copeia 1967(3):604-609.

Savage, J.M., & J. Villa (1986). Introduction to the herpetofauna of Costa Rica. Contributions to herpetology 3. Society for the Study of Reptiles and Amphibians, Oxford, Ohio. 207 pp.

Savage, J.M., & M.H. Wake (1972). Geographic variation and systematics of the Middle American caecilians, genera *Dermophis* and *Gymnopis*. Copeia 1972:680-695.

Schmidt, A. A., & G. Köhler (1996). Zur Biologie von *Bolitoglossa mexicana*: Freilandbeobachtungen, Pflege und Nachzucht. Salamandra 32(4):275-284.

Scott, N.J., Jr. & A. Starrett (1972). An unusual breeding aggregation of frogs, with notes on the ecology of *Agalychnis spurrelli* (Anura: Hylidae). Proc. So. Calif. Acad. Sci. (73):86-94.

Solórzano, A. (2014). Nature Notes. *Gymnopis multiplicata*. Size. Mesoamerican Herpetol. 1(2)281-282.

Starrett, P.H., & J.M. Savage (1973). The systematic status and distribution of Costa Rican glass-frogs, genus *Centrolenella* (family Centrolenidae), with description of a new species. Bull. So. Calif. Acad. Sci. 72:57-78.

Sunyer, J., S. Lotzkat, A. Hertz, D.B. Wake, B.M. Alemán, S.J. Robleto & G. Köhler (2008). Two new species of salamanders (genus *Bolitoglossa*) from southern Nicaragua (Amphibia, Caudata, Plethodontidae). Senckenbergiana Biol. 88(2):319-328.

Sunyer, J., J.T. Townsend, D.B. Wake, S.L. Travers, S.C. Gonzalez, L.A. Obando & A.Z. Quintana (2011). A new cryptic species of salamander, genus *Oedipina* (Caudata: Plethodontidae), from premontane elevations in northern Nicaragua, with comments on the systematic status of the Nicaraguan paratypes of *O. pseudouniformis* Brame, 1968. Breviora 526:1-16.

Sunyer, J., D.B. Wake, & L.A. Obando (2012). Distributional data for *Bolitoglossa* (Amphibia, Caudata, Plethodontidae) from Nicaragua and

Costa Rica. Herpetol. Rev. 43(4):560-564.

Taylor, E.H. (1952a). The salamanders and caecilians of Costa Rica.Univ. Kansas Sci. Bull. 34(12): 695-791.

Taylor, E.H. (1952b). A review of the frogs and toads of Costa Rica. Univ. Kansas Sci. Bull. 35:577-942.

Taylor, E.H. (1954). Additions to the known herpetological fauna of Costa Rica with comments on other species. No. I. Univ. Kansas Sci. Bull. 36:597-639

Taylor, E.H. (1955). Additions to the known herpetofauna of Costa Rica, with comments on other species. No. II. Univ. Kansas Sci. Bull. 37:499-575.

Tuttle, M.D., & M.J. Ryan (1981). Bat predation and the evolution of frog vocalizations in the tropics. Science 214:677-678

Valerio, D.C. (1971). Ability of some tropical tadpoles to survive without water. Copeia 1971: 364-365.

Vaughan, A., & J.R. Mendelson, III (2007). Taxonomy and ecology of the Central American toads of the genus *Crepidophryne* (Anura: Bufonidae). Copeia 2007(2):304-314.

Vial, 1968. The ecology of the tropical salamander *Bolitoglossa subpalmata*, in Costa Rica. Rev. Biol. Trop. 15:13-115.

Villa, J.D. (1977). A symbiotic relationship between frog (Amphibia, Anura, Centrolenidae) and fly larvae (Drosophilidae). J. Herpetol. 11:317-322.

Villa, J.D. (1984). Biology of a Neotropical glass frog, *Centrolenella fleischmanni* (Boettger), with special reference to its frogfly associates. Milw. Public Mus. Contrib. Biol. Geol. 55:1-60.

Vinton, K.W. (1951). Observations on the life history of *Leptodactylus pentadactylus*. Herpetologica 7:73-75.

Wake, D.B., J.M. Savage & J. Hanken (2007). Montane salamanders from the Costa Rica-Panamá border region, with descriptions of two new species of *Bolitoglossa*. Copeia 2007(3):556-565.

Wassersug, R. (1971). On the comparative palatability of some dry-season tadpoles from Costa Rica. Am. Midl. Nat. 86:101-109.

Whitfield, S., K.E. Bell, T. Philipp, M. Sasa, F. Bolaños, G. Chaves, J.M. Savage & M. Donnelly (2007). Amphibian and reptile declines over 35 years at La Selva, Costa Rica. PNAS 104(20):8352-8356

Wijngaarden, R., van & S. van Gool (1994). Site fidelity and territoriality in the dendrobatid frog *Dendrobates granuliferus*. Amphibia-Reptilia 15: 171-181.

Wilkinson, M., D. San Mauro, E. Sherratt & D.J. Gower (2011). A nine-family classification of caecilians (Amphibia: Gymnophiona). Zootaxa 2874:41-64.

Young, B. 1979. Arboreal movement and tadpole-carrying behavior of *Dendrobates pumilio* in northeastern Costa Rica. Biotropica 11:238-239

Zweifel, R.G. (1964a). Distribution and life history of the Central American frog, *Rana vibicaria*. Copeia, 1964:300-308.

Zumbado-Ulate, H., F. Bolaños, B. Willink & F. Soley-Guardia (2011). Population status and natural history notes on the critically endangered stream-dwelling frog *Craugastor ranoides* (Craugastoridae) in a Costa Rican tropical dry forest. Herpetol. Cons. Biol. 6(3):455-464.

Zweifel, R.G. (1964b). Life history of *Phrynohyas venulosa* in Panama. Copeia1964:201-208.

Photo Credits

All illustrations are by Twan Leenders. All photographs except for the following are by Twan Leenders:

Victor Acosta Chaves (26t, 42r, 49, 78, 79, 168, 210, 235ID-cr, 266ID-c, 271ID-l, 273ID-r, 275ID-cl, 283ID-r, 288t, 290[2x], 321ID-r, 324[2x]); Juan Abarca (119, 162[2x], 163, 163ID-l, 169ID, 172, 173[2x], 174, 175[2x]); Greg Basco, *Hypsiboas rosenbergi* (x); Abel Batista (374, 376t, 380ID-c, 382ID-l); César Barrio-Amorós (22, 23, 504r); Eduardo Boza Oviedo (45, 46, 58b, 80, 104, 105, 254, 256ID-c, 512ID-cr); Roberto Brenes (127, 128); Luis A. Coloma & Elicio E. Tapia (Centro Jambatu de Investigación y Conservación de Anfibios) (466, 468c); William Duellman (171, 230, 232ID-l, 232ID-c, 242ID-r, 243ID-r, 245, 245ID-c, 245ID-cr, 250, 250ID-c, 258ID-l, 275ID-r, 277, 279ID-l, 399ID-l, 399ID-cr, 404, 405t, 405ID-l, 413, 413ID-l); Tobias Eisenberg (308br, 345ID-l, 349t, 351ID-cl, 351ID-r, 357, 358[2x], 376ID-l, 380ID-l, 382t), Danté Fenolio (334, 335, 336b); Don Filipiak (34c, 84, 98, 161b, 219, 258br, 301t, 303ID, 419, 493l, 494b, 499bl, 499ID-r); Michael & Patricia Fogden (48, 96, 99t, 100, 106, 108, 110, 129, 130, 138, 139, 153b, 155[2x], 170, 181[2x], 229, 232, 237, 245ID-l, 250ID-l, 275ID-c, 279ID-r, 289b, 378t, 394, 395c, 411bl, 491l); Jason Folt (224, 262b, 356, 392b, 468ID, 470, 471b, 473ID-r); Brian Freiermuth (306); Luke Owen Frischkoff (24, 420); Brian Gratwicke (83, 359, 384, 385, 387ID-l, 395ID, 439t, 447t; Tammy Grella (167br); Marcos Guerra-Creative Commons (153t, 154ID, 158ID-l, 203, 205ID-l, 207ID-l, 209ID-c, 211ID-c, 223ID-r, 376ID-r, 378ID, 380t, 382ID-r); Andreas Hertz (52, 53, 67, 68l, 68r, 85, 169b[2x], 258bl, 378c, 380ID-r, 399t, 403t, 403b,409ID-cr, 413ID-r,415, 503ID-c); Fabio Hidalgo (32, 123); Peter Janzen (26b); Cesar Jaramillo (503); Donald Jimenez (116, 117, 178); Richard Dennis Johnston (25, 501b), Karl-Heinz Jungfer (264, 405b); Gunther Köhler (101, 107); Bill Leonard (140, 141); Karen Lips (268); Stefan Lotzkat (61); Jimmy Mata (378b, 401bl, 411br, 414); Alex Monro & Eduardo Boza Oviedo (87, 88); Piotr Naskrecki (81[2x]); Joachim Nerz (14t, 75b[2x], 99, 135t, 136, 142, 143); Kenji Nishida (385ID-r, 395t, 399ID-cl, 409ID-cl, 411ID-c); Tim Paine (41, 134, 135, 223br, 330[2x], 349b, 507ID-r); Seth Patterson (149c, 418ID-r, 431[2x], 435ID-l, 439ID-cr, 441ID-cr, 448ID-l, 471ID-r, 473ID-c, 487, 488t, 488c); Todd Pierson (30, 182, 328, 332b, 432bl, 436); Marcos Ponce (395b, 468t, 468b, 471ID-l, 473ID-l); Robert Puschendorf (194, 403ID, 415); Ira Richtling (477); Ignacio de la Riva (397, 399b, 401ID-l, 407ID-r, 409ID-l); Sean Rovito (47, 57, 62, 63); Jodi Rowley (401); Paddy Ryan (488); Stanley Salazar (202, 203ID-c, 205ID-r, 206, 207, 376b, 382b, 491r, 492t, 505ID); Rob Schell (261, 293, 386, 387t); Alex Shepack (200cl, 371ID-r, 373ID-cr, 385ID-c, 387ID-r, 392cl, 421); Angel Solís (91, 93t, 94bl, 160, 163ID-r, 169t, 171ID, 177ID-r, 179[3x], 183, 195, 197ID-l, 218ID-l, 223ID-l, 258t, 277ID-r, 294, 302, 392cr, 393, 397ID, 401t, 406, 492b, 503ID-r); Jason Straka (132); Javier Sunyer (58, 60, 69, 75t, 102, 133, 148b, 177t, 239, 242ID-cr, 262c, 286ID-c, 305, 332t, 391br, 428, 432t, 432br, 443, 451[2x], 469, 471ID-cr, 473[2x], 488, 493, 494, 507ID-cl); Roy Toft (350); Filip Uyttendaele (64); Andrés Vega (71, 471t); David Wake (55, 56, 59, 86, 125); Jon Wedow (82); Peter Weish (13t, 16, 17, 18); John Williams (426ID-l, 427); and Brad Wilson (vi, 147b, 254ID-r, 256ID-r, 336t, 510, 512[2x]).

(ID = photographs used in the Similar Species sections of the book. Location on the page is indicated as follows: t = top, b = bottom, l = left, r = right, c = center.)

Index

529

Notes

Notes

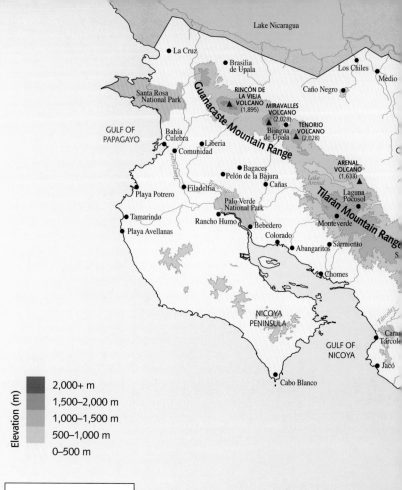

Lake Nicaragua

La Cruz

Brasilia
de Upala

Los Chiles

Medio

Santa Rosa
National Park

RINCÓN DE
LA VIEJA
VOLCANO
(1,895)

Caño Negro

GULF OF
PAPAGAYO

Bahía
Culebra

Liberia

Comunidad

MIRAVALLES
VOLCANO
(2,028)

Bijagua
de Upala

TENORIO
VOLCANO
(2,028)

ARENAL
VOLCANO
(1,633)

Bagaces

Pelón de la Bajura

Cañas

Laguna
Pocosol

Playa Potrero

Filadelfia

Palo Verde
National Park

Tamarindo

Rancho Humo

Bebedero

Monteverde

Playa Avellanas

Colorado

Abangaritos

Sarmiento

Chomes

NICOYA
PENINSULA

GULF OF
NICOYA

Caras
Tárcoles

Jacó

Cabo Blanco

Guanacaste Mountain Range

Tilarán Mountain Range

Lake
Arenal

Elevation (m)

2,000+ m
1,500–2,000 m
1,000–1,500 m
500–1,000 m
0–500 m

National parks
Volcanoes
Rivers
Cities

Costa Rica